AF352406

FUEL CELLS FOR PUBLIC UTILITY

AND INDUSTRIAL POWER

FUEL CELLS
FOR PUBLIC UTILITY
AND INDUSTRIAL POWER

Edited by Robert Noyes

NOYES DATA CORPORATION
Park Ridge, New Jersey, U.S.A.
1977

Copyright © 1977 by Noyes Data Corporation
No part of this book may be reproduced in any form
without permission in writing from the Publisher.
Library of Congress Catalog Card Number: 77-89632
ISBN: 0-8155-0676-7
Printed in the United States

Published in the United States of America by
Noyes Data Corporation
Noyes Building, Park Ridge, New Jersey 07656

FOREWORD

Fuel cell based power plants offer one of the most interesting possibilities for future power generation. The fuel cell is potentially more efficient than a conventional steam-fired plant. Since the fuel reacts electrochemically rather than by combustion, there is less of a pollution problem in terms of air, thermal, and noise pollution. They can be air-cooled and need not be adjacent to a body of water.

A most important consideration for a public utility is the concept of modularity. Fuel cell units in conjunction with a large power plant can be utilized for peaking and intermediate service. Because of efficiency considerations, other steam-driven units can be kept at their most efficient rated loads. Substation sized fuel cells scattered throughout the plant can also cut transmission costs.

The concept of modularity also has capital considerations since fuel cell units can be added to a power plant system incrementally over a period of time and built rapidly. The large maximum-efficiency power plants required to be built today tie up considerable capital in unused initial capacity, and take seven to ten years to build. Also, the advantages of small-scale fuel cell power units for smaller municipalities, large office complexes and shopping centers are obvious.

Fuel cells can be classified by the type of electrolyte used. Initial development work has been done with phosphoric acid; however, alkaline electrolytes are also being investigated. The second generation of fuel cells will operate at higher temperatures. A study of molten carbonate fuel cells was prepared by J.M. King, Jr. of United Technologies Corporation with Burns & Roe, Inc. and the Institute of Gas Technology (NASA CR 134955, FCR-0237; also available as EPRI-EM-335 from NTIS). The third generation will probably have solid oxide electrolytes operating at high temperature, high efficiency, and compatible with coal gasification processes.

The leader in fuel cell technology today is United Technologies Corporation. They will complete a 4.8 MW module demonstration plant to be financed by UTC, ERDA, and the Electric Power Research Institute. This could lead to expediting the introduction of commercial 26 MW power plants in the early 1980s. The TARGET program of UTC instituted in 1967 in conjunction with a number of gas companies, has resulted in the placement of 65 units of 12.5 kW for field testing, and the development of a 40 kW prototype. General Electric and Westinghouse are also conducting research in this field.

This book, based on information derived from U.S. government contracted studies, contains considerable practical, down-to-earth technical information relating to fuel cells for power plants.

PROPOSED 26 MW FCG-1 INSTALLATION

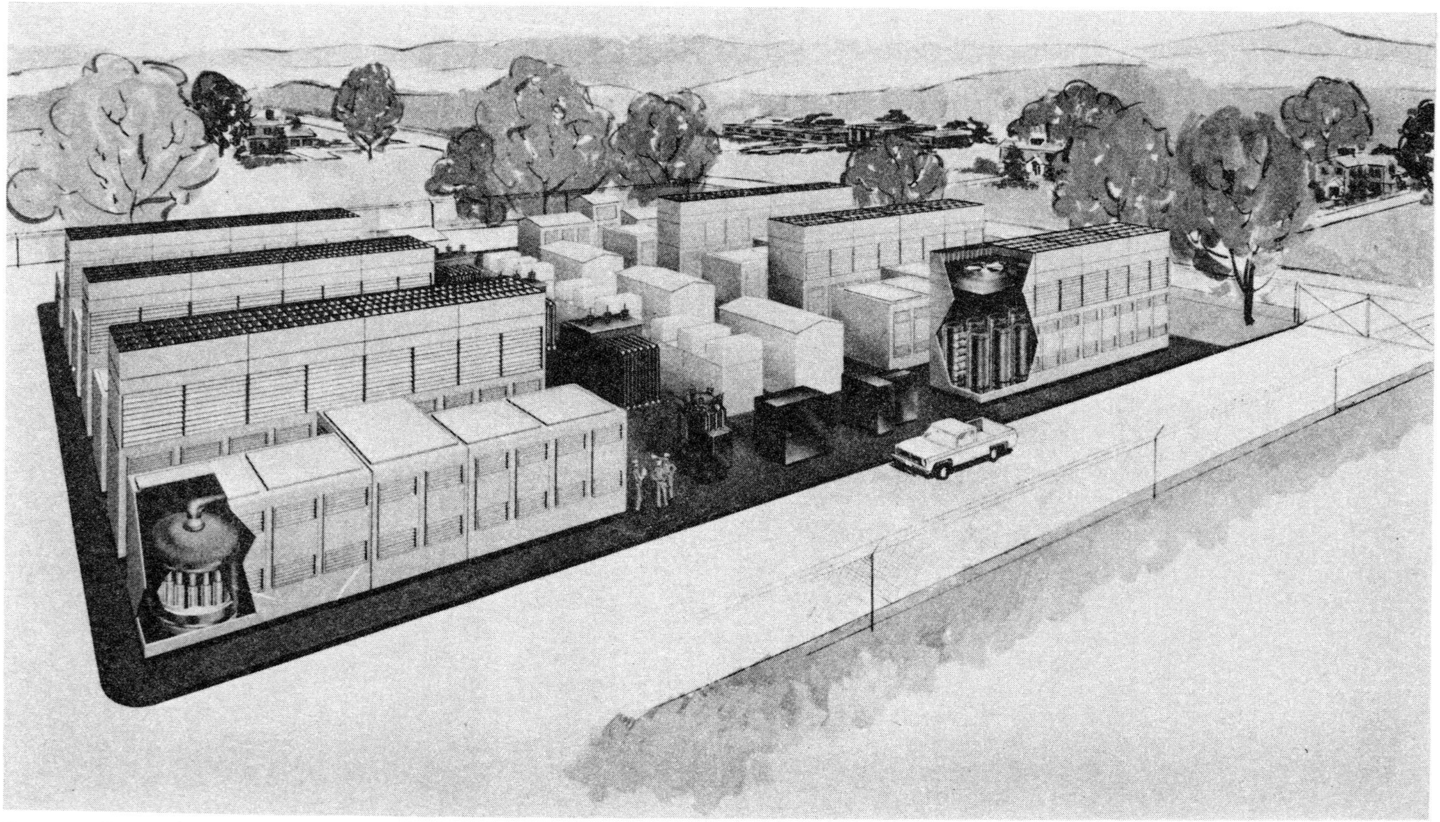

Courtesy Power Systems Division, United Technologies Corporation

CONTENTS AND SUBJECT INDEX

INTRODUCTION 1
Advantages of Fuel Cells 1
Application of Fuel Cells 1
Choice of Fuels 3
Status of Fuel Cells 4

TYPES OF FUEL CELLS–THEIR OPERATION AND USE 8
Introduction 8
Fuel Cell Operation 9
Fuel Cell Types 11
Direct Fuel Cells 11
Indirect Fuel Cells 12
Regenerative Fuel Cells 13
Fuels 14
Advantages 16
Efficiency 16
Pollution 18
Complexity 18
Scale 18
Special Applications 18
Space 19
Naval Propulsion 20
Electric Vehicles 20
Communication Systems 20
Commercial Systems 20
Other Applications 21
Fuel Cell Power Plants 21
System Characteristics 21
Applications 22
Future Systems 23
Economics 23
Major Commercial Research and Development Activities 25
P&WA/Gas Utilities 25

P&WA/Electric Utilities. 26
Esso-Alsthom 27
Westinghouse. 27
Status 28
Near-Future Systems. 29
Conclusions. 31
References. 31

**ASSESSMENT OF FUELS FOR POWER GENERATION BY ELECTRIC
UTILITY FUEL CELLS**. 33
Conversion Technology. 33
Summary. 33
Initial Process Screening 33
Later Process Screening. 37
Secondary Fuels 40
Gasification Technology 41
Central Alternatives. 41
On-Site Alternatives 41
Conceptual Fuel Supply Systems 42
Summary. 42
Modular Approach 48
Basis for Evaluation. 48
Central Fuel Conversion for Dispersed Fuel Cells 50
Synthesis Gas Treatment and Conversion Systems 51
Coal Gasification Systems. 51
Solid Waste Gasification 53
Heavy Oil Partial Oxidation. 53
Central Fuel Conversion Economics for Dispersed Fuel Cells 53
On-Site Fuel Conversion for Dispersed Fuel Cells. 53
Process Integration Opportunities (Central/On-Site). 53
Dispersed Fuel Cell/Processor Integration Analysis. 54
Capital Investment for On-Site Processors. 66
Central Fuel Conversion for Central Fuel Cells 67
Process Integration Opportunities. 67
Central Fuel Cell/Processor Integration Analysis and Economics 67
Recommendations for Future R&D Efforts 75

**FUEL CELLS FOR PUBLIC UTILITY APPLICATIONS—WESTINGHOUSE
STUDY**. 76
Summary. 76
State of the Art. 77
Aqueous Acid Fuel Cells. 78
Alkaline Fuel Cells 79
Molten Carbonate Fuel Cells. 81
Stabilized Zirconia Fuel Cells 83
Description of Parametric Points. 84
Phosphoric Acid Fuel Cell Power System 85
Alkaline Fuel Cell Power System. 88
Molten Carbonate Fuel Cell Power System 90
Solid Electrolyte Fuel Cell Power System. 94
Approach to Efficiency Calculations. 100
Phosphoric Acid Fuel Cell Power System 100
Alkaline Fuel Cell Power System. 103

Molten Carbonate Fuel Cell Power System 104
High-Temperature Solid Electrolyte Fuel Cell. 105
Capital, Site-Labor, and Operation and Maintenance Costs. 107
Phosphoric Acid Fuel Cell Power System 108
Alkaline Fuel Cell Power System. 110
Molten Carbonate Fuel Cell Power System 110
Solid Electrolyte Fuel Cell Power System. 112
Results of Parametric Assessment 113
Phosphoric Acid Fuel Cell Power Systems. 114
Alkaline Fuel Cell Power System. 119
Molten Carbonate Fuel Cell Power System 121
Solid Electrolyte Fuel Cell Power System. 124
Conclusions and Recommendations 127
Appendixes. 136
Appendix 1—Fuel Processing for Low-Temperature Fuel Cell Power
Plants 136
Appendix 2—Power-Conditioning Subsystem 143
Appendix 3—Oxygen Plants for Fuel Cell Power Systems. 147
References. 149

FUEL CELLS FOR PUBLIC UTILITY APPLICATIONS—GENERAL
ELECTRIC STUDY 152
Fuel Cells—Low Temperature 152
Description of Cycle 152
Analytical Procedure and Assumptions. 154
Design and Cost Basis 156
Results 162
Fuel Cells—High Temperature 167
Description of Cycle 167
Analytical Procedure and Assumptions. 167
Design and Cost Basis 169
Results 169
References. 173

FUEL CELL POWER PLANT EVALUATION 174
Introduction 174
Low Temperature Fuel Cells. 175
General Electric Treatment. 175
Westinghouse Treatment. 176
High Temperature Fuel Cells. 178
General Electric Treatment. 178
Westinghouse Treatment. 179
Overall Comparison. 181
Assessment of Low Temperature Fuel Cells 182
General Electric Solid Polymer Electrolyte Systems 182
General Electric Phosphoric Acid Systems 183
Westinghouse Phosphoric Acid Systems 183
Westinghouse Alkaline (KOH) Systems. 184
Assessment of High Temperature Fuel Cells 184
General Electric Zirconia Solid Electrolyte System. 184
Westinghouse Zirconia Solid Electrolyte System. 184
Westinghouse Molten Carbonate Systems 185
Significant Trends. 185

Detailed Evaluation. 190
   Solid Polymer Electrolyte. 190
   Phosphoric Acid Electrolyte 192
   Molten Carbonate Electrolyte. 195
   Zirconia Solid Electrolyte. 195
   Commercial Availability 196
Special Features of Fuel Cell Power Plants 196
Conclusions. 198
   Low Temperature Fuel Cell Systems. 198
   High Temperature Fuel Cell Systems. 199

MARKETING CONSIDERATIONS 200
   Summary. 200
   Historical Overview of Fuel Cells and Program .201
   Fuel Cell Characteristics 201
      Operating Characteristics. 201
      Environmental and Siting Characteristics .202
      Response to Varying Load 203
      Siting and Installation. 203
      Operation and Maintenance. 203
      Waste Heat 203
   Areas of Application. 203
      Electric Utility Uses 203
      Heat Recovery and Integrated Energy Systems Applications. 204
   Assumptions 204
   Benefit Analysis 205
      Utility Applications. 205
      Integrated Energy Systems 210
      Process Steam 211
      Export Market. 213
   Conclusions. 214
   Appendixes. 215
      Appendix 1–Characteristics of Advanced Fuel Cell Types 215
      Appendix 2–Effect of Electric Capacity on Fuel Cell Deployment. 215
      Appendix 3–Regional Fuel Prices 1985 216
      Appendix 4–Regional Capacity Shares 216
      Appendix 5–Annual Generating Cost per kW Based on 1987 Installation.217
   References. 223

PROPRIETARY PROCESSES 224

INTRODUCTION

The *Status and Outlook for Energy Conversion via Fuel Cells* was presented by John P. Ackerman of the Argonne National Laboratory, to the UMR-MEC Energy Conference at Rolla, Missouri on April 25, 1974, and is summarized below. Fuel cells have the potential of providing good solutions to a variety of energy-related problems. As supplies of fossil fuels are depleted, their cost will rise, and there will be increasing difficulty in obtaining certain premium fuels at any price. It is necessary, then, to use remaining reserves as efficiently as possible. Energy conversion via fuel cells represents one of the best ways to achieve this goal, because it is possible, simultaneously, to obtain more work and less pollution from a dollar's worth of fuel with a fuel cell than with any other device.

ADVANTAGES OF FUEL CELLS

Although much of the interest in fuel cells is due to their efficient use of fuel, there are considerable pollution control advantages to be gained as well. Because the fuel reacts electrochemically rather than by burning in air, no nitrogen oxides are formed. For the same reason, emissions of unburned and partly burned gaseous and particulate products are essentially nil. The only moving parts in fuel batteries are fuel pumps and, perhaps, electrolyte pumps, so operation is inherently very quiet. There is relatively little thermal pollution because less energy is lost as heat.

While there are a number of different kinds of fuel cell systems whose efficiencies vary from somewhat more to considerably less than the 60% that has been estimated for fuel cells, the message remains that more useful energy can be extracted from fuel with fuel cells than with any other energy conversion device.

APPLICATION OF FUEL CELLS

The uses to which fuel cells may most profitably be applied are electric power generation and transportation. Most of the nonelectrical energy in the industrial sector, and nearly all in the commercial and residential sector is used for heating. Conversion of fuel to heat usually proceeds with high efficiency, so relatively little application of fuel cells in these sectors is seen. Because the fuel cells convert chemical energy directly to electrical energy, electrical power generation is probably their most natural application. While the output of each cell is low voltage dc power, cells may be connected in various series and parallel arrangements to give whatever voltage is desired, and large highly efficient inverters are available for conversion to ac.

In this application, fuel cells must compete with large steam turbines, which are remarkably efficient devices. (At rated load, a large modern unit can approach 40% efficiency.) However, the demand for electrical energy is far from constant. Over the course of a year, the actual power output of a large utility may vary by nearly a factor of four, and the daily variation in load can be almost a factor of three. To adjust to this changing demand, either the large base load plants must sometimes operate at part power, or smaller cycling or peaker units must be used during periods of high demand. Either way, efficiency suffers and pollution increases. The fuel cell system not only has a greater efficiency at full load, but this efficiency is retained and even increases as load diminishes, so that inefficient peaking generators may not be needed.

A fuel cell system, unlike a heat engine, need not be big to be efficient. This characteristic, taken together with two others, low emissions and capability of operation on a variety of fuels, allows fuel cell systems to be operated almost anywhere. A small community power company can operate a power plant on the optimum fuel available locally with nearly the same efficiency achieved by a large central power station. A large metropolitan utility can disperse a number of generators throughout its area and match capacity to local demand, substantially reducing the expense and other problems associated with transmission and distribution of electricity.

Cost and other problems involved with local distribution of electrical energy are likely to be greater, especially as more utilities go to underground lines in urban areas. Also, the cost of transmission should be reduced by a factor of two rather than three for comparison with fuel cell generating systems, because fuel cells require less fuel per kilowatt hour of electricity generated than do conventional generating stations.

In the transportation industry, the same virtues of efficiency and low pollution make the fuel cell attractive. Here there are at least two other major requirements which must be met. These are the needs for a relatively high available energy/weight ratio (so-called energy density) and for a large power/weight ratio (power density). Fuel cells may be expected to meet the first criterion handily, since the amount of energy available is determined by the size of the fuel tank. A fuel cell powered vehicle can have a good long range without refueling, and can be refueled rapidly, just as can present day internal combustion vehicles. This represents a substantial advantage over battery powered vehicles, which are the competition for efficient, low pollution personal transportation.

The criterion of high power density is considerably more difficult to meet. It is very much worse for small personal vehicles than for large buses, trucks, trains, and ships. To propel a vehicle of weight comparable to an intermediate car with speeds and accelerations useable in present traffic conditions, it is probably necessary to achieve a power density of about 100 W/lb, which is equivalent to about 13 lb/hp. It may be possible to meet that goal by hybridizing a fuel battery with one of several high power energy storage devices, such as one of the new generation of flywheels.

Two factors will act to mitigate the necessity for high power densities. One is increasing cost and decreasing availability of fuel, which is even limiting the speeds and hence the power required for road vehicles. The other is that the presence on the road of low power vehicles will tend to change driving patterns in the same direction of decreased speed and acceleration requirements.

Fuel cell systems of adequate performance to propel railroad trains, barges, and ships can probably be built with existing technology, at least, as far as cells themselves are concerned. The detailed engineering necessary to actually build the power plant and ensure reliability and control is another matter. Although essentially all of the basic technology is available, considerable effort would have to be expended to develop a viable system. The power plant would be very smooth and quiet, virtually pollution-free, and could operate on conventional fuels. A detailed economic analysis would have to be undertaken to determine the break-even point where increased fuel costs would balance against lifetime and initial cost consideration.

CHOICE OF FUELS

Fuel cells have been made using a wide variety of fuels: hydrogen, hydrazine, ammonia, hydrocarbons of various sorts, alcohols, natural and synthetic gas, and others. The pragmatic truth of the matter is that the only fuel which performs nearly as well as hydrogen is hydrazine, and hydrazine is both toxic and very expensive. Unfortunately, the hydrogen economy is not yet developed, and hydrogen is not widely available in large quantities. Technology does exist for conversion of a variety of other fuels to hydrogen where tank or pipeline hydrogen is not available.

Natural gas and petroleum distillates are relatively easy to convert to hydrogen by several processes. One of the best for fuel cell uses is catalytic steam reforming at high temperature (900°C). The raw gas stream contains carbon monoxide, a notorious catalyst poison, which can be removed by the shift reaction with steam to form carbon dioxide and more hydrogen. Sulfur must be removed from the feed stream or raw gas stream because it ruins the reforming catalyst, the shift catalyst and the fuel cell catalyst. This is actually somewhat of an advantage, since now there can be no sulfur oxides in the fuel cell exhaust. Sulfur removal technology is well proven and in wide use in the petroleum industry. The pressure and temperature requirements imply that hydrocarbon fuels will be better suited to fixed than mobile uses.

Ammonia can be easily cracked in a simple reactor to provide a very suitable fuel stream containing only hydrogen and nitrogen. The small equilibrium amount of residual ammonia in this stream is easily removed in a trap. The simplicity of the cracker lends itself to easy control and thus, to mobile applications. Ammonia is relatively easy to store, and has a reasonable energy density (2.5 kWh/lb vs 2.76 kWh/lb for methanol and about 5.0 kWh/lb for gasoline.

Methanol has most of the virtues of ammonia, and in addition, can be converted to hydrogen at relatively low temperature (near 250°C) and a pressure near 1 atm. It is also likely to be less expensive than ammonia (about 10 to 15 cents per gallon at the plant if one has a coal gasification plant that makes high Btu gas at $1.50/million Btu or less, or if naphtha is available. Methanol is fairly reactive electrochemically, and there is a possibility that it can be used directly in a fuel cell without reforming. It is sufficiently involatile that it can be handled in the present gasoline distribution system without basic changes. All these factors combine to make it the most promising fuel for mobile applications.

As coal gasification technology matures, very satisfactory feed streams for fuel power plants will be available. Present processes convert coal to a carbon monoxide/hydrogen mixture which is scrubbed of sulfur-containing gases and converted to methane. For direct fuel cell use, the carbon monoxide in the sulfur-free stream could be shifted with steam via the water-gas shift reaction to carbon dioxide and hydrogen. The carbon dioxide could be scrubbed from the stream, if necessary, but it would probably be satisfactory to leave it in if the fuel cell station were nearby.

If the H_2 were pumped to a remote location it would be removed because of the pumping cost. This stream could also be used in ammonia synthesis. In this case, air would probably be used in the gasification processes to give the correct amount of nitrogen in the gas stream. For synthesizing methanol, carbon monoxide would be left in the stream, since it is one of the reactants in methanol synthesis.

STATUS OF FUEL CELLS

In 1974 there were essentially no commercial uses of fuel cell power plants in the field of transportation. Dr. Karl Kordesch of Union Carbide Corp. has had a small economy car converted to operate on a gaseous hydrogen fuel battery/lead-acid battery hybrid system for several years, but this is a hobby project, undertaken, perhaps, to demonstrate that it can be done. Six hundred cubic feet of hydrogen gas store 33 kWh of energy, and give the car a range of about 200 miles at 40 mph. General Motors had a fuel cell program which was active and making progress, especially on the air electrodes, but they had not announced any plans for putting fuel cells in even an experimental vehicle in the near future.

When sufficient progress has been made that fuel cells of high power density can be constructed, there will doubtless be much more interest from the transportation industry, but in 1974 there was not sufficient incentive for the automobile makers to launch the large research and development effort that would be required to construct an economically competitive vehicle.

In the utilities field, the situation was considerably brighter, Pratt and Whitney (United Technologies Corp.) contracted to place 26 MW fuel cell power plants in the field for a group of utilities. These plants would operate on a variety of fuels: natural gas, methanol, naphtha, or possibly even No. 2 fuel oil, depending on reformer technology. These would be the first commercial units to be developed, and successful application of these plants would mark the beginning of widespread use of fuel cells for power generation and the beginning of a new era in national use of energy.

Included below are illustrations from a report (*National Benefits Associated with Commercial Application of Fuel Cell Powerplants*) prepared by United Technologies Corp. for ERDA and published February 1976.

FIGURE 1.1: THE FUEL CELL

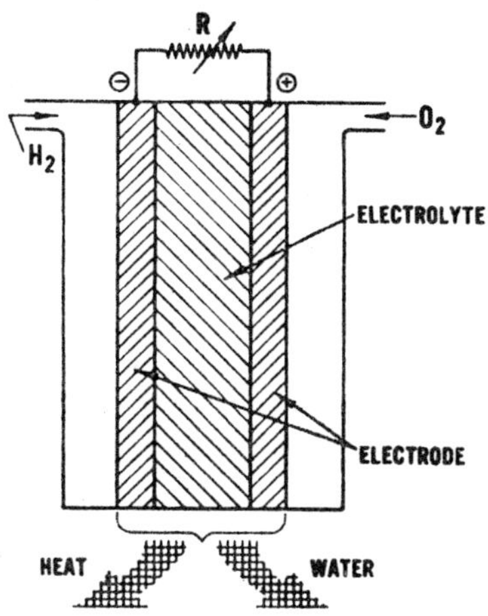

FIGURE 1.2: THE FUEL CELL POWER PLANT

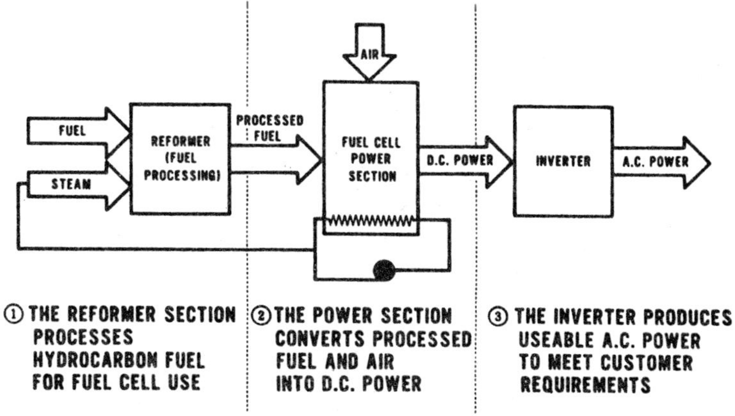

FIGURE 1.3: FUEL ECONOMY/EFFICIENCY FOR ALL SIZES

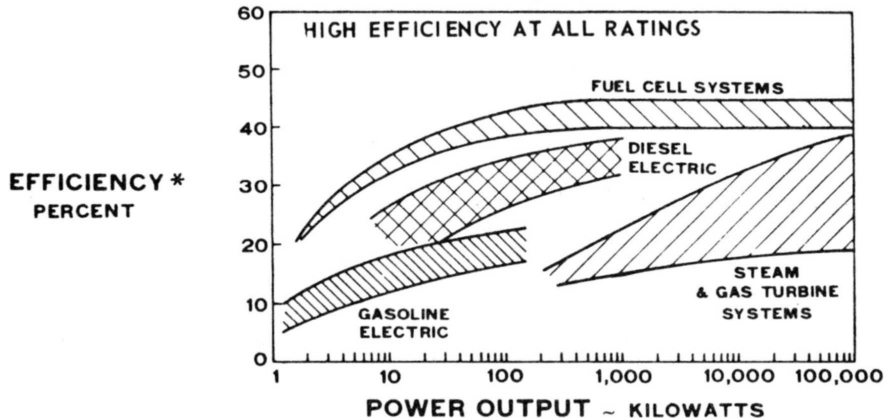

***BASED ON LOWER HEATING VALUE**

FIGURE 1.4: FUEL ECONOMY/EFFICIENCY AT PART LOAD

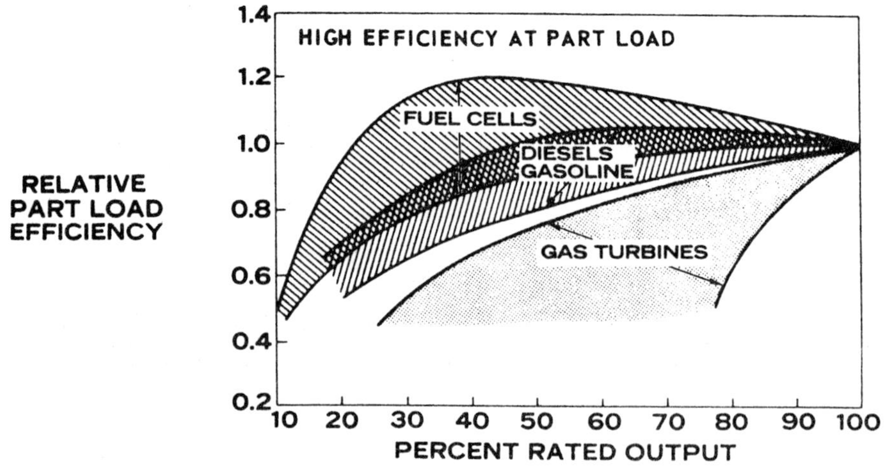

FIGURE 1.5: ENVIRONMENTAL IMPACT

POUNDS OF POLLUTANTS PER MILLION BTU HEAT INPUT

|-------- FEDERAL STANDARDS * --------|

	GAS-FIRED CENTRAL STATION	OIL-FIRED CENTRAL STATION	COAL-FIRED CENTRAL STATION	EXPERIMENTAL FUEL CELLS**
PARTICULATES	0.1	0.1	0.1	0.0000029
NO_X	0.2	0.3	0.7	0.013-0.018
SO_2	NO REQUIREMENT	0.8	1.2	0.000023
SMOKE	20% OPACITY	20% OPACITY	20% OPACITY	NEGLIGIBLE

*FEDERAL STANDARDS EFFECTIVE 8-17-71
**YORK RESEARCH CORP., Y-7309 APRIL 1970

FIGURE 1.6: WASTE HEAT RECOVERY POTENTIAL

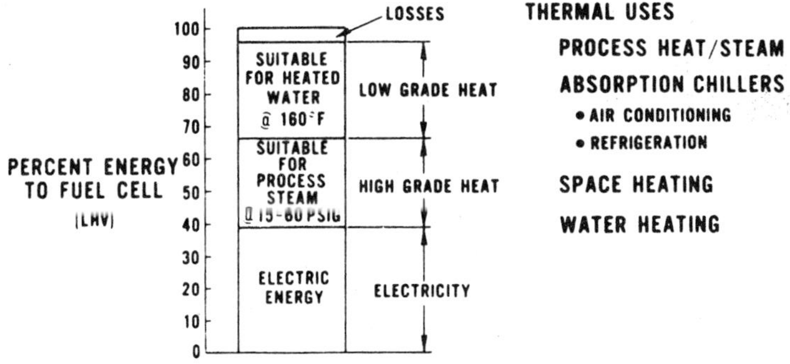

TYPES OF FUEL CELLS
THEIR OPERATION AND USE

This chapter is based upon a report prepared by Ralph T. Johnson, Jr. of the Sandia Laboratories of Albuquerque, New Mexico in August 1974, under the title of *Direct Conversion of Electrochemical Energy into Electricity* and is summarized and editorialized below.

INTRODUCTION

A fuel cell is an electrochemical device in which the chemical energy of a conventional fuel is converted directly and efficiently into low-voltage, direct current electrical energy. Fuel cells as well as conventional batteries (e.g., dry cells and lead-acid batteries) work by virtue of electrochemical reactions.

In a battery the chemical energy is stored within the cell, and the capacity of the battery is governed by the size and weight of the electrodes. Fuel cells are unlike batteries in that the reactants are supplied from outside the cell and the cell itself does not undergo an irreversible chemical change. Thus, it can continue to operate as long as fuel and oxidant are supplied and products are removed (or at least until the electrodes fail because of mechanical or chemical deterioration). In many respects a fuel cell can be thought of as a primary battery in which the fuel and oxidizer are stored external to the device and are fed to it as needed.

The history of fuel cell development as well as the present status of this technology has been reviewed by many authors (1)-(15). Initially the ultimate aim of much of the research and development was to develop new energy conversion devices which would run on conventional fuels (for example, gasoline and coal) and air. However, within the past decade some of the most useful and interesting systems which have been built and demonstrated are those which operate on more reactive fuels such as hydrogen, alcohol or hydrazine and on oxidizers such as oxygen or hydrogen peroxide.

The early development of fuel cells for electrical energy generation was overshadowed by the steam powered dynamo. The biggest boost to fuel cell development has been space power applications where high power density and low weight are important. Fuel cells have been an integral part of the Gemini and Apollo manned space flight system.

Fuel cells offer a number of important advantages over heat-engine/generator systems. These include high efficiency (no Carnot cycle limitation), no significant air or water pollution (less by a factor of 10 compared to other methods), high reliability, few moving parts (only auxiliary systems have moving parts), and quiet (as well as unattended) operation. Despite these obvious advantages, fuel cells have not been widely used or developed as terrestrial energy sources because of their high costs (partly due to expensive catalysts) and short operating lifetime (due primarily to degradation of electrode materials). It is only recently that it appears that some of these problems may be overcome and fuel cells may be commercially viable within this decade.

In this chapter, an attempt is made to acquaint the reader with the fuel cell, its operating principles and characteristics, its applications, and the present status of the technology. Specifically, the next section describes the operation of fuel cells and uses the hydrogen-oxygen cell (the most highly developed fuel cell) as an example. The various types of fuel cells and fuels are described in the next sections. This is followed by a discussion of the advantages and special applications of fuel cells.

The utilization of fuel cells for large-scale power plants and the economics of fuel cells are then discussed. The major commercial research and development activities are described in a subsequent section and final sections outline the present status of this technology and draw general conclusions.

FUEL CELL OPERATION

The essential features of a fuel cell (1)(2)(4)(5) are shown in Figure 2.1. The main components are a fuel electrode (anode), an oxidant or air electrode (cathode) and an electrolyte. In a typical application, the reactants are fed through the electrodes which are porous and are brought into contact with the electrolyte. Reactions take place which produce voltages at the electrodes. When an external load is connected, electrons are conducted through the load and perform useful work; whereas in the electrolyte, ions travel from one electrode to the other, completing the electrical circuit. Fuel cells continue to operate as long as the current flows through the load, as long as fuel and oxidant are supplied to the cell, and as long as structural and chemical integrity is maintained.

The principal reactions which occur in a hydrogen-oxygen fuel cell (2)(4) are shown in Figure 2.1. In this fuel cell hydrogen is fed through the anode where it reacts in the presence of a catalyst forming electrons and H^+ ions, the latter passing into the electrolyte. At the other side of the fuel cell, the oxygen which is fed through the cathode acquires electrons and reacts with water from the electrolyte to form hydroxyl ions (OH^-). In this type of fuel cell, a solution of potassium hydroxide (KOH) is often used as the electrolyte. The H^+ and the OH^- ions are in equilibrium with water in the electrolyte. Electrons released at the hydrogen electrode pass through the external load circuit and are captured at the oxygen electrode. The electrical output from an individual cell is typically a direct current of 100 to 200 mA/cm^2 at a voltage of about 1 volt (de-

pending upon the reactants employed) (3)(11)(12). These cells can be connected
in series and/or parallel to obtain a unit which will provide the appropriate voltage and current.

FIGURE 2.1: HYDROGEN-OXYGEN FUEL CELL

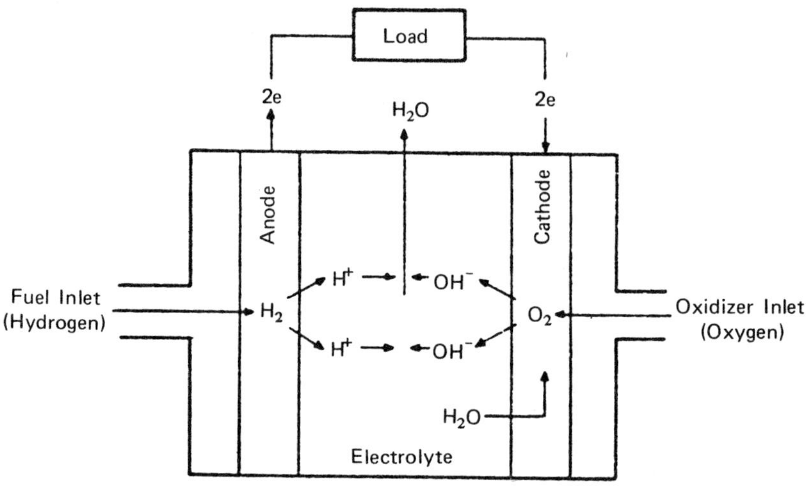

Source: Reference (2)

The reactants in a fuel cell should be as cheap and readily available as possible;
for example, one could use air for the oxidant and natural gas or petroleum derivatives for the fuel. This does not, however, rule out the importance of some
useful devices (probably of limited application) which use more expensive reactants.

Considerable research has gone into the design and choice of materials for electrodes since these are the heart of the fuel cell (2)(6)(7)(8). They must conduct
electricity well and be resistant to the highly corrosive environment. Since the
reactants are stored externally the electrodes may be made thin and light weight.
They must also have a high, catalytically-active area to promote the electrochemical reactions. Electrode catalysts may be made up of many materials (2)(6)(8).
However, most fuel cells work best with platinum or similar precious metals.
The use of these catalysts is highly restricted for economic reasons. There has
been much research into making more efficient use of restricted amounts of
precious metals and into the search for alternative cheaper materials.

A variety of electrolytes are utilized in various fuel cells (2)(6). In principle,
the electrolyte is simply an ionically conducting medium which prevents the two
electrodes from coming into electronic contact, and which allows the passage of
ions from one electrode to the other. Many substances, either acid or alkaline,
liquid or solid, may be used as electrolytes.

FUEL CELL TYPES

There are many different types of fuel cells, each differs in detail depending upon the fuel-oxidizer-electrolyte combination, and to a lesser extent upon the temperature and pressure at operation (1)(2)(5)-(8). All the various types will not be listed here, but there will only be an attempt to formulate a general classification for fuel cells. Some information is given in the next section regarding various fuels, but even the discussion of fuels will be almost exclusively restricted to hydrogen and hydrogen-containing fuels.

There are several ways of classifying fuel cells. The method used here is to distinguish between direct, indirect, and regenerative fuel cells (1)(5). Some examples are given in Table 2.1. The direct fuel cell is one in which the hydrogen or hydrogen-containing fuel is fed directly to the anode. The indirect fuel cell utilizes a reforming process outside the cell (or internally at the anode) to convert the fuel to hydrogen (or a hydrogen-rich gas) which is then fed to the anode. The regenerative fuel cell is one in which the fuel cell product is reconverted into its reactants by one of several methods (see Table 2.1) and is then recycled.

TABLE 2.1: CLASSIFICATION OF FUEL CELLS

. . . . .Direct. . . . .

Low Temperature	Intermediate Temperature	High Temperature
Hydrogen-oxygen	Hydrogen-oxygen	Hydrogen-oxygen
Organic compounds-oxygen	Organic compounds-oxygen	Carbon monoxide-oxygen
Nitrogenous compounds-oxygen	Ammonia-oxygen	
Hydrogen-halogen		
Metal-oxygen		

. . . . Indirect . . . .

Reformer　　　　　　Biochemical

Regenerative

Electrical
Thermal
Chemical
Photochemical
Radiochemical

Source: Reference (5)

Direct Fuel Cells

Probably the best known, and certainly the most highly developed direct fuel cell is the hydrogen-oxygen cell (see Figure 2.1). This cell has had extensive use in the manned space program. In this cell, hydrogen in gaseous form is used as the fuel; and oxygen, either pure or extracted from air, is used as the oxidizer. This cell is attractive because it requires only a very small amount of noble metal electrocatalyst. Other attractive features associated with this cell include the fact that the only chemical-reaction product is pure water, and excess heat generation is minimal.

Direct hydrocarbon fuel cells (illustrated in a of Figure 2.2) are still in the early stages of development but already have been shown to be more efficient than conventional engines using the same fuels. This success has been achieved only with the use of large amounts of precious metal catalysts. Until this requirement can be overcome, direct hydrocarbon cells will be too expensive for common use.

FIGURE 2.2: TYPES OF HYDROCARBON-AIR FUEL CELLS

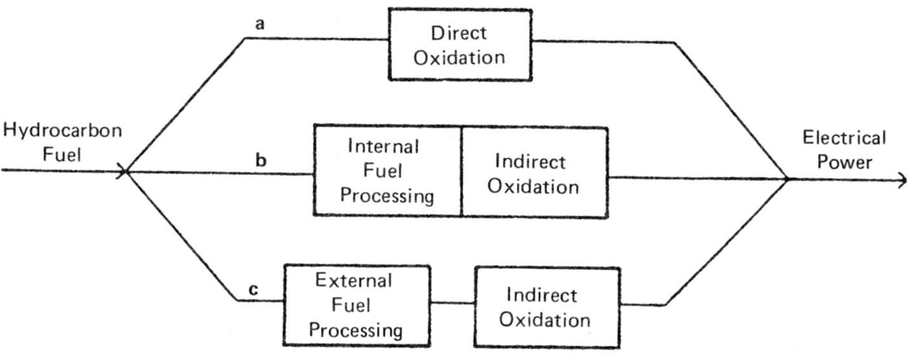

Source: Reference (1)

The direct fuel cell which shows the highest performance on a liquid fuel is the hydrazine cell (6). Of the direct fuel cells, this cell has received more engineering attention than any other except for hydrogen-oxygen. Unfortunately, hydrazine is an expensive, poisonous fuel, and thus its usefulness is limited.

Indirect Fuel Cells

Indirect fuel cells involve a process by which the fuel undergoes a reforming process either internally at the anode (**b**, Figure 2.2) or outside of the cell (**c**, Figure 2.2) to obtain hydrogen or a hydrogen-rich gas (1). When external fuel processing is used, the system can be arranged so that: (a) only very pure hydrogen is fed to the anode (most expensive), or (b) hydrogen of almost any degree of purity (including the untreated reformer exit gas) is fed to the anode (6).

The indirect systems which use very pure hydrogen are the most well developed and make high performance fuel cells. Furthermore, because of the high reactivity of hydrogen and the absence of chemical and electrochemical complications produced by impurities, a great deal of progress has been made toward the elimination of expensive electrocatalysts for these cells. For operation with low-purity hydrogen, fouling of the anode and electrolyte is a problem.

The most practical hydrocarbon fuel cell systems to be developed are indirect fuel cells in which the fuel is first converted to hydrogen and then the hydrogen is fed to the cell. This means that an additional subsystem has to be attached to the fuel cell system. The technology for the manufacture of hydrogen from hydrocarbons has been well established in large chemical plants, but there are serious problems involved in scaling down the plant size. In contrast to large chemical plants, the compact fuel cell processor (reformer) operates at high thermal efficiency and is responsive to rapidly varying outputs as well as to start and stop operations.

The main advantage of the indirect hydrocarbon fuel cell is that it does not require any (or at least as much) costly platinum electrode catalyst but can operate with less expensive materials. (In this case, however, the electrode area is generally larger and this increases other material costs.) The primary disadvantage of the indirect cell is that the reforming process requires energy which decreases the efficiency of the system.

A promising candidate for an indirect system is the molten carbonate electrolyte cell (1)(2)(4)(6). This cell operates at temperatures above 500°C and the hydrocarbon fuel is reformed with water to give hydrogen and CO. The cell is capable of consuming reformer gases containing relatively large amounts of CO with excellent performance. These cells can also operate on rather impure air and utilize relatively inexpensive electrocatalysts such as nickel at the anode and copper oxide at the cathode, making this an economically attractive system. There are still technical problems related to a complex recirculating fuel exhaust system and to materials stability which need to be overcome.

An indirect solid-electrolyte fuel cell system is the coal-gasification system being developed by Westinghouse (8)(16). This is a high temperature (1000°C) system which utilizes a stabilized zirconia (ZrO_2) solid electrolyte which conducts oxygen ions. The high temperature is required to enhance the oxygen ion transport through the zirconia. The excess heat is used in the coal-gasification process. The high temperatures in these cells pose many materials problems. Further discussion of this system is given in a later section.

Regenerative Fuel Cells

A regenerative fuel cell reprocesses and recycles the reactants (1)(4)(5). It is similar to a rechargeable (secondary) battery in the sense that the reactants can be regenerated and used over and over again. In these fuel cells there are two stages of operation: [1] conversion of reactants into products while producing electrical energy, and [2] reconversion (or regeneration) of the products into reactants. The regeneration is usually performed electrically by driving a current through the cell in the reverse direction. Heat may also be used to perform this function in certain special "thermally regenerative" cells. Such a system is illustrated in Figure 2.3. Other means of regeneration are listed in Table 2.1 and are discussed in Reference (5).

Much of the current research with the regenerative cells is with electrical regeneration (1). In this special case, the electrochemical energy converter passes over into an electrochemical energy storer. In comparison with secondary battery systems, these fuel cells offer the potential advantages of lighter weight, higher energy potential of the reactants, and longer recycle life. Fuel costs are less significant and more expensive fuels can be used in regenerative fuel cells because the reactants are recycled. The high overall efficiency of fuel cell systems makes them attractive for storing electricity at "off peak" periods because the amount of electricity used to regenerate the reactants, and hence the cost of regeneration, is reduced. Such cells might, therefore, find application in large-scale storage systems at generating sites by smoothing the load demand variations (1).

FIGURE 2.3: THERMALLY-REGENERATIVE FUEL CELL

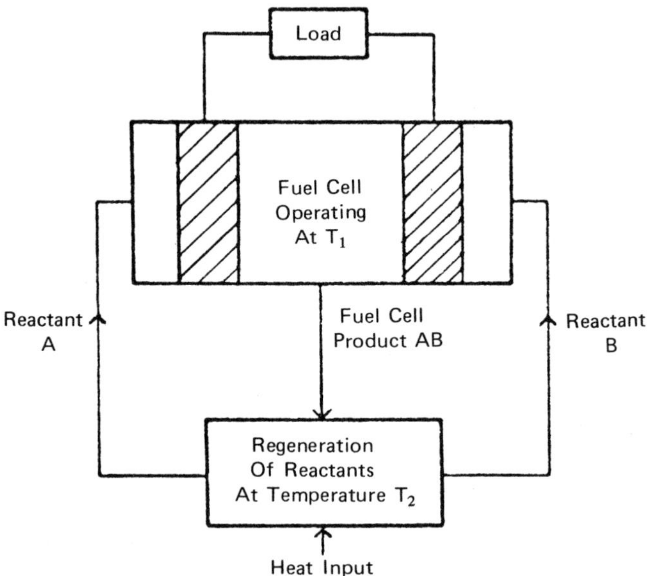

Source: Reference (5)

FUELS

An ideal fuel is dependent upon factors such as cost, availability, storageability, volume, and transportability. One of the factors which may seriously limit the use of fuel cells as a major (base line) source of electrical energy is the availability of appropriate fuels. However, even with this limitation there are still many applications for which fuel cells are ideally suited. This is discussed further throughout this report.

Hydrogen is an important fuel because hydrogen and oxygen are capable of releasing more energy, pound for pound, than most other fuel-oxidizer combinations (1). This is one of the reasons these reactants were selected for spacecraft power supplies. Hydrogen is in a class of its own because of the simplicity with which it reacts and because of its relatively high reactivity. Presently, hydrogen is a relatively expensive fuel.

Fuels other than hydrogen can be classified as hydrocarbons and so-called "compromise" fuels (1). Some of these fuels are listed in Table 2.2. The hydrocarbon fuels are generally cheaper and much easier to handle than hydrogen. They are, however, much less reactive than hydrogen, much more difficult to oxidize, and generally produce undesirable by-products. When a hydrocarbon is burned, oxides of carbon are produced. This usually leads to the use of acidic electrolytes, which can cause material compatibility problems (2). Also, large amounts of expensive catalysts and/or high temperatures must be employed to make the oxidation reactions proceed at acceptable rates. Some of these difficulties have led to fuel cells which convert the fuel into hydrogen outside the cell. Even with these difficulties, the hydrocarbon fuel cells will probably receive the widest application.

Compromise fuels (1)(2)(5)(7) have also been used in fuel cells. The reactivity of these fuels lies between that of hydrogen and the hydrocarbons, and they are generally easy to use. Both hydrazine and ammonia can be separated into hydrogen and nitrogen so that acidic electrolytes are not needed. That is not the case for methanol, however, where the oxidation process produces carbon dioxide.

TABLE 2.2: SOME POTENTIAL FUELS

Hydrocarbons

Methane	Cyclohexane
Ethane	Benzene
Ethylene	Heptane
Acetylene	Toluene
Propane	Octane
Propylene	Nonane
Butane	Decane
Butene	Hexadecane
Pentane	Kerosene
Hexane	

Compromise Fuels

Methanol	Hydrazine
Ammonia	

Note: Other possible fuels are indicated in
Table 2.1.

Source: Reference (1)

Hydrazine is highly reactive at normal temperatures and does not require an expensive catalyst (8). These desirable features are more than counterbalanced by the cost of hydrazine (15 to 20 times that of hydrogen) and the fact that it is poisonous. Despite these problems, hydrazine-powered fuel cells have been developed and used for military systems and for submersibles (5).

The cost of ammonia is similar to that of hydrogen (5) and it is more readily available and easier to handle than hydrazine. However, its reactivity is low and it is generally disassociated into hydrogen and nitrogen before being fed to the fuel cell. This results in increased complexity and reduced efficiency. Ammonia fuel cells are only considered suitable for specialized remote low-power applications.

Methanol fuel cells have received considerable attention (1)(2)(5). Methanol costs about the same as hydrogen, is made from hydrocarbons, and is readily available in pure form. Methanol can be reacted with steam (in an indirect cell) at temperatures lower than for hydrocarbons to produce hydrogen, or it will undergo direct electrochemical oxidation at the electrode which produces CO_2 contamination of alkaline electrolytes. Therefore, materials employed in direct cells are restricted to those which can withstand an acid's corrosive effects. Even with these problems and with other problems related to catalyst poisoning, methanol fuel cells have been under development by a joint French-American enterprise (11).

ADVANTAGES

Fuel cells offer a number of important advantages over other electrical energy conversion schemes. Specifically, fuel cells promise to be efficient, quiet, reliable, and nonpolluting. It is for these reasons that fuel cells are (and should be) considered for applications which call for the efficient utilization of available fuels in an environmentally acceptable way.

Efficiency

One of the chief advantages of fuel cells is their high conversion efficiency (1)(2)(4)(7)(11) (especially at low power levels). Fuel cells are not subject to the Carnot limitation on efficiency as are other conventional means (heat engines) of energy conversion. They are isothermal devices and the theoretical efficiency is determined by the thermodynamics of the cell reaction.

The energy conversion scheme for a conventional heat engine-generator is compared to that for a fuel cell in Figure 2.4 (1). The heat engine-generator has both heat loss and losses due to mechanical motion whereas a fuel cell is a one-step process (1)(2)(13). Although heat is given off in a fuel cell, it does not constitute an essential link in the energy conversion chain. The fuel cell, unlike a heat engine and generator, has no moving parts (except for possible auxiliary pumps).

All of this indicates that these systems have the potential for higher reliability and silent operation as well as unattended operation. Unfortunately, one of the general problems still facing fuel cells is materials degradation which limits life time. This, of course, affects reliability, costs, and increases maintenance. These are major problems which need to be overcome.

The efficiency of electrical production for fuel cells (hydrocarbon fuels) is compared with that of other types of generators in Figure 2.5 as a function of power output (3)(6)(11)(17). The best steam and gas turbine generators now operate at a thermal efficiency of ~39% at power levels of 100 MW. At this power level fuel cells are only slightly more efficient, but the efficiency remains high to power levels as low as 25 kW where other methods are much less efficient.

It is clear from Figure 2.5 that hydrocarbon fuel cell systems operate at much the same efficiency for small units in the 100 kW range to large multi-megawatt units. Efficiency is further increased to ~55% if pure hydrogen is substituted for the processed fossil fuel, and again to nearly 60% when oxygen is additionally substituted for air (3)(12).

FIGURE 2.4: COMPARISON OF ENERGY TRANSFORMATION PROCESSES

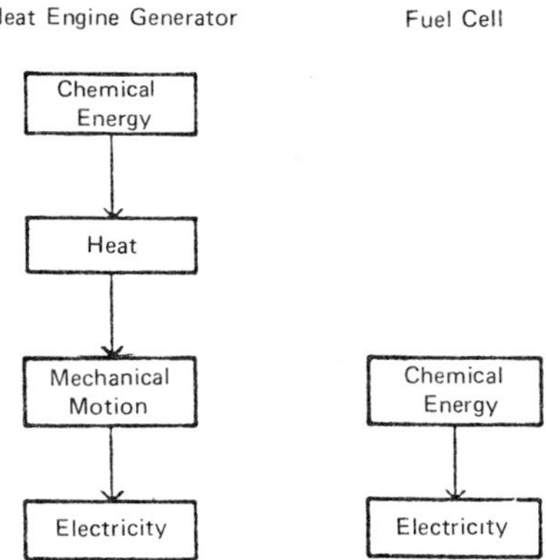

Source: Reference (1)

FIGURE 2.5: POWER SYSTEM EFFICIENCIES

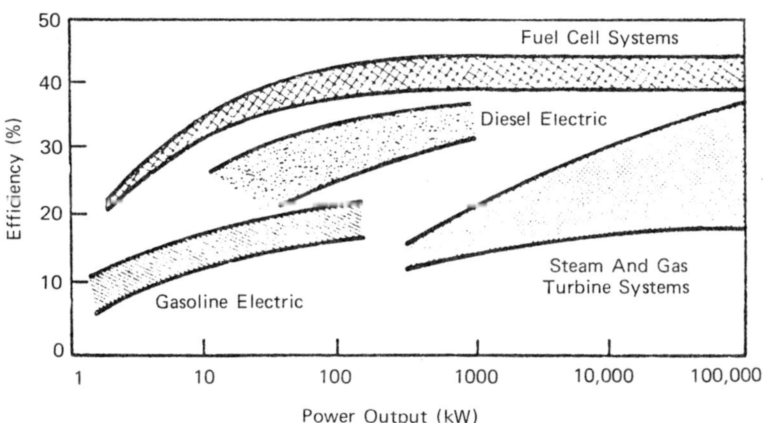

Source: References (3)(6)(11)(17)

Besides being highly efficient at peak load, fuel cells are even more efficient at partial load (12). While the efficiency of conventional systems declines, the efficiency of the fuel cell increases as the power level is decreased to about 40% load. In contrast to conventional systems, maintenance requirements are unaffected by extended operation at low or high power levels.

Pollution

Another important advantage of fuel cells over conventional power sources is that emissions are negligible because fuel cell operation is not based on combustion. In conventional power plants, a considerable quantity of nitrogen oxides, sulfur dioxide, hydrocarbons, and particulates are produced, whereas fuel cell systems emit an exhaust mostly of air, carbon dioxide, water vapor, and heat (3)(12)(17).

Studies have shown that emission of air pollutants from fuel cell systems are reduced by more than a factor of 10 over that of gas-, oil-, and coal-fired utilities (3)(11)(12)(13)(17). Thermal pollution of waterways is not a problem because excess heat is released directly to the atmosphere. Since fuel cell power plants are air cooled and are not dependent upon a source of cooling water, there are few restrictions on site locations (12)(13). Noise is also low because only the fuel and cooling systems have moving parts (11).

Complexity

The fuel cell itself is a very simple device with no moving parts. The complete (hydrocarbon) fuel cell power plant, however, does have some moving parts, but even these can be kept to a minimum. There is still an advantage in simplicity over the heat engine generator. Controls are necessary but again these can be kept fairly simple and automated.

Scale

Since fuel cell systems are efficient over a wide range of power levels, they are rather insensitive to scale (12). In a given power unit, however, there is a practical limit to the size of the thin fuel cell electrodes (probably several square feet at the most) (11). Scale-up to higher power systems is accomplished by interconnecting of cell stacks (modules). Since scale-up is by multiplication of identical components, one can expect some cost savings using mass production techniques.

SPECIAL APPLICATIONS

There are many potential applications for fuel cells since they can operate with high efficiency over a large range of power levels. Some possible applications are given in Table 2.3; only a few are discussed here. Each application generally imposes different requirements on the environment and power level, and this often dictates the need for a different type of cell which increases development time and costs. Some of the characteristics of the fuel cells used for (and considered for) special applications are outlined here (1)(2)(6)(8)(12)(15). Applications to large scale power systems are considered in the next section.

TABLE 2.3: POTENTIAL APPLICATIONS

Special Purpose

Remote Power

Communications system	Pipeline cathodic corrosion protection
Weather and oceanographic stations	Remote signal and beacon devices

(continued)

TABLE 2.3: (continued)

Military applications	Off-shore platforms

Emergency and Auxiliary Power

Hospitals	Police stations
Civil defense installations	Fire and intrusion protection
Radio, television, and telephone services	Aircraft

Portable Power

Two-way communications	Search and rescue
Lighting	Battery charging
Heating	Military systems

Propulsion

Underwater rescue and exploration	Forklift trucks
Silent power for search and rescue vessels	Off-road vehicles
Special low-pollution vehicles	Military vehicles

Other

Biomedical	Desalination
Oxygen for hospitals and aircraft	Dialysis (artificial kidney machines, etc.)
Environmental control	

General Purpose

Utility Power

Domestic power supplies	Power distribution networks
Commercial and industrial installations	Central generating plant
Peak power storage	

Propulsion

Urban automobiles	Boats
Hybrid power plant highway vehicles	Locomotives
Mass transit vehicles	

Recreation

Portable power for beach, mountains, etc.	Transportable power for camping, trailers, etc.
Silent power for naturalists, fishermen	

Source: Reference (1)

Space

The single application which has proved the fuel cell to be a useful power source is the space program. The most important criteria for a space craft power source are minimum weight and reliability (1)(11). These criteria guided the selection and development of special fuel cells for the Gemini (General Electric system) and Apollo [Pratt & Whitney Aircraft (P & WA) system] space missions. [Skylab uses 2 Apollo type fuel cell units (1)(2)(11).]

The General Electric system and the Pratt and Whitney Aircraft system were of different designs (1)(2)(5) but both used hydrogen and oxygen stored cryogenically. The Gemini system utilized two fuel cell units each capable of delivering 1 kW, whereas the Apollo system used three units capable of 2 kW each; both systems were designed for 14 day continuous operation. These fuel cells provide higher specific energy than batteries and higher efficiency than engine-generators. An additional mission weight savings is obtained in that the reaction product is water which is used for the astronauts' drinking supply. Unfortunately, the fuel cells developed for space are not directly adaptable to terrestrial commercial applications because of their specialized fuels, high intrinsic cost, and limited life. An important difference between the space package and commercial fuel cells is that the space power plant needs no reformer, a component which processes fossil fuels into hydrogen (12). Pure hydrogen and oxygen reactants are supplied from the space craft's cryogenic fuel supplies.

Naval Propulsion

Fuel cells have application to powering submersibles where low volume and high-power density are required (5)(6)(11)(15). An example of a fuel cell developed for this type of application is Alsthom's hydrazine/hydrogen peroxide cell which utilizes a modular construction and achieves power levels of up to 30 kW (11). Fuel cells are also of interest for surface vessels. Nuclear power is economical only in very large ships. In small vessels fuel cells may offer cost advantages over steam-turbine-powered systems.

Electric Vehicles

Batteries, fuel cells, and combinations of both [hybrid systems (6)] have been used to electrically power vehicles. Fuel cells have been utilized in off-road vehicles (fork lifts, agricultural and construction machinery) and in short-range delivery vehicles (1)(2). The extension to private automobiles is a serious challenge but appears to be possible.

The most practical systems will probably be hybrid power plants in which batteries provide peak power and the fuel cell acts as a charging unit during low-power periods. Several vehicular fuel cell and fuel cell/battery power systems have been demonstrated for military (6) and special purpose vehicles (1). For passenger cars, a fuel cell powered van and a fuel cell/battery powered car have been tested. Although these vehicles are far from general use, it appears that they may provide practical, low-pollution vehicles with useful performance and range.

Communication Systems

Special fuel cell systems have been designed for powering remote relay stations and for advanced military communication systems (2)(4)(6)(15). Monsanto has developed a hydrazine-air fuel cell system for the Army which provides 60 w, weighs 12 lb, and can operate for 12 hr on 1 lb of fuel (2). This unit is almost completely silent in operation and can be recharged quietly. This system is probably not suitable for commercial use because of high costs and the toxicity of hydrazine. However, it does appear that with further development other fuel cells may find increasing application for powering communication systems especially for remote area stations requiring unattended operation. This is discussed further in the next section(s). A discussion of the overall U.S. Army fuel cell development program is given in References (6), (8) and (15).

Commercial Systems

Economic considerations and fuel availability are the dominant factors influencing commercial applications (11). The advantages of the fuel cell over the gas-turbine or piston engine are low pollution, quietness, high efficiency, and reliability (2). Although these factors could swing the balance in favor of fuel cells, the relatively high development costs involved means that substantial markets must be sought. Discussions of commercial systems, applications, and R & D programs are given in following sections.

Other Applications

Special fuel cells are being considered for artificial heart power supplies (1)(5)(8). For this application, they must be small and light weight, must be made from blood- and tissue-compatible materials, must operate on reactants available within the body, and must not give off significant quantities of heat or undesirable waste products. Tests have proven the concept feasible, but there are still problems with low power levels and by-product contamination. Some of these problems can be overcome. Fuel cells are also being considered for other biomedical applications (1).

The previous examples have outlined some uses of fuel cells in power-conversion schemes. There are other applications to which this technology can be applied. Specifically, improvements in ion-exchange and electrocatalysis technologies could be used in purification and separation of liquids and gases (1). This would be useful for oxygen generators and concentrators in hospitals and in aircraft, for dialysis membranes in artificial kidney machines, and for environmental control systems (air cleaners, dehumidifiers, water purifiers, and possibly as desalination plants). The important point to note is that fuel cells are being developed for a number of diverse applications.

FUEL CELL POWER PLANTS

General purpose commercial fuel cell power plants are being considered for homes, commercial and industrial installations, power distribution sites, central generating plants, and peak power storage. Some of the characteristics and possible applications of these systems are discussed in this section.

System Characteristics

The commercial fuel cell power plant generates electricity from fossil fuels through three major subsystems (see Figure 2.6) (1)(3)(12):

[1]　Reformer, which processes fossil fuels such as natural gas, kerosene, diesel fuel, and coal into a more reactive form—a gaseous mixture of hydrogen with some carbon dioxide and/or carbon monoxide. Fossil fuels are processed through steam reforming, where fuel and steam are catalytically reacted with the addition of heat to produce a hydrogen-rich gas. This processed gas flows to the fuel cell where most of it is consumed in the electrochemical reaction. In gas and liquid hydrocarbon systems a small amount ($\sim$ 6%) is circulated back to the reformer and burned to sustain the reaction. (In high temperature fuel cells the excess heat is used to sustain the reaction.) Most of what little pollution (thermal, gaseous, and particulate) that the power plant emits generally takes place at the reformer.

[2]   Fuel cell, which reacts the processed fuel with oxygen
        from air to produce dc power and water, a by-product.
        The fuel cell power section is made up of single cells
        assembled into a stack (or cell banks) to produce the
        desired voltage and power. A single fuel cell generates
        100 to 200 W of electricity at approximately 1 V dc
        for each square foot of electrolyte cross section.
        Stacking of cells permits power levels from 1 kW to
        the multimegawatt range.

[3]   Inverter, which converts the dc power into useful ac
        power, and a transformer to step up voltage to power-
        line voltage.

Removal of both the heat and the by-product water generated in the power
plant is accomplished with a radiator, which rejects the heat to the atmosphere
and condenses the water for reuse in the fuel processor. The power plant is air-
cooled. Its only moving parts are fans used for the air supply and for cooling,
providing an almost noiseless operation.

FIGURE 2.6: MAJOR SUBSYSTEMS FOR (NATURAL GAS) FUEL CELL POWER PLANT

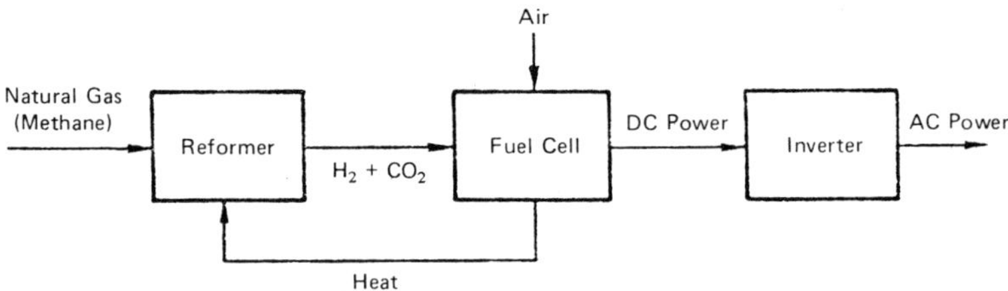

Source: Reference (1)

Applications

Fuel cell power plants have shown an inherent adaptability toward a broad range
of applications. The direct nature of the electrochemical process minimizes the
effects of both scale and load level on operating characteristics. The power
plant's modular construction and the ease with which units can be linked permits matching of capacity to demand. Increased capacity can be added quickly
and simply. Because the power plant is air-cooled, it is not dependent on a
source of cooling water and thus has few location restrictions (12). This is an
important consideration for many New Mexico applications.

For individual homes, apartments, commercial buildings and industrial sites, independent on-site power plants for utility purposes could meet power demands
from 10 to several hundred kW. They can be integrated into total energy pack-
ages that not only generate electricity, but provide control of temperature,
humidity and air cleanliness.

Other applications include the use of fuel cells as generators throughout the existing electric utility system. Operating on low-cost liquid fuels, power plants in the 10 to 50 MW class could augment central stations to cover peak loads. They could also decentralize the generation of electricity and trim the expense of transmission over networks by serving as substations. Some of the special features of coal-gasification plants which have been proposed for central power generating stations are discussed under Major Commercial Research and Development Activities and in References (6), (8) and (14).

The extent to which these applications may be realized, based on economic factors, etc., is discussed in later sections.

Future Systems

Besides being used as energy conversion devices, fuel cells can also be used as energy storage systems (12). Interesting applications here include nuclear power plants and solar-powered communities. Current nuclear technology indicates that breeder reactors must operate continuously at full power to achieve maximum efficiency (12). This would favor using breeder reactors to supply base-load requirements, augmented by supplemental and peaking non-nuclear (e.g., fuel cell) plants.

The excess electricity generated by nuclear power plants during periods of low demand could be used to reactivate regenerative fuel cells or could be used to power electrolysis units, which separate water into its elemental components of hydrogen and oxygen. These gases subsequently would be used to fuel the cells when power is needed. A possible limiting factor here is the relatively large amount of electricity presently required by electrolysis; however, the high efficiency and pollutionless character of fuel cells is a plus factor in such applications. Many of the same arguments and applications can be advanced for energy storage systems in relation to solar-powered communities.

One of the other energy schemes being considered for the future is that associated with a "hydrogen economy" (18)(19). In this case, all of the major energy sources would be used to produce hydrogen, which would then be distributed as a nonpolluting multipurpose fuel.

In such an economy, the hydrogen-air fuel cell would become an extremely important electrical generation device (19). Here, generating units would be dispersed to the immediate vicinity of using facilities. Individual fuel cells would be located in homes or neighborhoods. For an individual residence one could envision such devices mounted on the roof using the hydrogen which is piped to the residence. The technology for these fuel cell devices exists. The future of the hydrogen economy is more dependent upon the economics of producing hydrogen and the storage and transmission of hydrogen than it is upon the generation of electricity.

ECONOMICS

Successful large-scale use of fuel cells will depend more upon resolving economic (and fuel availability) problems than technical ones. The economics of these power generating systems are not firmly established, and cost figures are hard to obtain (11). One set of figures is available in Reference (1) for aerospace, military, and commercial systems in the 20 to 200 kW range. Other figures which compare fuel cell power plants to large turbines for large-scale power plants are available in Reference (12).

The most expensive fuel cell systems are the aerospace fuel cells which range from $100,000 to $400,000/kW. The military systems in 1974 cost about $30,000/kW, with a figure of $10,000/kW as a feasible future goal. These high costs are partly due to the limited quantities produced and the requirement for lightweight, compact units. For stationary hydrocarbon fuel cell power systems in the 10 to 15 kW class, the goal is for costs ranging from $100 to $300/kW. [Indications are that these costs range from $800 to $1,000/kW (20).] Some estimates even indicate that these costs should be about $10 to $30/kW to be competitive for commercial applications in the 20 to 200 kW range (1). A reduction of this magnitude will be very hard to achieve.

The most expensive elements in fuel cell systems are generally the catalysts and the electrodes [and also possibly the power regulator system (20)]. In order to achieve a lower cost, improvements are needed in all of these areas. The development of more stable materials able to withstand the electrochemical environment is also required to extend the operating period, which also affects cost.

Cost estimates for large-scale fuel cell power plants have been quoted in *Aviation Week and Space Technology* (12):

> The present market cost of the fuel cell power plant manufactured on a production basis is about $350 per kW, and the economic operating period is approximately 16,000 h. Pratt and Whitney estimate a specific cost of $200 per kW and a 40,000 h operating period would establish the fuel cell power plant as an economic generator for a variety of duties. In comparison, large (gas) turbines for central utility stations cost about $150 per kW and operate for 100,000 h between major overhauls.

From these estimates, it appears that substantial progress has been made in the past several years in commercial hydrocarbon systems, with specific material costs reduced nearly ten- to twenty-fold and operating life increased by a factor of five (3). It has been estimated that the break-even point for commercial systems may be reached within a reasonable time.

Although these cost estimates and projections are optimistic, it must be remembered that the economics for these systems has been (and still is) one of the biggest problems. Being realistic, one must also remember that much of the detailed information on costs is of a proprietary nature with the developing companies, and until all of the cost analysis figures are available, one must view final figures and cost estimates with caution.

There is some indication that some cost estimates on commercial systems may be off by as much as a factor of three (20). Furthermore, crude estimates indicate that, based on the present economics and on costs of $800 to $1,000/kW, cost reductions of 30 to 100 times are needed to make these units attractive for gas utility applications in individual homes, and reductions of 4 to 7 times are needed for large-scale electric utility use.

On the positive side, however, it seems clear that (in the future) as other systems and fuels become more expensive, fuel cells which are more efficient (and clean) will become much more attractive from an economic point of view for some applications.

MAJOR COMMERCIAL RESEARCH AND DEVELOPMENT ACTIVITIES

Many companies were involved in fuel cell research in the 1960s since it appeared that these devices might solve many social problems (1)(2)(4)(5)(8)(11) (15). But fuel cell development was plagued with many problems which had to be solved before the full potential could be assessed. The early enthusiasm turned to cautious optimism in the 1970s (1). The period in the late 1960s saw a drop in funding for fuel cell development. Government research funds which were granted almost exclusively for space and military applications dropped from nearly $16 million in 1963 to about $3 million in 1970 (3).

Unwilling or unable to assume the substantial investment required for commercialization, companies that had so eagerly rushed into fuel cell development quietly abandoned their research programs or reduced them to token operations. The Pratt & Whitney Aircraft Division of United Technologies Corporation, (P&WA), East Hartford, Connecticut is one of the few companies pursuing a full-scale commercial fuel cell program. About 43 United States and three foreign utilities and Pratt & Whitney Aircraft have invested more than $50 million to prove the technical feasibility of commercial fuel cells. They will probably invest twice that amount during the next 3 years in an attempt to demonstrate the economic viability of fuel cells (3).

In this section, there is described the Pratt & Whitney program(s) pursued in conjunction with the gas utilities (TARGET program) and the electric utilities (1)(3)(6)(11)(12)(13), as well as the Esso-Alsthom program in France (11), and the Westinghouse program in the U.S. (3)(8). There are probably other programs of major interest now being pursued (5) (e.g., by automobile companies, etc.) which are not discussed here. Rather than attempt a survey of all current R&D activities, an attempt will be made to give an indication of the diversity of the activities and to outline some of the programs.

P&WA/Gas Utilities

A major fuel cell commercial development project was undertaken in 1967 called TARGET (Team to Advance Research for Gas Energy Transformation) (1)(3)(6) (11)(12)(13). The TARGET program is a joint development by 32 gas utilities (including Southern Union Gas Company) in conjunction with P&WA (17)(21).

This program was in its third phase in 1974 after more than $50 million of initial research and field testing. The original motivation was for the gas utilities to be able to compete with the electric utilities in the total energy market. The electric utilities utilize electric energy for heating, cooking, as well as lighting, etc. The gas utilities sought to develop a practical fuel cell system which could also offer a total-energy alternative.

This program first led to the development in about 1970 of a power plant denoted as PC11. These power plants have been used in extensive field testing in a variety of environments and locations (60 units in 37 locations in 19 states and the District of Columbia) including residential homes, commercial buildings and industrial plants.

Information obtained from this test program led to the design of an improved (12.5 kW) power plant, PC16 (11)(12). This unit is considered by P&WA to be the first genuine commercial fuel cell power plant, and is expected to compete with the diesel generator for remote site generation and portable or emergency power. The ultimate goal for fuel cells is a total energy and environmental conditioning system (1)(21). That is, such systems should provide not only electrical energy but should also provide heat and air conditioning, should humidify and purify the air, and should also be used as a waste processor (3).

One obvious concern is the uncertain future supply of gas for natural-gas powered cells. Although the gas supply may be limited, the inherently greater efficiency of the fuel cells should make them the preferred outlet for many of the natural gas applications (13). [Some estimates indicate that fuel cell power plants would produce about 20 to 25% more useful energy per unit of fuel (3).] A greater efficiency advantage holds regardless of the fuel since the fuel cell can utilize hydrogen derived from any fossil fuel and even from nuclear or solar processes.

P&WA/Electric Utilities

Electric utilities are now interested in extending the fuel cell power range of interest from tens of kilowatts up to the multimegawatt range (11)(17). These utilities are becoming increasingly concerned with air pollution, heat rejection to water, land use, aesthetics, radioactive waste disposal, power plant siting, and transmission losses and rights-of-way. Some of these concerns could be alleviated by utilizing fuel cell systems in dispersed power generation networks (11) (12). In this way, pollution sources would be both minimized and dispersed, transmission line requirements lessened, existing substation and generating sites utilized, area protection provided and central power plant reserve requirements reduced.

In other words, the use of fuel cell power plants as dispersed generators would provide the electric utilities with an environmentally acceptable system. Furthermore, the modular construction of fuel cell systems would result in shorter construction times. The electric utilities would principally use these plants to augment central power stations to cover peak loads, and would use them in rapidly growing areas and in areas where siting or transmission is a problem for conventional plants.

A program was undertaken by P&WA to develop a 26 MW fuel cell system (designated the FCG-1 Fuel Cell Power Plant) to generate electrical power (13) (17). The first phase of this program was financed by nine electric power utilities ($28 million) and by P&WA ($14 million). The utilities were also to contribute $7 million for a demonstration unit. This project was expected to produce $250 million in initial fuel cell deliveries beginning in 1978. There were orders for about 56 of these power plants.

These fuel cell plants would operate on fossil fuels such as natural gas, distillate fuel oil, or gas from coal. A major selling point for these fuel cell systems is size and flexibility. The P&WA 26 MW unit (enough power for 20,000 people) would be about 18' high and cover less than half an acre of ground (13). The modular construction would allow use of individual or multiple units in almost any location.

Although the goals and applications of fuel cells by the electric utility and the TARGET companies are different, they rely on the same related technologic progress.

Esso-Alsthom

Alsthom and Esso combined in a joint development program with funding of about $10 million over a 5 year period (1)(11). Alsthom had previously developed the hydrazine/hydrogen peroxide fuel cell for submersibles. The aim of this new development program was to exploit the cell design and engineering concepts previously developed. The initial objective of this program was to provide a practical fuel cell system which would compete with engine generators and batteries in remote power applications. Their program also looked toward the possibility of fuel cells for automotive propulsion.

Alsthom chose to develop a methanol-air cell (5)(11) rather than utilizing hydrocarbon fuels as in the TARGET programs. Although methanol is more expensive than hydrocarbons, it can be derived from coal. Methanol also has a lower specific energy than hydrocarbons but it can be used directly in a fuel cell without having to use a fuel reformer, as is necessary with present hydrocarbon fuels. Much of the effort in this program was aimed at developing new and improved (cheap) catalysts (1).

Westinghouse

Central-station electrical generation using conventional gas and liquid hydrocarbon fuel cells appears to be impractical because such cells are no more efficient than the best large turbines. Westinghouse Electric Corporation, however, investigated high-temperature (1000°C), solid-electrolyte fuel cells (5)(8)(10)(16) to be used in conjunction with coal-gasification plants. These systems could have thermal efficiencies greater than 60% (8)(16) which makes them of interest for central generating stations. Westinghouse has worked on these systems in conjunction with the Office of Coal Research, Department of the Interior (3)(6)(8)(16).

A schematic diagram of the Westinghouse coal-gasification fuel cell power system (8)(16) is given in Figure 2.7. This is a complete power system in that it includes a means for gasifying the coal as well as a means for producing electrical power. The heat released in the fuel cell is used to supply the heat required for the gasification process. For efficient heat transfer the fuel cells are immersed within the reactor. In operation, coal is fed to a fluidized reactor where it reacts with a partially oxidized fuel stream coming from one bank of fuel cells. The resulting fuel gas, after a cleaning process, is rich in H_2 and CO. This fuel gas is fed to the fuel cells. The second bank of fuel cells is used to completely oxidize the fuel.

In the fuel cells, each individual cell is a thin-film solid-state device which uses a zirconium oxide electrolyte. Oxygen ions, obtained from the air electrode, are transported through the solid electrolyte and react with the fuel. The system operates at 1000°C in order to enhance the transport of oxygen ions through the electrolyte, to provide sufficient heat transfer for the coal-gasification reaction, and to enhance the rate of this reaction.

By utilizing coal-gasification fuel cell power systems, nearly one and one-half times more electrical energy can be extracted from coal than can be extracted using a conventional generating station. There is also a decrease in heat waste since the excess heat is used in the reformer (gasification reactor), and pollution of the environment is minimal. These power systems do not require plentiful water supplies since cooling water is not needed. These power systems are also very compact (a one megawatt generating capacity would fit in a drum that has the dimensions of seven feet in diameter and six feet in height) (16).

FIGURE 2.7: COAL-GASIFICATION FUEL CELL POWER PLANT

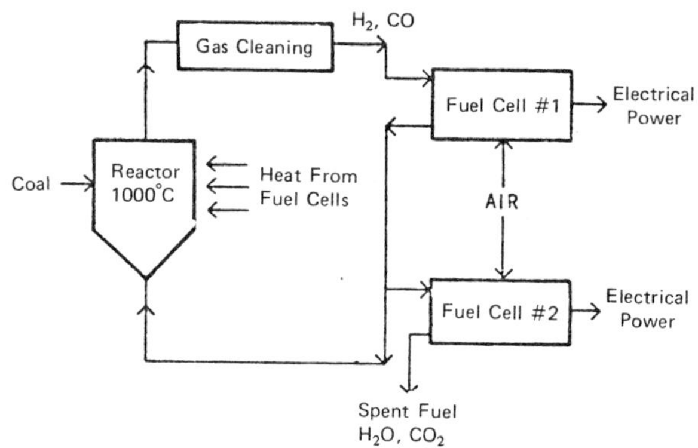

Source: References (8)(16)

The technical feasibility for such coal-gasification fuel cell power plants has been demonstrated (8)(16). A program for establishing the commercial feasibility of this new power system is underway. However, it appears that it will be several years before 100 kW prototype power plants will be available (3), and thus, it will be quite some time before such systems are fully evaluated and commercially available.

STATUS

The technology for fuel cell power systems exists. The principal barriers to the utilization of fuel cells as power sources relate to their relatively high cost and their relatively short lifetime (1). These are both serious problems and unfortunately it is difficult to obtain a thorough, accurate, and objective assessment of these problems. In order to increase lifetime, which would also affect cost, problems such as materials degradation in acid and high-temperature environments must be overcome. It is hoped, of course, that additional development will lead to reliable, trouble-free (and thus maintenance-free) systems.

In considering the general areas where improvement in fuel cells is needed, the three most prominent are (6): (a) electrocatalysts, primarily as they affect cost and performance, (b) electrode materials, as they affect lifetime, costs and weight, and (c) engineering, as it affects endurance, weight and volume. None of the problems is insoluble; all will take further research and development, which also translates into time and money. There is likely to be a gradual opening-up of the field, with wider and more common uses of fuel cells, as they become more economically and environmentally attractive.

The availability of suitable fuel at a reasonable cost is an important factor which must be considered for widespread future utilization of fuel cell power generation systems. This can impose severe limitations on the applications for fuel cells. With the present technology and fuels, it seems clear that these systems will not be a major (base line) source for electrical power. However, they appear to be ideally suited for substations and on-site generation for providing peak load requirements, standby power, and remote site electrical power generation.

Fortunately, for some possible future applications the fuel cell is not tied solely to the dwindling supply of fossil fuels (12). All the fuel cell requires is air and a source of hydrogen, one of the most abundant elements. Although hydrogen gas is a very desirable fuel, it is not clear that a sufficient quantity can be produced economically in order to make a significant impact upon the energy crisis. Coal, on the other hand, is readily available and reserves are such that it should be available for several hundred years.

Fossil fuels are likely to remain the dominant source of energy for the rest of this century, but hydrogen from water is receiving considerable mention as the fuel of the future. If the cost of electrolysis can be reduced or an economical method to crack water by another means can be devised (e.g., by thermo-chemical decomposition), fuel cell operations will be even more economical.

Near-Future Systems

The use of fuel cell power systems will become more widespread in the future, but the growth will be gradual. The advantages for many applications over other power systems (high efficiency, high reliability, clean exhaust, silent operation, modular construction, and others) almost ensure a continued effort to develop these energy sources. Speculations on future trends for some specific applications are as follows (2)(5)(15):

[1]　The space program will continue to use fuel cells for missions requiring about 2 to 20 days. For longer missions solar and nuclear generation will be used (possibly in conjunction with regenerative fuel cells).

[2]　The future of the total-energy fuel cell systems developed for supplying the total energy (utility) requirements for single family homes, apartments, commercial buildings, and industrial sites seems uncertain. The intent was for these systems to provide heating and cooling as well as electrical generation and environmental conditioning. Unfortunately, economics and fuel availability may limit these systems to only special applications, such as to remote sites where other power sources may not be available or are equally expensive. The technology for these systems has been largely developed through the TARGET program. Depending upon need, commercial systems could probably be available within several years.

[3]　A principal focus for commercial fuel cell power plants will be for units in the hundreds of kilowatts to multi-

megawatt range which would be used as dispersed generators throughout the electric utility network. These units would augment large central fossil fueled and nuclear power plants to cover peak loads and to provide auxiliary power in rapidly developing areas (or at remote sites). These systems are under development and are scheduled for delivery beginning in 1978. The competitiveness of these systems should be determined by 1979 to 1980 (17), and widespread utilization should be possible by the early 1980s.

[4] Commercial communication systems which require reliable uninterrupted power and are located in remote sites will eventually be powered by fuel cell systems in the 200 W to 10 kW range. Such systems will also be useful for providing emergency or stand-by power for many applications. These systems will be based on much of the technology developed for the total-energy fuel cell systems (and also on the technology developed in the Esso-Alsthom program). The extent to which these systems will be used will depend principally upon economics.

[5] Advanced communication systems for the military will require silent power. Fuel cells in the 300 W to 10 kW range will be developed for this application. Some special purpose systems have already been developed.

[6] Special purpose low power (10 W to 1 kW) hydrogen-air and ammonia-air fuel cells will be used for a variety of applications including marine electronic equipment, off-shore navigational aids, mobile power systems, recreation, etc. Several of these units are now marketed by Engelhard Industries Division (Engelhard Minerals and Chemicals Corporation, East Newark, New Jersey).

[7] Regenerative fuel cells which store their reactants external to the electrodes will eventually be developed as energy storage devices. They will be used in conjunction with solar and nuclear electrical generating systems to handle peak loading. The time scale for these systems appears to be far into the future.

[8] Coal-gasification fuel cell power plants also appear to be well into the future. It has been estimated that it will be some time before prototype units are available.

[9] Hybrid fuel cell/battery systems probably will be developed for electrically powered vehicles. They will be first applied to larger vehicles such as buses and military vehicles before they appear in the family automobile. The advantages of quietness and clean exhaust will stimulate this development. However, the extent of the possible impact on transportation systems in the near future seems very questionable.

The potential for applications is enormous, but the extent to which fuel cells will be used in the future will be determined principally by economics and the availability of appropriate fuels. Research and development should be aimed at reducing costs and increasing lifetime, and should be directed toward those applications for which fuel cells are uniquely suited or where they provide distinct advantages over other methods of electrical power generation (especially with regard to the efficient utilization of fuels).

CONCLUSIONS

Fuel cells have been proven as practical power sources in certain specific applications, such as space missions and remote site operation. The development of fuel cells for widespread commercial applications is now underway. The technical feasibility of commercial power plants has been largely proven (11), but much work needs yet to be done in lowering costs and improving lifetime.

The utilization of fuel cells for practical power sources will grow in the future. Recent moves toward multimegawatt power generation systems is an important example of this growth. Future growth will be increasingly influenced by social and strategic factors, such as pollution and fuel costs and availability. Fuel cells should be seriously considered for some of our future power generation requirements.

REFERENCES

(1) B.J. Crowe, *Fuel Cells: A Survey*, NASA SP-5115, (Technology Utilization Office, NASA, Washington, DC, 1973).

(2) D.P. Gregory, *Fuel Cells* (Mills and Boon Limited, London, 1972).

(3) A.L. Hammond, W.D. Metz and T.H. Maugh II, *Energy and the Future* (American Association for the Advancement of Science, Washington, DC, 1973) p. 109: see also *Science* 178, 1274B, 1972.

(4) L.G. Austin, *Scientific American* 201, 72 (1959).

(5) J.O'M. Bockris and S. Srinivasan, *Fuel Cells: Their Electrochemistry* (McGraw-Hill Book Co., New York, 1969) p. 18 to 33, 167, 516 to 627.

(6) *Fuel Cell Systems II* (Edited by R.F. Gould, American Chemical Society Publications, Washington, DC, 1969).

(7) R. Roberts in *The Primary Battery* (Edited by G.W. Heise and N.C. Cahoon, The Electrochemical Society, John Wiley and Sons, Inc., New York, 1971, Vol. 1) p. 293.

(8) *From Electrocatalysis to Fuel Cells* (Edited by G. Sandstede, Battelle Seattle Research Center, University of Washington Press, Seattle, 1972).

(9) J.H. Renken, *Fuel Cells*, Sandia Laboratories Memorandum, September 1973.

(10) R.T. Johnson, Jr. and B. Morosin, *Fuel Cells, Batteries, and Thermoelectric Generator Technology: Solid Electrolytes*, Sandia Laboratories Memorandum, July 1973; and *Solid Electrolyte Technology*, Sandia Laboratories Proposal, 1973.

(11) A.D.S. Tantram, *Energy Policy*, March 1974, p. 55.

(12) C. Martin, *Aviation Week and Space Technology*, p. 56, January 1, 1973.

(13) "Fuel Cell Research Finally Paying Off", *Chemical and Engineering News*, p. 31, January 7, 1974.

(14) J.L. Wirth, Sandia Laboratories Memorandum, July 1973.

(15) *Proceedings of the 25th Power Sources Symposium* (May 23 to 25, 1972, published and distributed by the PSC Publication Committee, P.O. Box 891, Red Bank, N.J. 07701).

(16) *1970 Final Report Project Fuel Cell*, Research and Development Report No. 57; prepared for Office of Coal Research, Department of the Interior; prepared by Westinghouse Electric Corporation (Contract No. 14-01-0001-303, Superintendent of Documents, U.S. Government Printing Office, Washington, DC).

(17) Information provided by Pratt and Whitney Aircraft, East Hartford, Connecticut, and by Southern Union Gas Company, Dallas, Texas.
(18) D.P. Gregory, *Scientific American* 228, 13 (1973).
(19) E. Fein, *A Hydrogen Based Energy Economy*, (prepared for Northeast Utilities Service Company; The Futures Group, 124 Hebron Ave., Glastonbury, Connecticut, 06033, October 1972).
(20) S.C. Levy, Sandia Laboratories, private communication.

Acknowledgments

Helpful discussions were conducted with and reference material was provided by B. Morosin, J.H. Renken, D.M. Haaland, S.C. Levy, B.H. Van Domelen and R.S. Claassen of Sandia Laboratories; S.D. Neblett of Southern Union Gas Company; and L.M. Weaver of Public Service Company of New Mexico.

ASSESSMENT OF FUELS
FOR POWER GENERATION
BY ELECTRIC UTILITY FUEL CELLS

In October 1975, R.P. Stickles, E. Interess, G.C. Sweeney, P.E. Mawn and J.M. Parry of Arthur D. Little Inc. prepared a report for the Electric Power Research Institute assessing various fuels for utility-sized fuel cells. Most of the report relates to cost analysis, however since this book is primarily concerned with technology, only Sections 3 and 4 of the report are editorialized here. The complete report can be ordered from NTIS (PB 247 216).

CONVERSION TECHNOLOGY

SUMMARY

A broad list of potential technologies for central and on-site processing of raw fuel sources to fuels for fuel cells were screened. After a final screening three central conversion alternatives, namely, coal gasification with oxygen, partial oxidation of crude oil, and solid waste gasification, and six on-site conversion alternatives including steam reforming of natural gas, SNG, naphtha, low sulfur distillate, or methanol, and partial oxidation of distillate fuel oil were retained for further consideration.

Representative economics for central fuel conversion plants producing syngas are summarized in Table 3.1. Product gas costs range from $2.25 to $3.65 per MM Btu. The economics for the integrated on-site and central base load systems are developed later.

INITIAL PROCESS SCREENING

This study was begun with a broad list of energy sources shown in Table 3.2 for both central and on-site processing. Initially there was an evaluation of the conversion technology associated with these energy sources for applicability to furnishing fuel to fuel cells.

TABLE 3.1: SUMMARY OF ECONOMICS FOR CENTRAL FUEL CONVERSION TO SYNTHESIS GAS (COSTS IN 1975 DOLLARS)

Process	Capital Investment ($/kw)	Fuel Cost ($/MM Btu)[1]	Capital Charge ($/MM Btu)[1]	Operating & Maintenance ($/MM Btu)[1]	Product Gas Cost ($/MM Btu)
Lurgi[2]	275	2.02	1.05	0.26	3.33
Koppers-Totzek[2]	285	1.91	1.03	0.72	3.66
Texaco P.O.[2]	163	2.07	0.64	0.56	3.27
Agglomerating Fluid Bed[2]	172	1.84	0.61	0.28	2.73
Entrained Bed[2]	141	1.64	0.44	0.25	2.33
Solid Waste[3]	414	(0.72)	1.53	1.44	2.25
Asphalt Partial Ox.[4]	131	2.76	0.51	.26	3.53

[1]$/MM Btu of product gas.
[2]Based on 81.5×10^6 MM Btu/yr coal feed.
[3]Based on 10×10^6 MM Btu/yr MSW feed.
[4]Based on 24.7×10^6 MM Btu/yr asphalt feed.

TABLE 3.2: EARLY PROCESS SCREENING

Raw Fuel Source	Primary Conversion Process	Status	Rationale for Elimination
Central			
Coal	Gasification	Retained	---
	Liquefaction	Eliminated	No advantage over gasification[1]
	Solvent Refining		
	Pyrolysis		
Crude Oil and Refined Products	Standard Refining	Retained	---
	Steam Reforming		---
	SNG		---
	Partial Oxidation		---
	Flexicoking		---
Nuclear	Electrolysis	Retained	---
	Thermochemical	Eliminated	Beyond time period of study
Oil Shale	To Syncrude	Eliminated	Future development uncertain, syncrude price comparable to imported crude
Tar Sands	To Syncrude	Eliminated	Subject to Canadian export policy
Solid Waste	Gasification	Retained	---
	Liquefaction	Eliminated	No advantage over gasification[1]

(continued)

TABLE 3.2: (continued)

Raw Fuel Source	Primary Conversion Process	Status	Rationale for Elimination
Onsite			
Natural Gas	Reforming	Retained	---
Naphtha	Reforming	↓	---
Distillate	Partial Oxidation		---
Low-Sulfur Distillate	Reforming		---
Methanol	Reforming		---
Nuclear	Electrolysis		---
Cell Hydrogen	---		---

---

[1] This technology might be utilized to supply nonpolluting utility grade fuels, however, the deciding factors would not be associated with fuel cells.

Gasification of coal to produce any of a range of fuel cell fuels via the synthesis gas route is an obviously strong candidate because of the large coal reserves available in the United States and the uncertainty clouding the supply of other fuels. Even though it is a liquid, methanol is obtained via the gasification route and is therefore considered herein.

Unlike direct gasification, coal liquefaction, solvent refining of coal and pyrolysis produce various quantities of gaseous, liquid, and solid coproducts. Generally, the quantity of gases is quite small. The liquid products from liquefaction and solvent refining are as heavy or heavier than natural crude oil; the liquid from pyrolysis, which is also quite heavy, requires hydrotreating.

In all cases the liquid must be upgraded to fuel cell quality. The solid product must be either gasified or otherwise converted for electric utility use. Because of the considerable downstream processing required for the liquid and solid products, liquefaction, solvent refining, and pyrolysis were eliminated from consideration as techniques for captively producing fuel cell fuels.

Despite the political and economic uncertainties affecting the price and availability of crude oil and refined products, these materials will clearly play a significant role in the United States' energy picture for the remainder of this century. Therefore, all aspects of oil processing including a dedicated refinery for the production of a slate of electric utility fuels were retained for further study.

Two methods of producing hydrogen from nuclear electric power were considered. The first, electrolysis of water, was retained pending the findings of an EPRI-sponsored study (RP320-1) on the cost of producing industrial hydrogen from off-peak nuclear power. The second, thermochemical decomposition of water has been dropped. Numerous chemical cycles to utilize the high level heat available from nuclear reactors have been proposed and studied. However, it is not expected that this technology will be available before 1990 or economically attractive before 2000.

Up to 1990, oil shale and tar sands development is likely to proceed modestly; still, synthetic crude oil produced from these resources will be high priced, probably competitive with imported crude oil. Therefore, these materials are seen as augmenting the available oil supply rather than resources to be directly exploited by electric utilities for fuel cell purposes.

Solid waste gasification and liquefaction were retained and eliminated, respectively. The rationale for these decisions is analogous to that for coal. All of the proposed on-site options including steam reforming of naphtha and natural gas, partial oxidation of distillate fuel oil, gasification of methanol, electrolysis of water, and the use of caustic/chlorine cell hydrogen were retained in the initial process screening.

LATER PROCESS SCREENING

A subsequent screening narrowed the list of potential candidates for detailed economic evaluation (See Table 3.3). A preliminary analysis of coal gasification indicated that the use of air as the oxidant in the gasification reactions is impractical for this application.

First, the economics of using air rather than oxygen is virtually a standoff. Larger piping, heat transfer equipment, and compressors practically offset the capital cost of an air separation plant. Not only are the large air compression requirements of the same magnitude as those for producing oxygen, but the product, a hydrogen-carbon monoxide-nitrogen mixture, is of questionable value. It cannot be used economically for synthesis of another secondary fuel cell fuel, and the relatively high transportation and storage costs contribute to its inferiority. Therefore, air-blown coal gasification is eliminated from consideration as a possible central conversion process.

A dedicated refinery for the production of fuel cell fuels with coproduction of other fuels for captive utility use has limited advantages as a new venture business for electric utilities. Several scenarios were studied using a linear program refinery model.

In developing the refinery model representative economics for modular process units were used with the assumption that (1) heat recovery among the units can be optimized in a detailed design, and (2) that the capital investment is linearly proportional to the throughput. The latter assumption can be modified by adjusting the cost per unit capacity to be approximately correct in the expected range of actual throughput.

The processes included in the model were:
- (a)   atmospheric and vacuum distillation,
- (b)   gas liquids separation,
- (c)   desulfurization of heavy and light fractions,
- (d)   hydrocracking,
- (e)   solvent deasphalting,
- (f)   flexicoking,
- (g)   partial oxidation,
- (h)   steam reforming,
- (i)   shift conversion,
- (j)   methanol synthesis, and
- (k)   methanation.

Analysis utilizing this program led to a seemingly obvious conclusion—that generally the wider the slate of refinery products, the better the profitability. For example, in an extreme situation where only two products are produced, naphtha and residual oil for example, there is no outlet for the middle distillates other than to produce naphtha by expensive hydrocracking. If a third middle distillate product is permitted and valued at prevailing premium prices, the profitability improves dramatically.

However, a dedicated refinery predicated on fuel cells will generally be quite small. For example, 20 fuel cells operating at 26 MW for an average of 9 hours per day would require about 7,000 bbl/d of naphtha. For Arabian Light where the natural content of C_5-375°F is about 27%, a total crude run of 26,000 bbl/d could satisfy the entire naphtha need.

This is far below the average size refinery being built today. If only the C_5-300°F cut is taken to naphtha, the total crude run could be increased to about 40,000 barrels per day.

TABLE 3.3: LATER PROCESS SCREENING

Raw Fuel Source	Primary Conversion Process	Status	Rationale for Elimination
Central			
Coal	Gasification with O_2	Retained	---
	Gasification with air	Eliminated	Economics, inferior product
Crude Oil and Refined Products	Standard Refining	Eliminated	Not a fuel cell concept
	Steam Reforming	Eliminated	Inexpensive naphtha transport to FC
	Partial Oxidation	Retained	---
	Flexicoking	Eliminated	LP model chose other routes
Nuclear	Electrolysis	Eliminated	Exposure
Solid Waste	Gasification	Retained	---
Onsite			
Natural Gas	Reforming	Retained	---
Naphtha	Reforming	Retained	---
Distillate	Partial Oxidation	Retained	---
Low-Sulfur Distillate	Reforming	Retained	---
Methanol	Reforming	Retained	---
Nuclear	Electrolysis	Eliminated	Transport and storage

The only practical means to get the refinery capacity into a higher, more economic size range is to find other uses for the excess products through product marketing not familiar to electric utilities. Clearly, fuel cells alone will not provide the impetus for electric utilities to enter the dedicated refining business, since the fuel cell fuel requirements will be a small fraction of the refinery's total output.

In effect, the price of naphtha that is produced by a dedicated refinery is not expected to be significantly different than the price of purchased naphtha. Therefore, this concept has been eliminated as a general approach for meeting fuel cell needs.

Naphtha is an excellent fuel cell fuel because it is easily transported and stored and because the fuel cell integration possibilities provide a relatively good heat rate in the on-site conversion to electric power. Thus, there is no advantage to converting naphtha centrally to produce products such as hydrogen, syngas, methanol, or SNG.

These processes would result in additional capital requirements for the conversion plant, energy losses in conversion, and the costs associated with the transportation and storage of another product. In the case of methanol and SNG, on-site processing would still be required. Consequently, central steam reforming and SNG production from naphtha are dropped from consideration.

Partial oxidation was retained as a proven, economically viable process for producing syngas from a range of petroleum products. Flexicoking was indirectly eliminated from consideration by the refinery linear programming model used in the study of dedicated refineries. Typically, partial oxidation was chosen as a preferable route to producing syngas, as opposed to the flexicoker which is geared toward light hydrocarbon production.

Of the on-site conversion concepts, only electrolysis was eliminated in the second process screening. Central and on-site electrolysis require virtually the same equipment. However, transportation and storage are the determining factors. In the central concept, hydrogen is transported and may be stored centrally and in part by line packing in the transport pipeline. In the on-site approach, electric power is sent to the fuel cell and hydrogen must be stored on-site creating siting problems with respect to safety and land area requirements. Furthermore, assuming an all new transportation link, energy transmission in the form of hydrogen will be less expensive than that for electric power.

It has been indicated that the projected price of electrolytic hydrogen will be in the range of $15/MM Btu by 1985. This cost does not include an allowance for transportation or storage which is significant for this fuel. Consequently, it is concluded that electrolysis of hydrogen is economically unattractive in the time frame of interest.

SECONDARY FUELS

Six fuels have been identified as secondary fuels for feed either directly to a fuel cell or to an on-site fuel conditioner. These are pure hydrogen, impure hydrogen, syngas, SNG, methanol and naphtha. Of these, naphtha would not be produced in a central conversion facility since both solid fuel liquefaction and dedicated refineries have been eliminated. Nor would impure hydrogen be produced centrally because raw gasifier products, which can typically be classified as synthesis gas, would require water-gas shift to meet the impure hydrogen specification. Since 1 mol of carbon dioxide is produced for each mol of hydrogen produced by shifting, the total volume of dry gas product is increased by this step.

If subjected to acid gas removal, the stream would be close to pure hydrogen quality and, if not, would require considerable additional expense for transportation and storage. Therefore, unshifted syngas and pure hydrogen are always lower cost central conversion alternatives than impure hydrogen. In summary, then, there are five secondary fuels to be considered--hydrogen, syngas, methanol, SNG, and naphtha. All but the latter are produced centrally.

GASIFICATION TECHNOLOGY

Central Alternatives

Based on the results of the second screening, several raw fuel conversion processes were evaluated as central alternatives. These included:

- (a) Existing Coal Gasification Technology:

    Lurgi Gasification
    Koppers-Totzek Gasification
    Texaco Partial Oxidation

- (b) Advanced Coal Gasification Technology:

    Agglomerating Bed
    Entrained Bed

- (c) Partial Oxidation of Heavy Oil
- (d) Solid Waste Gasification

The evaluation included preparation of flow diagrams, material balances, and economics for each of the above candidates. In addition the comparative economics of using air or oxygen as an oxidant were determined for the coal gasification and oil partial oxidation options.

On-Site Alternatives

Design criteria limit the number of processes suitable for on-site fuel conversion with dispersed fuel cells because of logistical, operational, and environmental factors. The constraining factors are listed below:

- (a) No visible plumes
- (b) Unattended operation
- (c) Rapid start-up capability (30 minutes)
- (d) Siteable (insurable, meets local codes)
- (e) Nonpolluting
- (f) Load following capability
- (g) Low noise
- (h) No cooling water requirements

Due to the characteristics of the raw fuel, and the complexity of the process, on-site coal and solid waste conversion would have difficulty meeting most of these criteria. These systems also possess unattractive means of raw fuel storage and high capital investment cost per unit of output for the capacities required. Partial oxidation of heavy oil fuels, as commercially practiced, is equally difficult for this application. However, certain modifications of this process incorporating a catalyst, such as are being developed by Jet Propulsion Laboratories, may have potential for application to fuel cells.

In general, the fuel conversion process which comes closest to meeting the above design criteria for on-site processing is steam reforming of low-sulfur fuel such as SNG, methanol, and low-boiling petroleum fractions. In fact, United Technologies Corporation, under their TARGET progam, field tested a 12.5 kW power plant based on converting natural gas to hydrogen-rich fuel using steam reforming.

CONCEPTUAL FUEL SUPPLY SYSTEMS

SUMMARY

In this section the cost of producing fuel cell fuels at central conversion facilities for subsequent transportation to on-site fuel cells is discussed. Further, there is a discussion of the cost of integrating fuel processors with fuel cells to improve the overall system heat rate by using fuel cell waste energy and water, where appropriate, within the processing unit. Included in the integrated systems are both on-site processors for dispersed, peak to intermediate power generation and coal gasification for larger central, base load generation. Fuel costs as produced at a central processor are summarized in Table 3.4 through Table 3.7.

The advantages of integrating the fuel cell with the fuel processor are illustrated in the Table 3.8 summary of the more promising on-site and central converter-fuel-cell combinations. In general, integration significantly enhances the overall thermal efficiency because of the need to continually vent the fuel, thereby wasting energy in the form of combustible gases. If the gases can be utilized effectively as fuel in the processor, the overall efficiency of the processor-fuel-cell system will improve. This generalization, of course, does not pertain to systems requiring no processor such as those taking hydrogen or syngas as feed. For each of the six secondary fuels there is a near optimum fuel cell relative to minimizing the overall heat rate. These are as follows:

Fuel	Fuel Cell	Fuel to ac Power Heat Rate (Btu/kWh)
Hydrogen	Alkaline	6,500
Impure hydrogen	Acid	7,750
Syngas	Molten carbonate	8,200
Methanol	Acid	6,500
SNG	Molten carbonate or acid	7,500
Naphtha or low sulfur distillate	Molten carbonate or acid	7,500

TABLE 3.4: FUEL COSTS EXIT CENTRAL FUEL CONVERSION PLANT—HARTFORD
[$/MM Btu (1975 DOLLARS)]

Product	Syngas			H_2			Methanol			SNG		
	1980	1985	1990	1980	1985	1990	1980	1985	1990	1980	1985	1990
100 Mwe Dedicated Plant[1]												
Coal – Commercial	4.99	5.03	5.17	5.79	5.39	5.99	7.25	7.37	7.49	6.62	6.74	6.85
Coal – Advanced	–	3.22	3.30	–	4.87	4.97	–	5.57	5.68	–	4.30	4.40
Solid Waste	2.09	2.09	2.09	2.75	2.75	2.75	3.30	3.30	3.30	2.96	2.96	2.96
Partial Ox–Residual	3.90	3.81	3.93	4.80	4.71	4.84	5.32	5.21	5.36	5.40	5.29	5.45
Partial Ox–Vacuum Residual	3.24	3.16	3.26	4.10	4.02	4.13	4.52	4.42	4.55	4.57	4.47	4.60
1300 Mwe Multipurpose Gasifier[2]												
Coal – Advanced	–	–	–	–	–	–	–	5.31	5.42	–	–	–
1300 Mwe Central Station Fuel Cell[3]												
Coal – Advanced	–	2.63	2.70	–	–	–	–	–	–	–	–	–

[1]Product shown is dedicated solely to dispersed fuel cell use.
[2]100 Mwe of gasifier raw gas extracted as methanol for dispersed fuel cell use only.
[3]1300 Mwe fuel cell integrated with gasifier in base load central station.

TABLE 3.5: FUEL COSTS EXIT CENTRAL FUEL CONVERSION PLANT–DALLAS
[$/MM Btu (1975 DOLLARS)]

Product	Syngas			H_2			Methanol			SNG		
	1980	1985	1990	1980	1985	1990	1980	1985	1990	1980	1985	1990
100 Mwe Dedicated Plant[1]												
Coal – Commercial	5.14	4.51	5.52	–	5.24	–	–	–	–	4.32	5.35	4.65
Coal – Advanced	–	2.71	2.84	–	4.20	–	–	4.82	5.26	–	3.70	3.85
Solid Waste	2.35	2.35	2.35	3.05	3.05	3.05	3.64	3.64	3.64	3.29	3.29	3.29
Partial Ox–Residual	3.74	3.68	3.82	4.64	4.57	4.72	5.13	5.05	5.22	5.21	5.13	5.30
Partial Ox–Vacuum	3.48	3.47	3.67	4.36	4.35	4.56	4.82	4.80	5.04	4.88	4.86	5.11
Residual												
1300 Mwe Multipurpose Gasifier[2]												
Coal – Advanced	–	–	–	–	–	–	–	4.56	5.00	–	–	–
1300 Mwe Central Station Fuel Cell[3]												
Coal – Advanced	–	2.20	2.30	–	–	–	–	–	–	–	–	–

[1] Product shown is dedicated solely to dispersed fuel cell use.
[2] 100 Mwe of gasifier raw gas extracted as methanol for dispersed fuel cell use only.
[3] 1300 Mwe fuel cell integrated with gasifier in base load central station.

TABLE 3.6: FUEL COSTS EXIT CENTRAL FUEL CONVERSION PLANT—COLUMBUS
[$/MM Btu (1975 DOLLARS)]

Product	Syngas			H_2			Methanol			SNG		
	1980	1985	1990	1980	1985	1990	1980	1985	1990	1980	1985	1990
100 Mwe Dedicated Plant[1]												
Coal – Commercial	4.19	4.22	4.23	4.91	4.94	4.95	6.18	6.22	6.23	5.60	5.64	5.65
Coal – Advanced	–	2.48	2.50	–	3.89	3.91	–	4.47	4.49	–	3.43	3.44
Solid Waste	2.35	2.35	2.35	3.05	3.05	3.05	3.64	3.64	3.64	3.29	3.29	3.29
Partial Ox-Residual	3.68	3.55	3.59	4.57	4.43	4.47	5.05	4.89	4.94	5.13	4.96	5.01
Partial Ox-Vacuum Residual	3.17	3.14	3.24	4.03	3.99	4.10	4.44	4.39	4.52	4.49	4.44	4.57
1300 Mwe Multipurpose Gasifier[2]												
Coal – Advanced	–	–	–	–	–	–	–	4.21	4.23	–	–	–
1300 Mwe Central Station Fuel Cell[3]												
Coal – Advanced	–	2.00	2.00	–	–	–	–	–	–	–	–	–

---

[1] Product shown is dedicated solely to dispersed fuel cell use.
[2] 100 Mwe of gasifier raw gas extracted as methanol for dispersed fuel cell use only.
[3] 1300 Mwe fuel cell integrated with gasifier in base load central station.

TABLE 3.7: FUEL COSTS EXIT CENTRAL FUEL CONVERSION PLANT—LOS ANGELES
[$/MM Btu (1975 DOLLARS)]

Product	Syngas			H_2			Methanol			SNG		
	1980	1985	1990	1980	1985	1990	1980	1985	1990	1980	1985	1990
100 Mwe Dedicated Plant[1]												
Coal – Commercial	4.92	4.29	5.24	4.84	5.01	5.14	6.10	6.30	6.46	4.13	5.11	4.41
Coal – Advanced	–	2.54	2.64	–	3.97	4.10	–	4.55	4.70	–	3.49	3.61
Solid Waste	2.88	2.88	2.88	3.64	3.64	3.64	4.32	4.32	4.32	3.96	3.96	3.96
Partial Ox-Residual	3.59	3.41	3.47	4.47	4.28	4.35	4.94	4.72	4.80	5.01	4.79	4.87
Partial Ox-Vacuum Residual	3.04	2.85	2.93	3.89	3.68	3.77	4.28	4.04	4.14	4.32	4.08	4.18
1300 Mwe Multipurpose Gasifier[2]												
Coal – Advanced	–	–	–	–	–	–	–	4.29	4.44	–	–	–
1300 Mwe Central Station Fuel Cell[3]												
Coal – Advanced	–	2.04	2.13	–	–	–	–	–	–	–	–	–

[1] Product shown is dedicated solely to dispersed fuel cell use.
[2] 100 Mwe of gasifier raw gas extracted as methanol for dispersed fuel cell use only.
[3] 1300 Mwe fuel cell integrated with gasifier in base load central station.

TABLE 3.8: HEAT RATES FOR PROMISING COMBINATIONS

Cell Technology	Raw or Primary Fuel	Conversion Technology	Fuel Cell Fuel	Fuel Cell Heat Rate	
				Unintegrated[2] (Fuel Cell Only)	Integrated (Total System)
Central Fuel Cell:					
Molten carbonate	Coal	O_2 Gasification	Syngas	8,200	7,720
Acid	Coal	O_2 Gasification	Impure H_2	7,750	7,840
Onsite Fuel Cell:					
Molten carbonate	Naphtha[1]	Reforming	Syngas	8,200	7,500
	Distillate	Partial oxidation	Syngas	8,200	7,500
	Methanol	Reforming	Syngas	8,200	7,500
Acid	Naphtha[1]	Reforming	Impure H_2	7,750	7,500
	Methanol	Catalytic conversion	Impure H_2	7,750	6,500
	No. 2 distillate	Partial oxidation	Impure H_2	8,200	9,200
Alkaline	Naphtha	Reforming	H_2	6,500	8,700
	Methanol	Partial oxidation	H_2	6,500	8,700

[1] Low sulfur distillate also appropriate.
[2] Unintegrated heat rates are based on fuel to fuel cell; integrated heat rates are based on raw or primary fuel.

MODULAR APPROACH

In developing fuel cell fuel supply systems, three basic approaches were considered. The first, referred to as System A, consists of converting raw fuel at a central conversion facility and distributing a clean product fuel to dispersed fuel cell power plants. System B involves the delivery of purchased hydrocarbon feedstocks directly to the dispersed power plants where they are converted on-site to fuel cell grade fuels. System C involves integrating a central station fuel cell with a central coal gasifier to eliminate fuel transportation costs, to fully integrate all necessary conversion in a single plant, and to take advantage of the economics of scale of a large base load system. The general system schematic for these three options is shown in Figure 3.1.

The cost associated with these total fuel supply systems was developed modularly, e.g., many of the central conversion options initially produced synthesis gas which, depending on the final product, is subsequently shifted or methanated or treated for removal of carbon dioxide. Modular costs were estimated for the basic fuel conversion processes, various subsystems, and the transmission and storage elements in the system. The total system cost was obtained by assembling the modular costs appropriate to the product fuel. This approach offered flexibility in that many alternative systems were costed using the basic modules.

Although economics are projected for four U.S. locations, no attempt was made to estimate the relative capital investment differences among the locations or the differences in operating labor costs. Capital related costs in future years are based on 1975 investment estimates with no escalation, since the application of estimated inflation rate only adds uncertainty to these costs. Since all comparisons are in 1975 dollars, this eliminates the need to deflate the costs backward. Essentially, the only differences in overall economics among the locations stem from fuel cost differences.

BASIS FOR EVALUATION

In developing modular costs, a general basis for evaluation was defined. This is outlined below.

(a) Dispersed Fuel Cell Operation
   Peak to intermediate service - 3,000 hr/yr
   Dispersed location - 30 mile average distance to cell
   Cell technologies - acid, base, molten carbonate
   Integration with on-site processor when possible

(b) Central Fuel Cell Operation
   Base load service - 8,000 hr/yr
   Located at coal gasifier site
   Cell technologies - acid, base, molten carbonate
   Integration with gasifier when possible

(c) Central Conversion Facility
   Size - fuel output for ten fuel cell power plants: 51.8 MM scfd H_2;
   59.4 MM scfd syngas; 17.6 MM scfd SNG; and 780 tpd methanol

FIGURE 3.1: FUEL CELL POWER PLANT FUEL SUPPLY SYSTEMS

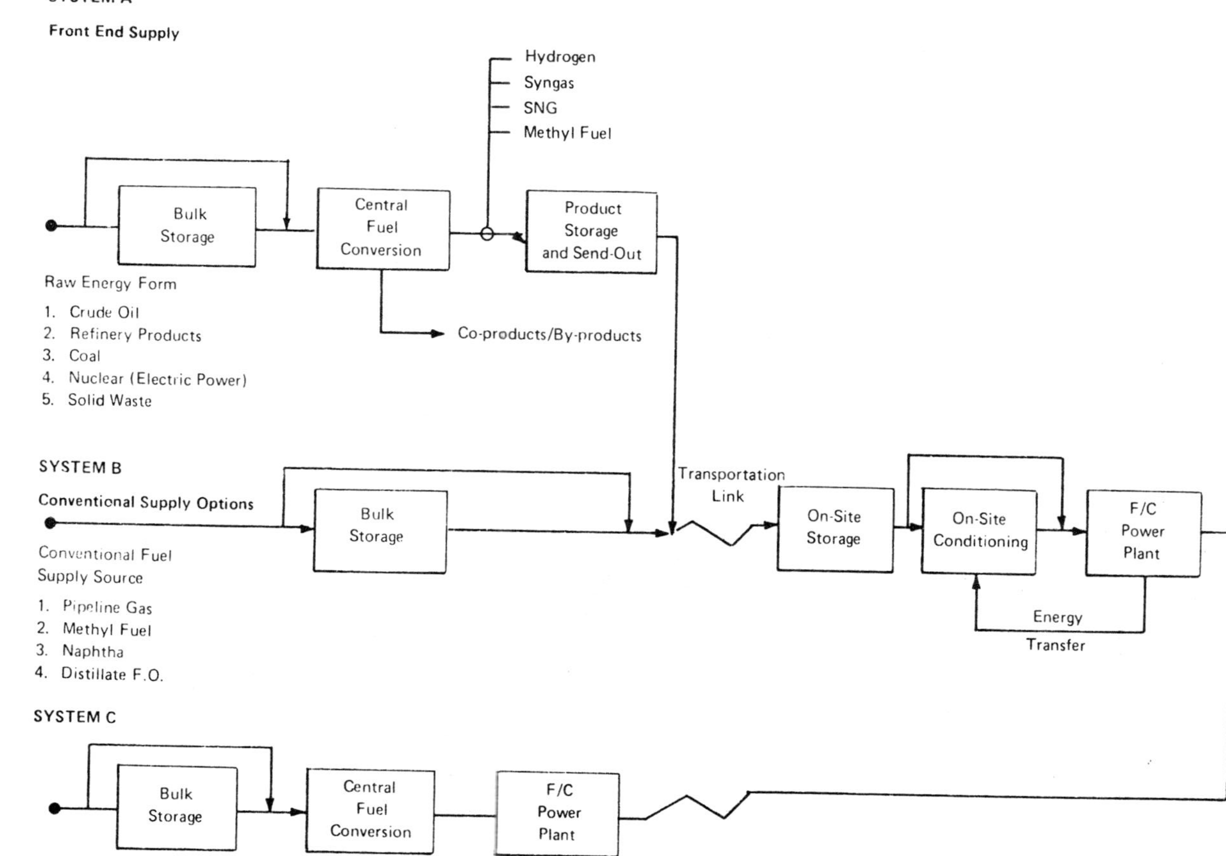

Operation - continuously, 8,000 hr/yr
Gaseous products dried for pipeline transmission

(d) Transportation and Storage
Pipeline, truck and rail systems
Short-term storage for daily load fluctuations
Longer-term storage for emergency shutdowns

Reasonable methods of transportation and storage of product fuels produced at the central conversion facility were identified.

CENTRAL FUEL CONVERSION FOR DISPERSED FUEL CELLS

In the central conversion concept a raw or primary fuel is fed to a single facility for the production of primary or secondary fuel for subsequent transportation to a series of dispersed fuel cells which may or may not have a fuel conditioner.

Like the total fuel supply concepts, the central conversion schemes were developed modularly. For example, as shown in Figure 3.2, SNG is produced from coal via Lurgi gasification followed by shift conversion, acid gas removal, methanation, and drying.

FIGURE 3.2: BLOCK FLOW DIAGRAM—SNG FROM COAL VIA THE LURGI PROCESS

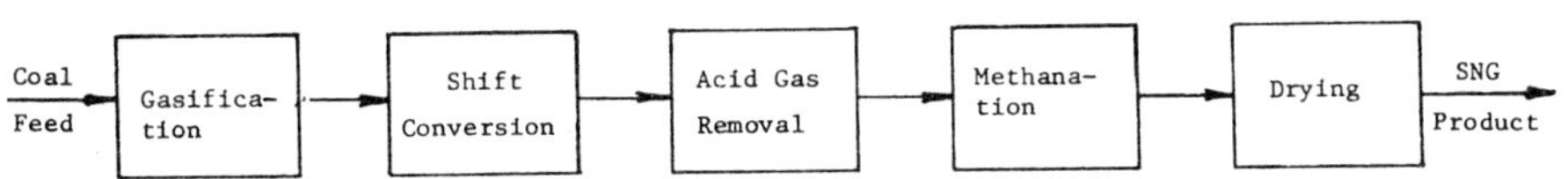

Capital investments and operating requirements were estimated for each of these modules and summed to give the overall economics estimates. Consideration was given to integration of energy utilization within the plant although no optimization was attempted. This approach was used for each central processing case to expedite the cost estimates for the many cases considered:

(a) Four locations
(b) Four conversion processes in each location
two (or more) coal processes
one solid waste process
one partial oxidation process with two possible feeds
(c) Four secondary products from each process/feed combination above:
syngas
pure hydrogen
methanol
SNG

Synthesis Gas Treatment and Conversion Systems

As developed earlier, the economics for central coal and oil gasification include
only the production of raw, undesulfurized syngas. In order to complete the
modular cost buildups for central facilities, typical economics for the necessary
process steps were prepared. These include:

- (a)　Acid gas removal
- (b)　Sulfur recovery
- (c)　Water gas shift
- (d)　Methanol synthesis
- (e)　SNG synthesis
- (f)　Methanation (to meet H_2 purity)

Note that the economics for these modules, particularly water gas shift, cannot
always be added linearly to give final overall process economics because of the
various opportunities for integration with other process units—also, in the case
of water gas shift the cost is importantly a function of the particular syngas composition and the product to be produced.

Therefore, when linear addition was not appropriate, in-house design models and
engineering judgement were used. In general, the systems were not optimized,
but it was assumed that they could be, with appropriate credit given.

Coal Gasification Systems

Dedicated Gasifiers for Fuel Cell Fuels: For central coal conversion to produce
hydrogen, syngas, methanol, or SNG, economics were developed for three commercially available processes—Lurgi, Koppers-Totzek (T-K), and Texaco—and two
advanced systems presently under development—agglomerating ash fluid-bed and
two-stage entrained-bed.

The K-T process is a low pressure, high temperature system which produces a
gas with very low methane content. Texaco, although a high pressure process
(>1,200 psi), operates above 2000°F and also yields a low methane gas. These
gases are particularly appropriate for hydrogen or methanol production where
methane is a troublesome inert material rather than a valuable fuel. For SNG
and syngas production the other three processes, which operate near 30 atm at
moderate temperatures (about 1600° to 1800°F), are appropriate as their significant methane yields are tolerable and even desirable in the case of SNG.

The Texaco system, actually straddles the line between commercial and advanced
in several ways. There are presently no commercial-scale systems operating or
announced, although the process is fully technically developed. Further, the
estimated economics for the system are more in line with the advanced systems.
For these reasons and the fact that there is no known advanced concept which will
produce a high pressure, low methane gas for methanol and hydrogen production,
the Texaco economics were chosen as typical, if conservative, of what may be
available in 1985-1990 for low methane gas, available at elevated pressure.

Of the other two advanced gasifiers, the two-stage entrained system was selected
to carry through the remaining economic analyses of SNG and syngas. The two-
stage entrained system appears to be very economic. In addition, it is a relatively
easy system to operate.

Multipurpose Nondedicated Gasifier: In the multipurpose gasifier concept a large 1,300 MWe gasifier would be installed to produce a variety of utility-related fuels, including 780 tpd (100 MWe) of methanol for distribution to ten 26 MWe dispersed fuel cells. Two approaches were considered to meet this methanol demand.

In the first, a slip stream of raw synthesis gas would be purchased from the central gasifier and the gas would be shifted, scrubbed for acid gas, and reacted in a conventional recycle system to produce methanol. Essentially, the only economic advantage of this scheme over a dedicated plant would be the economy of scale of the gasification section.

In the second, a larger syngas stream would be taken through the shift, scrubbing and reactions steps, the quantity being determined by the yield of methanol possible with no syngas recirculation. Analysis showed that virtually all the gas from the 1,300 MWe two-stage entrained gasifier would be required assuming synthesis at a pressure of 500 psig to produce the necessary once-through methanol.

Table 3.9 shows the relative costs of producing methanol from these schemes. For comparison purposes, the economics for dedicated methanol are included. As noted above, the recycle methanol saves modestly on capital investment, resulting in a saving of $0.26/MM Btu of product. The once-through scheme, however, suffers drastically from the need to handle and treat the large syngas stream and enjoys no benefit from the fact that the unused stream returned to the multipurpose gas header has a very low sulfur content (less than 1 ppm).

TABLE 3.9: ESTIMATED ECONOMICS FOR METHANOL FROM (DEDICATED AND MULTIPURPOSE) COAL GASIFICATION

Capital Investment (Mid-1975), MM$	Dedicated	Multipurpose Conventional Synthesis*	Multipurpose Once-Through Synthesis**
Gasification	28.5	22.2	20.1
Acid gas removal	1.8	1.8	19.9
Shift conversion	3.9	3.9	15.0
Methanol synthesis	6.3	6.3	4.4
Total estimated plant cost, MM$	40.5	34.2	59.4
Operating Requirements Annual, M$			
Coal at $1.26/MM Btu	10,420	10,420	9,234
Catalyst and chemicals	100	100	2,000
Water at $0.30/Mgal	70	70	189
Power at $0.02/kWh	2,784	2,784	4,503
Labor and overhead at $80,000	800	800	623
Maintenance at 4% of plant investment	1,620	1,368	2,376
Capital charge at 17% of plant investment	6,885	5,814	10,098
Total operating cost, M$	22,679	21,356	29,020
$/MM Btu	4.47	4.21	5.72

Note: Capacity, 780 tpd methanol; stream factor, 8,000 hr/yr; location, Columbus, Ohio

*A syngas stream sufficient for the necessary methanol production by conventional recycle is purchased from a product gas header.

**The entire multipurpose syngas is treated and passed, on a once-through basis, through the synthesis reactor. Unused gas is returned to the header for use by other customers.

Solid Waste Gasification

For the gasification of solid waste Union Carbide's Purox process was chosen as the basis for the economic analysis. The Purox gasifier is a fixed-bed, partial oxidation reactor capable of accepting as-received refuse. Therefore, there is a minimum of feed preparation required.

As a low pressure, low effluent temperature process, Purox produces a gas containing a significant yield of methane. This is fine for SNG and syngas but not for methanol and hydrogen, as discussed above. In these latter cases the cost of a steam reformer, to reduce the methane content to less than 1% was included.

Heavy Oil Partial Oxidation

Partial oxidation processes of the type licensed by Shell and Texaco are versatile processes capable of converting almost any hydrocarbon feedstock to syngas. The Texaco process further is developing to a point where it will soon be capable of accepting slurried coal feed for operation at elevated pressure (above 1,200 psig).

Since these processes are capable of high pressure operation, they are quite attractive for producing fuel cell fuels. Gaseous products do not require compression to attain pipeline pressures and in the production of methanol the partial oxidation reactor can be run at a pressure sufficiently above that in the synthesis loop.

Partial oxidation is most advantageously run on feedstocks which cannot be converted to synthesis gas by other routes; consequently, heavy hydrocarbons show favorable economics. Economics were developed for partial oxidation utilizing both residual fuel oil and high sulfur vacuum residue.

Central Fuel Conversion Economics for Dispersed Fuel Cells

The costs of producing fuels centrally in Hartford, Dallas, Columbus and Los Angeles are shown in Tables 3.4 through 3.7. These are based on modular analyses referred to above with typical example flowsheets for SNG via Lurgi. All costs are in 1975 constant dollars.

ON-SITE FUEL CONVERSION FOR DISPERSED FUEL CELLS

Process Integration Opportunities (Central/On-Site)

The degree to which the hydrogen-rich fuels are utilized in the fuel cell stack varies according to the type of fuel cell and fuel form. Typical heat rates for unintegrated fuel cells have been shown in Table 3.8 for three different 26 MW, fuel cell types and corresponding fuels. Solid oxide electrolyte fuel cells were not considered since their commercialization is likely to be beyond the time frame of this study.

When the hydrogen-rich fuel is produced at a central conversion plant and transported to a distant fuel cell, no improvement in the heat rates of Table 3.8 is possible. In order to utilize the cell's waste heat from both the unconsumed fuel in the anode vent and the heat emitted by the stack, the fuel cell must be integrated with an adjacent fuel conversion plant. Improvements in the overall heat rate require much regenerative heat exchange in addition to using the energy in the anode vent.

Dispersed Fuel Cell/Processor Integration Analysis

With the cooperation of United Technologies Corporation sufficient information was obtained on the operating characteristics of the various fuel cells to perform a fuel cell/processor integration analysis and derive representative heat rates for integrated fuel cell generators.

Several fuel/fuel cell combinations were considered. Both steam reforming and catalytic partial oxidation were evaluated as on-site conversion options. In the latter case, JPL provided ADL with conceptual system designs and capital investment estimates.

In performing this integration analysis, the cell voltage was fixed at 0.78 volt. Consequently, efficiency improvements associated with changes in cell operating conditions were not considered. In all cases, the cell operating pressure was less than 40 psia.

Steam Reforming: Complete mass and energy balances were developed for three of the more important integrated options involving steam reforming of naphtha. The first option is the use of impure hydrogen with an acid cell. A material balance and flowsheet are shown in Table 3.10 and Figure 3.3. For this case the overall heat rate is 7,490 Btu/kWh (HHV basis—naphtha to ac power).

The second case is for pure hydrogen with a base cell, shown in Table 3.11 and Figure 3.4. In this case the overall heat rate is 8,700 Btu/kWh (naphtha to ac power) including the auxiliary heat required for carbon dioxide removal. The last case shown in Table 3.12 and Figure 3.5 is for the syngas-molten carbonate combination. The heat rate in this case is 7,525 Btu/kWh (naphtha to ac power).

Integrated cell heat rates for these and other fuel/fuel cell combinations that were studied are summarized in Table 3.13. The unintegrated heat rates corresponding to the product fuel from the on-site processor are shown for comparison. As a result of the analysis of steam reforming, several interesting aspects concerning integration opportunities were identified. Briefly, they are as follows:

    (a)   The heat balance around the reformer is the key to reducing fuel requirements. The only other way to improve efficiency is to use vent stream energy to reduce parasitic load (i.e , increase kWh output). The heat load of the reformer can be reduced by preheating the reactants, combustion air, and fuel streams through utilization of sensible waste heat in the system—both from the reformer and cell stack.

    (b)   The reformer flue gas temperature required to achieve heat transfer to the reformer coil is set by radiant and convective heat transfer mechanisms and fixes the efficiency by which heat is transferred to the process.

FIGURE 3.3: ACID CELL INTEGRATED PROCESSOR

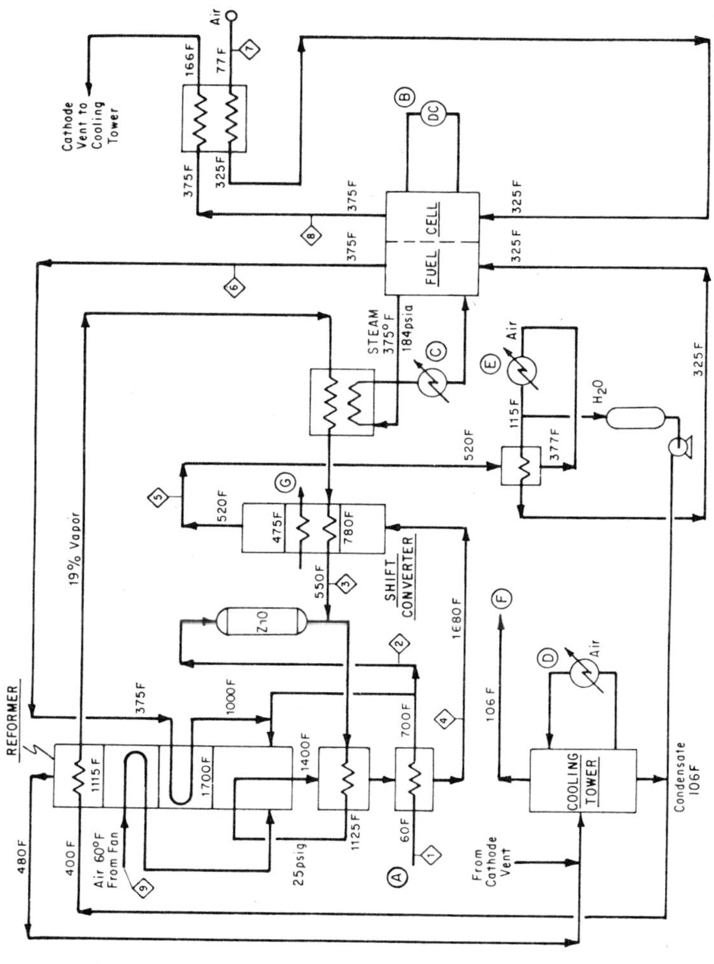

TABLE 3.10: ON-SITE INTEGRATION ANALYSIS STEAM REFORMING/ACID CELL

Product – Impure H_2

MASS BALANCE
(lb mols/hr)

Stream No.	1	2	3	4	5	6	7	8	9
Component									
H_2	---	---	---	1428.1	1655.0	165.5	----	----	---
CO	---	---	---	243.3	16.4	16.4	---	---	---
CO_2	---	---	---	288.6	515.5	515.5	---	---	---
H_2O	---	---	2153	1307.1	1080.2	115.5	---	1490	---
CH_4	---	---	---	6.2	6.2	6.2	---	---	---
O_2	---	---	---	---	---	---	1655.3	910.3	393.2
N_2	---	---	---	---	---	---	6227.1	6227.1	1475.3
TOTAL			2153	3273.3	3273.3	819.1	7882.4	8627.4	1868.5
lbs/hr	9,849	7,689	38,754	46,443	46,443	25,650	127,329	130,309	54,000

HEAT BALANCE
DATUM 77°F

MMBtu/hr

A	B	C	D	E	F	G
197.0	95.5	23.9	33.6	24.7	16.4	2.9

FIGURE 3.4: ALKALINE CELL INTEGRATED PROCESSOR

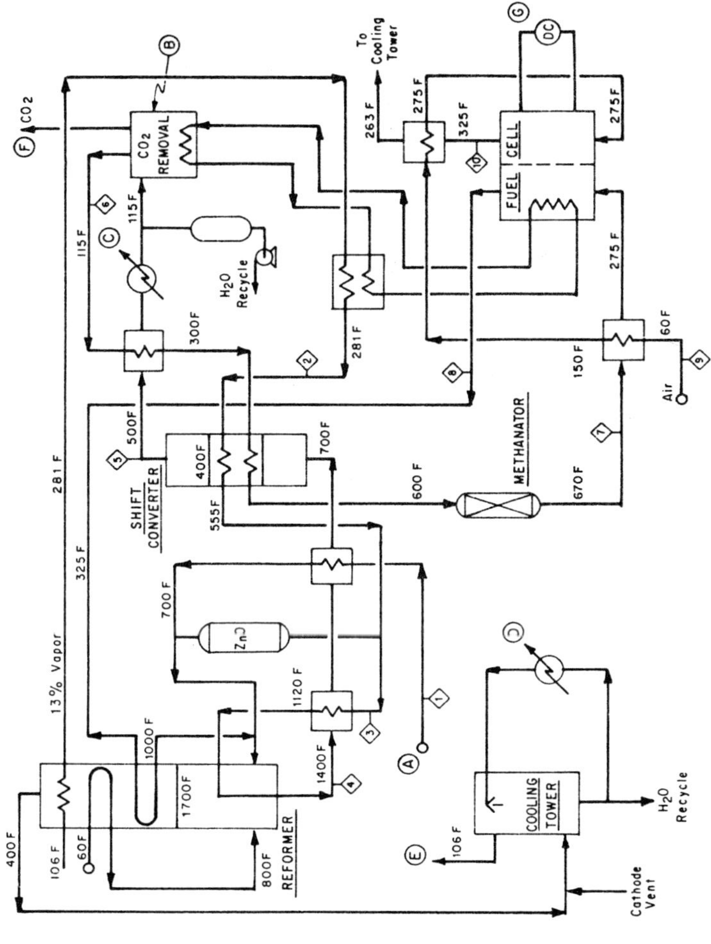

TABLE 3.11: ON-SITE INTEGRATION STEAM REFORMING/BASE CELL

Product - Pure H_2

MASS BALANCE
(lb mols/hr)

Stream No.	1	2	3	4	5	6	7	8	9
Component									
H_2	---	---	---	1420.4	1420.4	1420.4	1395.2	27.9	---
CO	---	---	---	207.9	8.4	8.4	---	---	---
CO_2	---	---	---	246.8	446.3	---	---	---	---
H_2O	---	1841	---	1117.4	917.8	99.0	107.4	107.4	---
CH_4	---	---	---	5.4	5.4	5.4	13.8	13.8	---
O_2	---	---	---	---	---	---	---	---	1395.2
N_2	---	---	---	---	---	---	---	---	5248.6
TOTAL	8,504	1841	---	2798.3	2798.3	1533.2	1516.4	149.1	6643.8
lbs/hr	8,504	33,138	6,574	39,712	39,712	5336	5336	2601	191,607

HEAT BALANCE
DATUM 77°F
MMBtu/hr

A	B	C	D	E	F	G
171.1	50.1	(21.6)	(32.1)	(13.1)	(58.9)	(95.5)

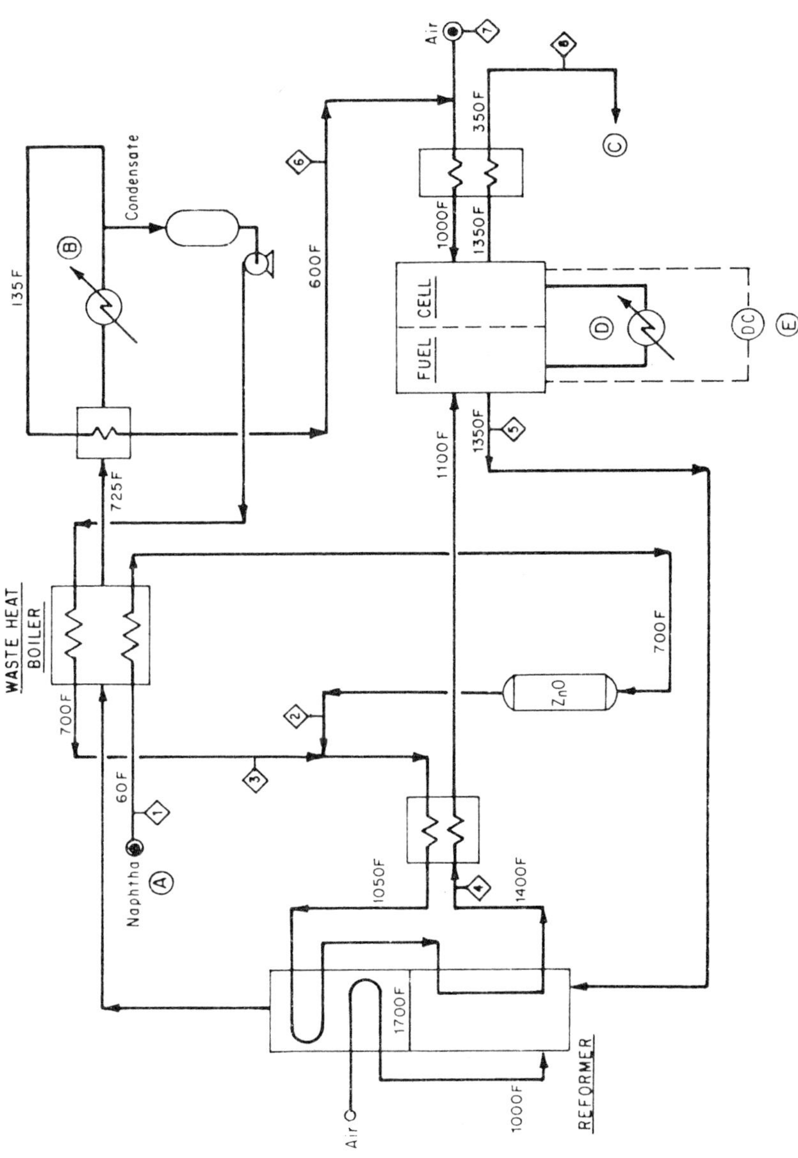

FIGURE 3.5: MOLTEN CARBONATE CELL INTEGRATED PROCESSOR

TABLE 3.12: STEAM REFORMING/MC-CELL

Product – Syngas

MASS BALANCE

(lb mols/hr)

Stream No.	1	2	3	4	5	6	7	8
Component								
H_2	---	---	---	1776.6	568.3	---	---	---
CO	---	---	---	398.5	127.5	---	---	---
CO_2	---	---	---	293.6	2060.8	2190.0	---	712
H_2O	---	---	2079.4	1090.6	2304.8	797.1	---	797
CH_4	---	---	---	5.3	1.7	---	---	---
O_2	---	---	---	---	---	55.1	1478	739
N_2	---	---	---	---	---	1519.8	5355	6873
TOTAL	---	---	2079.4	3564.6	5063.1	4562	6833	9121
lbs/hr	9,917	9,917	37,462	47,334	136,942	155,084	197,322	261,896

HEAT BALANCE
DATUM 77°F

(MM Btu/hr)

A	B	C	D	E
198	58.7	20.3	23.4	95.6

TABLE 3.13: ON-SITE INTEGRATION ANALYSIS SUMMARY OF HEAT RATES

Cell Technology	Primary Fuel	Onsite Conversion Technology	Fuel to Cell	Heat Rate, Btu/kwh [1]	
				Unintegrated	Integrated
Acid	Naphtha [2]	Reforming	Pure H_2	$7,200^3$	$8,660^{4,5}$
	Naphtha	Reforming	Impure H_2	$7,750^3$	7,490
	Naphtha	Reforming	Impure H_2	$9,000^6$	8,700
	Methanol	Catalytic Conversion	Impure H_2	$7,750^3$	6,500
	Methanol	Catalytic Conversion	Impure H_2	$9,000^6$	7,550
	No. 2 Distillate	Partial Oxidation	Impure H_2	$8,200^3$	9,200
Alkaline	Naphtha	Reforming	Pure H_2	$6,500^8$	$8,700^{4,5,7}$
	Methanol	Catalytic Conversion	Pure H_2	$6,500^8$	$8,650^{4,5,7}$
Molten Carbonate	Naphtha	Reforming	Impure H_2	$8,200^3$	7,525
	Naphtha	Reforming	Syngas	$8,200^3$	7,525
	Methanol	Catalytic Conversion	Syngas	$8,200^3$	6,500
	No. 2 Distillate	Partial Oxidation	Syngas	$8,200^3$	10,000
	No. 2 Distillate	Partial Oxidation	Impure H_2	7,750	9,200
	No. 2 Distillate (Low Sulfur)	Reforming	Impure H_2	7,750	7,490

[1] Unintegrated heat rates are based on fuel to fuel cell; integrated heat rates are based on primary fuel.
[2] Naphtha based heat rates also apply for low sulfur distillate.
[3] Fuel cell voltage 0.785 vdc.
[4] Includes heat for CO_2 removal (MEA System).
[5] Primary cell coolant used directly in MEA stripper.
[6] Fuel cell voltage 0.65 vdc (first generation).
[7] Potential for improvement with efficient CO_2 separation process.
[8] Fuel cell voltage 0.85 vdc.

(c) To derive maximum benefit from the combustion of vent streams, water removal before combustion should be considered. For the molten carbonate cell, an efficient, lower pressure CO_2 removal process would also be advantageous.

(d) To manufacture pure hydrogen on-site and achieve an attractive heat rate is difficult. With low pressure reforming, CO_2 removal becomes a real disadvantage due to the high energy requirements for an amine system which is the only acceptable process—vis-a-vis scrubbing efficiency at low pressure. A hot carbonate process such as Benfield can reduce this energy requirement by about 60% if the operating pressure can be increased to above 100 psig.

(e) The reaction and preheat energy requirements for low temperature (400°F) catalytic conversion of technical grade methanol to produce impure hydrogen can be supplied by waste heat. Methanol is required only to furnish the hydrogen and carbon values of the product gas.

United Technologies indicates that steam reforming of distillate fuel oil is possible with their conditioner design, provided a desulfurized fuel containing 300 ppm sulfur can be purchased. However, because distillate fuel oil contains considerably more aromatics than naphtha, carbon deposition on the reformer catalyst is more likely. To counteract this tendency, higher steam-to-carbonate ratios and possibly lower space velocities are required. Even with these modifications, it is believed that the catalyst life would not equal that achieved with naphtha. The heat rate for reforming of distillate fuel oil will be about the same as for naphtha.

Another technology suitable for on-site fuel conversion is steam reforming of methanol. In this process, which is a modification of conventional reforming, methanol reacts at low pressure between 375° to 425°F, and in the presence of a low-temperature shift catalyst to form hydrogen and carbon monoxide. By operating in this temperature range, the heat energy required for feed vaporization can be supplied by waste heat from a fuel cell with a 375°F outlet and the energy for the reaction can be supplied by the anode vent stream. Only the feed to produce hydrogen and carbon monoxide is required. Consequently, the heat rate approaches that of hydrogen cells.

Experiments conducted so far (Kurpit, S. and E. Gillis, *Methanol Fuel for Fuel Cells,* Ft. Belvoir) are based on technical-grade methanol. Manufactured methyl fuel will likely have higher alcohols present, since they do not affect the cost of transportation. The presence of higher alcohols would tend to raise the temperature required for reforming and hence lower the efficiency. However, these higher alcohols could be easily removed at the methanol production facility through an appropriately designed distillation unit.

Catalytic Partial Oxidation: Catalytic partial oxidation with air was also considered as an alternative to on-site steam reforming. The JPL portable hydrogen generator was evaluated in the context of a larger, stationary unit suitable for application to dispersed fuel cell power plants. For this evaluation, JPL was most cooperative in providing technical and cost information for a system scaled up to meet the requirements of a 26 MW power plant.

A process flow diagram for a catalytic partial oxidation fuel conditioner using distillate oil is shown in Figure 3.6 for synthesis gas production. No water is used in the process and carbon monoxide is not shift reacted to hydrogen, since the product gas is fed to a molten carbonate cell.

The process flow diagram for the partial oxidation system integrated with an acid cell is shown in Figure 3.7. In this system, the fuel is mixed with water and vaporized. Preheated air is mixed with the vaporized fuel/steam mixture and the total stream enters the reactor where catalytic partial oxidation and steam reforming take place. After heat exchange and sulfur removal, the carbon monoxide in the gas is shifted.

The advantage of the catalytic partial oxidation system is that it is inherently simple since no external heat is required to reform the feed. Furthermore, the fuel cell interacts relatively little with the processor; thus the need for an elaborate control system is eliminated. With fewer subsystems required, the overall system is reasonably reliable and relatively economical to maintain.

Another advantage afforded by this system would be the ability to handle heavy feedstocks not suitable for steam reforming. However, the use of a nickel reforming catalyst again introduces the potential for poisoning by feedstock impurities, particularly sulfur. This problem is minimized at operating temperatures above 1400°F, according to JPL.

This still leaves the problem of sulfur removal before the cell. With 0.2 weight percent sulfur oil, about one-half ton of sulfur per week must be removed. At a 0.2 pound per pound loading, 2.5 tons of zinc oxide are required for one week's operation in peaking service. Similarly, 10 tons are required for a month's operation. At this sulfur level the zinc oxide system might be tolerable; however, with residual fuel oil containing 1% sulfur, the quantity of ZnO required would be five times greater. At this level, the size and replacement frequency of the zinc oxide system is unacceptable, thereby limiting the feedstock options for catalytic partial oxidation to distillate fuel oil or lighter.

The major disadvantage of this conversion system is that there are few opportunities to make use of unconverted fuel and waste heat available from the cell. This reduces the initial cost of the system since few heat exchanges are required at the continuing expense of a significantly higher heat rate. Heat rates for the partial oxidation/molten carbonate system operating on distillate fuel oil are 9,200 and 10,000 Btu/kWh for impure hydrogen and syngas, respectively. These values are presented in Table 3.13.

Best Heat Rates: The best available heat rate for each of the secondary fuels is listed in Table 3.14.

These heat rates were used in the development of the final cost comparison. The heat rate shown for distillate fuel oil is based on steam reforming since the price differential for low-sulfur distillate fuel oil does not offset the much lower efficiency of the partial oxidation process.

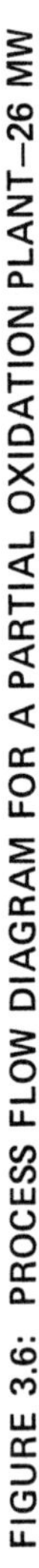

FIGURE 3.6: PROCESS FLOW DIAGRAM FOR A PARTIAL OXIDATION PLANT—26 MW
PARTIAL OXIDATION REACTORS
NO. 1
NO. 2
NO. 3
NO. 4
1800°F
600°F
REACTOR DIMENSION:
D = 4 ft
L = 4.4 ft
(NiO on Al₂O₃ CATALYST BED)
HEAT EXCHANGER NO. 1
HEAT EXCHANGER NO. 2
(VAPOR)
1400°F
A₁ = 300 ft²
A₂ = 100 ft²
MASS (5.2)
70°F
COMPRESSOR
AIR
MASS (1.0)
70°F
PUMP
FUEL (NO. 2 FUEL OIL)
PRODUCT SYNTHESIS GAS
(TO MOLTEN CARBONATE FUEL CELL)
1.5 psig
1150°F

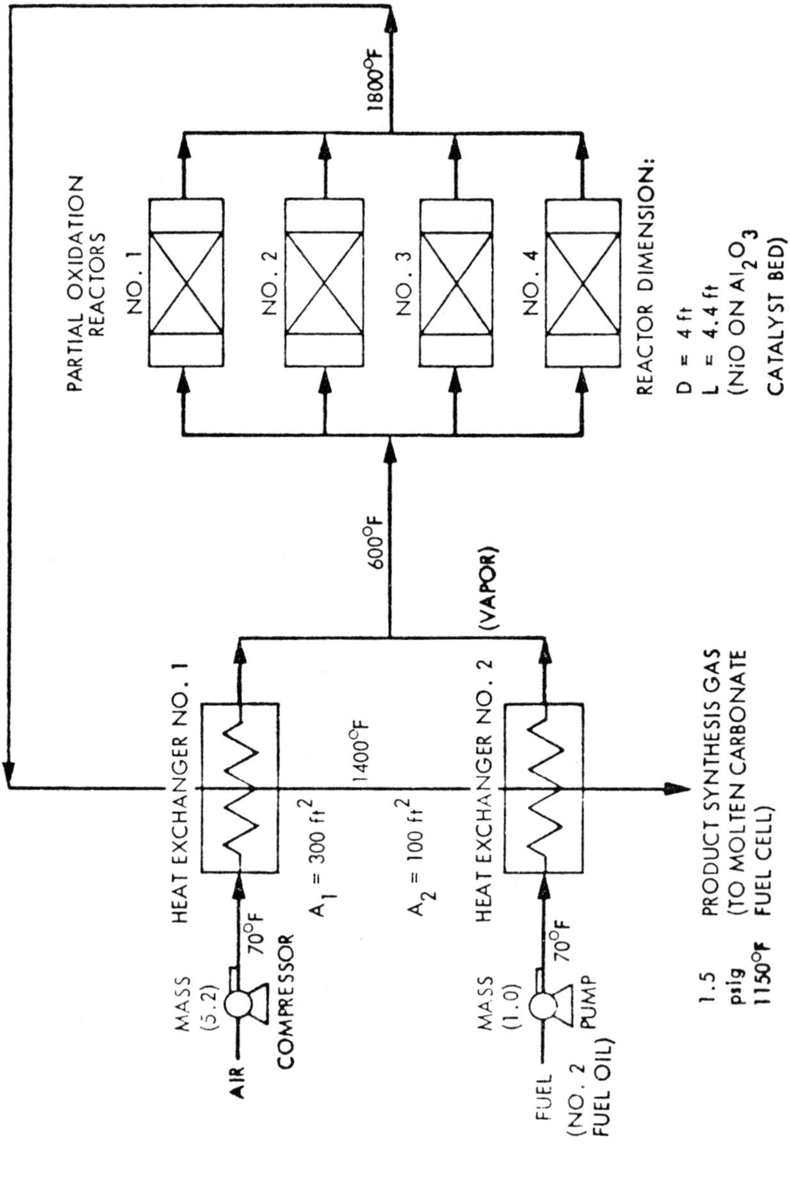

FIGURE 3.7: PROCESS FLOW DIAGRAM FOR A PARTIAL OXIDATION PLANT (WITH SHIFT CONVERTER)—26 MW

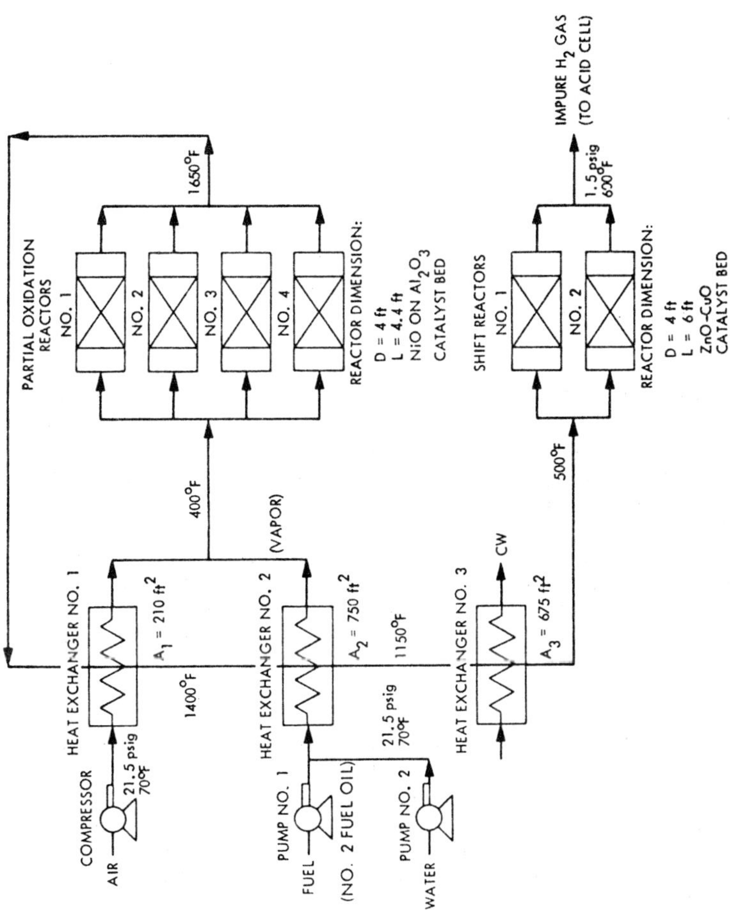

TABLE 3.14: BEST AVAILABLE HEAT RATES

Fuel Cell Fuel	Cell Technology *	Heat Rate Net Plant Btu/kwh
H_2	Base	6,500
Impure H_2	Acid	7,750
Syngas	MC	8,200
Methanol	Acid	6,500
SNG	Acid/MC	7,500
Naphtha	Acid/MC	7,500
Distillate F.O. (Low Sulfur)	Acid/MC	7,500

*Assumes 0.78 V dc for acid/MC; 0.85 V dc for base

Capital Investment for On-Site Processors

Although the concept of on-site processing provides an efficiency advantage over alternative supply concepts, the cost of the on-site processor cannot be ignored if the system is to be competitive with other advanced generator concepts.

Consequently, an effort was made to estimate the capital investment cost of on-site processors without a complete knowledge of the status of UTC's development effort in this area. Estimates are based on knowledge of commercially available technology and an intuitive feel for the kinds of improvements UTC has been able to incorporate into the system.

The capital cost pertinent to this study is that associated with second generation technology. The approach used to estimate this cost was to begin with the estimated cost of present day technology as perceived, and to apply appropriate learning curves to arrive at the cost of second generation technology in constant 1975 dollars.

Based on current information (Fickett, A.P., *An Electric Utility Fuel Cell: Dream or Reality,* 25 p., American Power Conference, Chicago, Illinois, April 1975) it is estimated that the cost of the on-site reformer is about $125/kW in 1975 dollars. Based on information supplied by JPL with adjustments for an on-site hydrodesulfurization system and reactor design uncertainties, the current cost of this system was placed at $85/kW. The estimated costs of advanced systems based on these technologies are shown in Table 3.15.

These figures represent the cost of the hundredth unit produced assuming a 0.9 learning curve, i.e., for every doubling of the units produced, the cost of the last unit in constant dollars is reduced 10%. The differential between the cost of the reformer and the catalytic partial oxidation unit reflects in part the limited integration opportunities for the latter. One hundred units is about twice the number projected for the low growth case in 1985. This also assumes that the design and fabrication of the fuel processor has been well optimized by the time the hundredth unit is assembled.

TABLE 3.15: ON-SITE PROCESSOR ESTIMATED CAPITAL INVESTMENT

Fuel	Process	Second Generation Technology – Capital Investment – $/kw[1] (1975 Dollars)	Capital Charge[2] $/MM Btu[3] (1975 Dollars)
Naphtha	Reforming	65	0.49
SNG	Reforming	60	0.45
Methanol	Cat. Conversion	55	0.43
No. 2 Fuel Oil	Part. Oxidation	45	0.28

[1] Installed capacity

[2] 17% of capital and 3,000 operating hr/yr

[3] Delivered to fuel cell processor

It is interesting to note that if a 0.8 learning curve is applied to the cost of a commercial package hydrogen plant which currently runs about $220/kW, the cost of the hundredth unit in constant dollars is $50/kW. In this case a steeper learning curve might be justified since the technology has never been adapted to the needs of the dispensed fuel cell generator. The capital investment required for the fuel cell generator is presented by type in Table 3.16.

TABLE 3.16: FUEL CELL CAPITAL INVESTMENT

Fuel Cell Type	Capital Investment[1] $/kw (1975 Dollars)	Heat Rate Btu/kwh	Capital Charge[2] $/MM Btu (1975 Dollars)
Acid Electrolyte	180	7,500	1.36
		6,500	1.57
Base Electrolyte	180	6,500	1.57
Molten Carbonate	145	8,200	1.00
		7,500	1.10
		6,500	1.26

[1] Source: EPRI; does not include fuel processor

[2] 17% of capital and 3,000 operating hr/yr

CENTRAL FUEL CONVERSION FOR CENTRAL FUEL CELLS

Process Integration Opportunities

The earlier discussion on improving the overall fuel cell/conversion system thermal efficiency also applies to central fuel cells since the cell is near the fuel conversion plant.

Central Fuel Cell/Processor Integration Analysis and Economics

At the outset of the central/central integration analysis, several guidelines were set:

(a) maximize the overall coal-to-power efficiency (i.e., minimize the heat rate);

(b) do not attempt to optimize in detail with respect to energy recovery versus capital cost;

(c) look at the acid, alkaline and molten carbonate fuel cell technologies to the extent that they appear to offer good heat rates; and

(d) consider only the two-stage entrained, pressurized, oxygen-blown coal gasifier as the primary conversion step because of its superior thermal efficiency and economics among the coal gasification processes that were studied.

The results of the analysis are shown schematically in the form of two flow-sheets, Figures 3.8 and 3.9 for the molten carbonate and acid fuel cells, respectively. Conspicuously absent is the base cell which suffers from the relatively high methane content of the raw gasifier effluent which would have to be reformed prior to entering the fuel cell.

To meet the specification for pure hydrogen, the gas would have to be steam reformed at low pressure with subsequent shift conversion and carbon dioxide removal, the latter operation being very costly in terms of energy requirements for a low-pressure operation (i.e., requiring an amine scrubbing system).

Molten Carbonate Fuel Cell: Handling the methane content of the gas is also apparent in the molten carbonate and acid systems, as reformers of one type or another are required. In the molten carbonate (m/c) system (see Table 3.17 for heat and mass balance), the fuel cell feed (syngas) passes through a reformer after sulfur removal, gas expansion for energy recovery, and appropriate heat exchange.

The energy for reforming is supplied to the burners by the anode vent, which along with unconsumed carbon dioxide and hydrogen contains some methane which is not converted in either the reformer or fuel cell. This level of methane in the vent was chosen by trial and error to match the energy requirements of the reformer with the energy availability in the vent. Finally, the flue gas, which contains all the carbon dioxide produced in both the cell and the reformer, passes to the cathode side of the fuel cell where most of the CO_2 is absorbed to regenerate the molten carbonate electrolyte.

Because the m/c cell operates at elevated temperature (1350°F), the cell waste heat is rejected at a high level. To recover this heat as electric power, a bottoming steam cycle was added which generates more than 10% of the total system output.

Acid Fuel Cell: The acid cell can utilize only hydrogen. Any fuel value as carbon monoxide must be shift converted to hydrogen (see Table 3.18 for heat and mass balance). Although the cell tolerates carbon dioxide well, there are two steps of CO_2 removal to afford the best possible driving force for the shift conversion steps to minimize the necessary energy consumption to generate intermediate and high level steam.

The anode vent contains the unconsumed hydrogen, some carbon dioxide and the unconverted methane which was carried through from the gasifier.

FIGURE 3.8: MOLTEN CARBONATE TECHNOLOGY

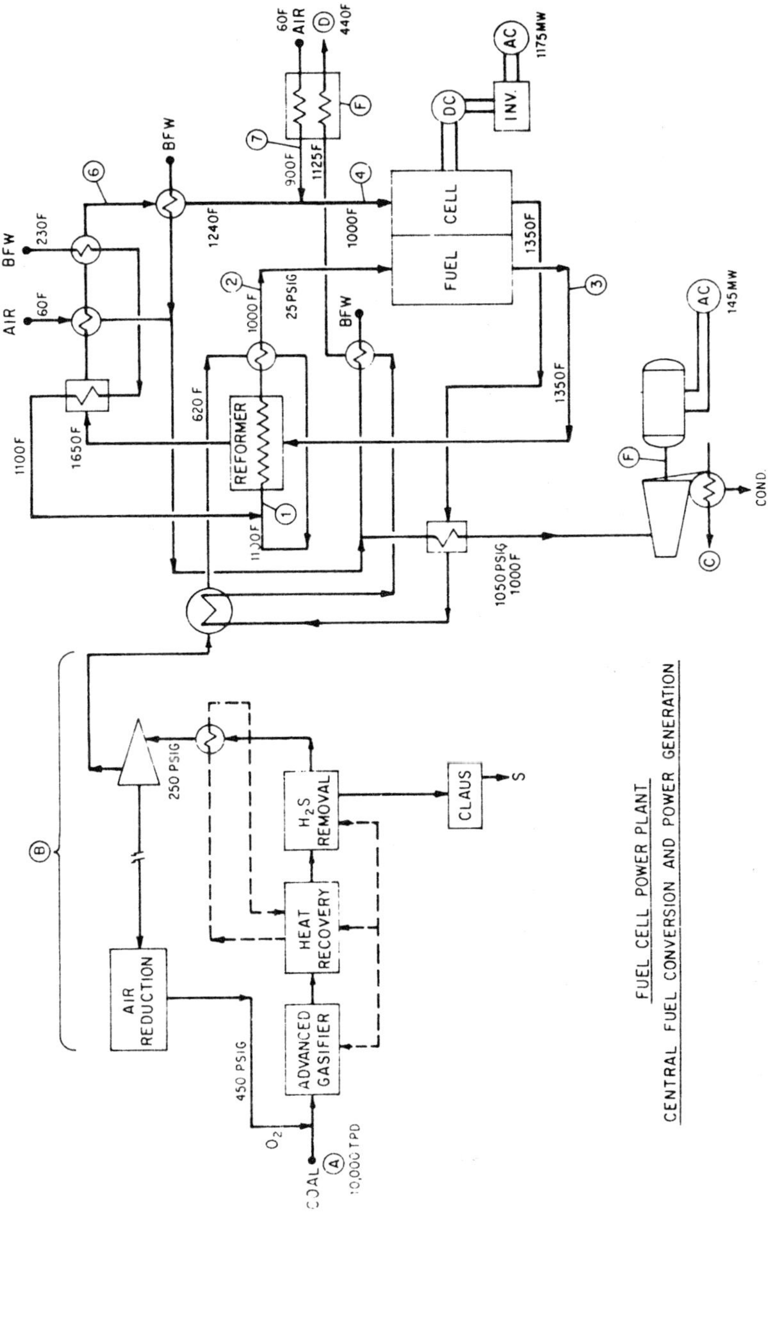

FIGURE 3.9: ACID TECHNOLOGY

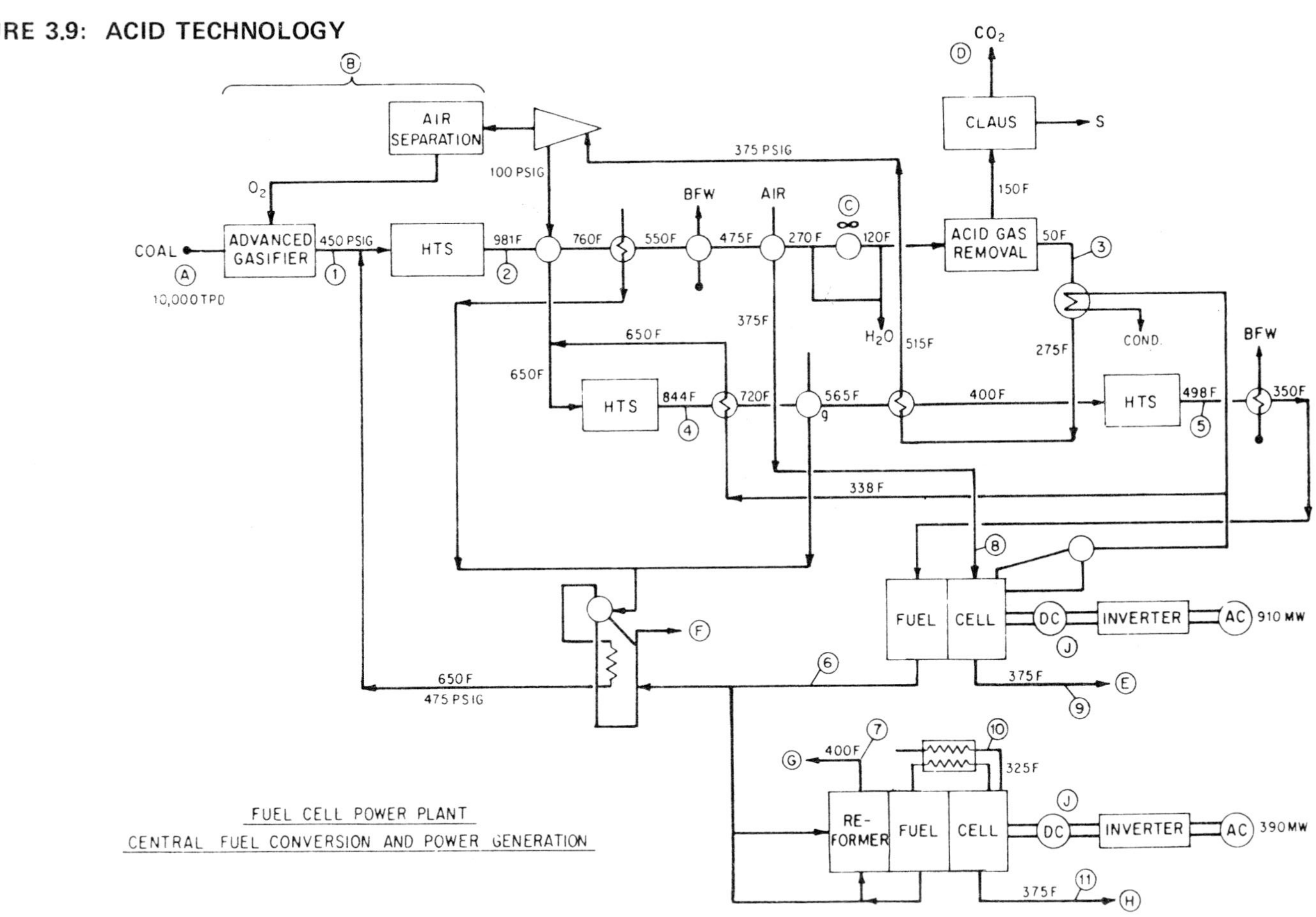

TABLE 3.17: FUEL CELL POWER PLANT GENERATOR CENTRAL FUEL CONVERSION AND POWER GENERATION MOLTEN CARBONATE CELL

MASS BALANCE
(Mols/Hr)

Stream No.	1	2	3	4	5	6	7
Component							
H_2	23,960	45,380	6,807	---	---	---	---
CO	34,730	32,950	4,943	---	---	---	---
CO_2	6,682	13,370	107,958	114,612	48,032	114,612	---
H_2O	25,000	13,4C0	51,973	62,202	62,202	62,202	---
CH_4	6,619	1,711	1,711	---	---	---	---
O_2	---	---	---	163,679	130,389	1,395	162,284
N_2	590	5S0	590	651,309	651,309	40,810	610,499
TOTAL	97,581	107,4C1	173,982	991,802	891,932	219,019	772,783

HEAT BALANCE
(MM Btu/Hr)

A	B	C	D	E	F
10,192.0	1,192.9	1,024.9	3,215.8	4,273.5	484.9

TABLE 3.18: FUEL CELL POWER PLANT CENTRAL FUEL CONVERSION AND POWER GENERATION—ACID CELL

MASS BALANCE
(Mols/Hr)

Stream No.	1	2	3	4	5	6	7	8	9	10	11
Component											
H_2	23,963	43,692	43,692	52,692	57,192	5,719	---	---	---	---	---
CO	34,729	15,000	15,000	6,000	1,500	1,500	---	---	---	---	---
CH_4	6,620	6,620	6,620	6,620	6,620	6,620	---	---	---	---	---
CO_2	6,683	26,410	300	9,300	13,800	13,800	18,109	---	---	---	---
H_2O	6,778	24,016	---	14,400	9,900	9,900	14,042	---	51,473	---	24,779
O_2	---	---	---	---	---	---	451	51,473	25,737	24,779	12,390
N_2	590	590	590	590	590	590	13,495	193,636	193,636	93,216	93,216
H_2S	941	941	---	---	---	---	---	---	---	---	---
NH_3	100	100	---	---	---	---	---	---	---	---	---
	80,404	117,369	66,202	89,602	89,602	38,129	47,098	245,109	270,846	117,995	130,385

HEAT BALANCE
(MM Btu/Hr)

A	B	C	D	E	F	G	H	J	Cell Heat Utilized	Cell Heat Unutilized
10,192	1,192.9	346.1	17.5	1,554.4	47.1	354.2	777.3	4,722.9	548.9	630.7

To utilize this stream, there was included a fully integrated fuel processor/fuel cell which, as discussed earlier, is capable of a heat rate of 7,500 Btu/kWh. This bottom cell produces almost one-third of the electric power from the overall system.

Heat Rates and Economics: The overall heat rates from the above described systems are 7,720 Btu (HHV of coal)/kWh (net power output) for the molten carbonate system and 7,840 Btu/kWh for the acid system. These have very encouraging potential for use as base load central systems.

Table 3.19 presents estimates of the capital costs of the molten carbonate and acid central/central concepts. The total investment costs of $370/kW and $405/kW for 1975 installation of these systems are extremely encouraging for base load operation, particularly in light of the project heat rates. Table 3.20 shows the estimated cost of power generation, in 1975 costs, for four U.S. locations, based on a forecast of 1985 coal prices.

TABLE 3.19: CENTRAL FUEL CELL POWER PLANT CAPITAL INVESTMENT (1975 DOLLARS) ($ MM)

CAPACITY 81.5×10^{12} Btu/yr coal input

STREAM FACTOR   8,000 hr/yr

FUEL CELL TYPE:	Molten Carbonate	Acid
MODULE COSTS		
Gasification	196.8	196.8
Acid Gas Removal (incl. Claus)	21.1	27.0
CO Shift	---	8.0
Reforming	34.0	---
Heat Exchange	23.1	27.0
Auxiliary Boiler	---	9.0
Turbo Generator	21.8	---
Fuel Cell Power Plant		
Stack and Ancillaries	191.4	234.0
Fuel Conditioning	---	25.4
TOTAL FACILITY	488.2	527.2
Generating Capacity, MW	1,320	1,300
Total Investment, $/kW	370	405

TABLE 3.20: CENTRAL FUEL CELL POWER PLANT ENERGY COSTS

Raw Fuel: Coal

FUEL CELL TYPE:	Molten Carbonate				Acid			
Capital Investment (1975), $MM			488.2				527.2	
Generating Capacity, MW			1,320				1,300	
LOCATION:	Hartford	Dallas	Columbus	Los Angeles	Hartford	Dallas	Columbus	Los Angeles
Cost of Coal, $/MM Btu[1]	1.93	1.47	1.26	1.31	1.93	1.47	1.26	1.31
Annual Operating Costs,[2] $MM/yr								
Fuel	157.3	119.8	102.7	106.8	157.3	119.8	102.7	106.8
O & M	25.4	25.4	25.4	25.1	27.8	27.8	27.8	27.8
Capital Charge	83.0	83.0	83.0	83.0	89.6	89.6	89.6	89.6
Total	265.7	228.2	211.1	215.2	274.7	237.2	220.1	224.2
Mils/kWh	25.2	21.7	20.0	20.4	26.4	22.8	21.2	21.6

[1] 1985 price in 1975 dollars.
[2] 8,000 operating hrs/yr.

Recommendations for Future R&D Efforts

To reiterate, the economics and energetics of a central gasifier/central base load
fuel cell appear excellent after this first cut analysis. Again, no optimization
was attempted with respect to the choice of a most compatible gasifier (oxygen
or air, pressurized or atmospheric, entrained-, fluidized- or fixed-bed, etc.)/fuel
cell combination. This represents an area in which EPRI will undoubtedly want
to expand its efforts in light of these results. Other issues which arose which
EPRI may wish to investigate further are:

(a)   the possibility of accomplishing methane reforming within the molten
      carbonate cell, and
(b)   the development of a nonenergy intensive carbon dioxide removal
      system for operation at atmospheric or slightly higher pressure.

FUEL CELLS
FOR PUBLIC UTILITY APPLICATIONS
WESTINGHOUSE STUDY

In 1976 the Westinghouse Electric Corporation prepared a report (NASA CR-134941, Vol. XII) outlining fuel cell use in a public utility power plant, with a parametric assessment of four systems: phosphoric acid, potassium hydroxide, molten carbonate and stabilized zirconia. The report was authored by C.J. Warde, R.J. Ruka and A.O. Isenberg, and is editorialized below. The complete report is presented with the exception of the computer output sheets.

SUMMARY

Four fuel cell power systems, differentiated by electrolyte type, have been investigated from cost and efficiency standpoints. For the phosphoric acid system, operating at 190°C, a power plant efficiency, an overall energy efficiency (based on the limiting value of the coal employed in the production of the power plant fuel), a capital cost and a cost of electricity were calculated for each of sixteen points in the parametric assessment. Similar calculations were performed for aqueous alkaline fuel cell power plants at 343°K [70°C (158°F)] (16 points), molten carbonate plants at 923°K [650°C (1202°F)] (17 points), and stabilized zirconia plants at 1273°K [1000°C (1832°F)] (20 points).

In parametric assessments, the following parameters were varied: useful life and rating of the fuel cell subsystem, fuel cell power density and electrolyte thickness, fuel and oxidant types, performance degradation over the useful life of the fuel cell subsystem, anode and cathode catalyst loadings in the acid and alkaline systems, and temperature of operation and use of waste-heat recovery systems in the molten carbonate and stabilized zirconia power systems. Four of these—fuel cell useful life and power density, use of a waste-heat recovery system, and fuel type—proved to be of particular importance in efficiency improvement and/or electricity-cost reduction.

Typical capital costs, overall energy efficiencies and electricity costs of fuel cell power plants were found to be as shown on the following page.

(a) Phosphoric acid $350–450/kWe, 24 to 29%, and 11.7 to 13.9 mills/MJ (42 to 50 mills/kWh)

(b) Alkaline $450–700/kWe, 26 to 31%, and 12.8 to 16.9 mills/MJ (46 to 61 mills/kWh)

(c) Molten carbonate $480–650/kWe, 32 to 46%, and 10.6 to 19.4 mills/MJ (38 to 70 mills/kWh)

(d) Stabilized zirconia $420–950/kWe, 26 to 53%, and 9.7 to 16.9 mills/MJ (35 to 61 mills/kWh)

Projections as to the lowest possible cost of electricity (in mills/kWh) for the acid, alkaline, carbonate and zirconia systems are mid to high 30s, low 40s, low 30s and high 20s, respectively.

Three types of fuel cell power plants are recommended for further study. These are: [1] solid-electrolyte plant with steam bottoming; [2] molten carbonate plant with steam bottoming; and [3] solid electrolyte plant with an integrated coal-gasification reactor for waste-heat recovery (the Westinghouse Fuel Cell Power System).

STATE OF THE ART

Fuel cell power-generation systems may be grouped in four distinct classes differentiated by electrolyte type. These are:

(a) Aqueous acid, which encompasses systems based on phosphoric and sulfuric acids, and the solid polymer electrolyte;

(b) Aqueous alkali, based on potassium hydroxide and buffered carbonate-bicarbonate solutions;

(c) Molten salt, specifically molten carbonates; and

(d) High-temperature solid electrolyte, specifically stabilized zirconia.

NASA has specified that the minimum power plant rating to be considered in this study should be 25 MWe. No fuel cell power system of even two orders of magnitude lower in rating has been constructed or operated, so there is no utility experience to guide this investigation. All that is available is the advocacy of their own systems by the corporations engaged in fuel cell research. While there are, of necessity, some constraints on the published estimates of system efficiencies, there appears to be no such restraint shown in the projections of the expected cost and performance of the fuel cell subsystems.

A fuel cell power plant will consist, in general, of fuel-processing and power-conditioning equipment in addition to the fuel cell subsystem. Most of the effort of this study was directed toward the fuel cell component because of the greater degree of uncertainty with respect to its cost and performance. Further, as no fuel cell power plant concept has been conclusively demonstrated as optimal from the standpoints of technical feasibility or economic desirability, four base cases are considered, rather than the minimum of two mandated by the contract work statement, with one base case selected from each of the four electrolyte-type categories described above.

The scientific literature pertaining to fuel cells was reviewed in order to facilitate this selection procedure and, particularly, to assist in establishing a meaningful framework of parametric values. Within each of the four general areas the fuel cell power systems were considered from the standpoint of their suitability for use in central-station power generation. Among the more important considerations were the voltage efficiency, demonstrated cell and battery useful lifetimes, life-limiting processes, quantities of noble-metal catalysts required, problems posed by the use of coal-derived fuels, and the state of the art with respect to the engineering of the overall power systems. The results of the literature survey in each area are outlined in the following subsections.

Aqueous Acid Fuel Cells

Systems based on three distinct electrolyte types have been considered in this area:

(a) Phosphoric acid fuel cells, which operate in the temperature range $423°$ to $463°K$ ($302°$ to $374°F$);

(b) Solid polymer electrolyte fuel cells, operating at $348°$ or $423°K$ ($167°$ or $302°F$), depending on whether air or oxygen is employed as the oxidant;

(c) Sulfuric acid fuel cells which operate at approximately $333°K$ ($140°F$).

The Power Systems Division of United Technologies Corporation (formerly Pratt and Whitney Aircraft Division of United Aircraft) is the undisputed leader in acid fuel cell systems. It has field-tested complete power systems of up to 40 kW capacity and is currently building 26 MW systems (FCG-1) for a number of utilities. The preferred system employs immobilized phosphoric acid electrolyte at temperatures between $423°$ and $463°K$ ($302°$ and $374°F$). The higher temperatures in this range are desirable mainly because of air electrode (cathode) activity, but they also serve to accelerate life-limiting processes such as platinum electrocatalyst recrystallization and degeneration of the phosphoric acid matrix.

One problem with this system is typical of all acid systems—fuel electrode deterioration due to carbon monoxide blockage of active sites on the anode electrocatalyst, thus denying them to the electrochemically more active hydrogen. This problem is a good deal less severe here than in other acid systems because the fuel electrode functions adequately, provided the carbon monoxide concentration in the fuel gas is maintained at 0.5% or below.

In a recent ERDA-directed assessment study of devices for the generation of electricity from stored hydrogen, conducted at Argonne National Laboratory (ANL), an efficiency of 38% (based on the higher heating value of the fuel gas) was quoted by United Technologies personnel for a fuel consisting of a typical reformer effluent gas, shifted (and perhaps methanated) to meet the carbon monoxide concentration specifications quoted above (1). The selling price (FOB factory) of the fuel cell power system, including the inverter, was given as $225/kW of installed capacity. If an allowance for the inverter system of $40/kW (2) is subtracted from this total, the final cell subsystem cost is $185/kW. The targeted (but as yet unachieved) useful life of the power system is 144 Ms (40,000 hr) of operation with a 5% loss in efficiency in that period.

Over the past 20 years, the General Electric Company has developed a fuel cell based on an electrolyte consisting of a porous film of a fluorocarbon with chemically-bound sulfonic acid groups (3). The solid polymer electrolyte (SPE), developed and marketed by Du Pont under the tradename Nafion, is reportedly extremely stable in the fuel cell environment. With near-ambient pressure air as oxidant, the optimum temperature of operation is approximately $348°K$ ($167°F$). Because the air (and fuel) streams must be presaturated to ensure membrane stability, higher temperatures cause excessive dilution of the oxygen in the air stream by water vapor. When oxygen is employed, operational temperatures up to $423°K$ ($302°F$) are possible.

State-of-the-art performances are 180 mW/cm^2 (167 W/ft^2) of active area at 0.66 V per cell (44% HHV) with air at $348°K$ ($167°F$) and 470 mW/cm^2 (437 W/ft^2) at 0.75 V per cell (51% HHV) with oxygen at $423°K$ ($302°F$). This work has also shown that at $348°K$ ($167°F$), with a fuel gas containing 0.3% carbon monoxide, the current noble metal catalyst will not tolerate this carbon monoxide level without a prohibitive performance loss, even at noble metal loadings in the anode of approximately 4 mg/cm^2 (5.69×10^{-5} lb/in^2) (4). This phenomenon, discussed above, is obviously an even severer limitation on this power system.

The state-of-the-art fuel cell manufacturing cost (not selling price) is estimated to be approximately $400/kW for the system operating on hydrogen and air. Of this total, approximately $260/kW is attributable to the noble-metal loadings and approximately $120/kW to the membrane. The economic viability of this system is seen, therefore, to depend on significant breakthroughs in the areas of noble-metal loading reductions and membrane substitution or cost reduction, as well as the development of methods for the lowering of the carbon monoxide content substantially below concentrations of 0.3%. A carbon monoxide concentration of 10 ppm is considered to be tolerable.

Sulfuric acid-based systems, in which the anode electrocatalyst is tungsten carbide, do not suffer from the carbon monoxide poisoning problem described above for the phosphoric acid and SPE systems (5). As is usual in sulfuric acid systems which operate typically at approximately $333°K$ ($140°F$), however, the major problem is the low activity of the cathode even when it is operated on oxygen. A further problem arising with long-time operation is the need to reject the water formed in the cathode reaction.

The phosphoric acid-based system was chosen as the best representative of the acid systems because it is the fuel cell which best addresses the problems of tolerance towards carbon monoxide and minimization of noble-metal loadings while maintaining high levels of anodic and cathodic activities. The confidence of the electrical utilities, as displayed by their support of the United Technologies fuel cell program, tends to support this judgment.

Alkaline Fuel Cells

The systems considered in this electrolyte-type group are those based on 30 weight percent aqueous potassium hydroxide solutions at approximately $343°K$ ($158°F$), 75 weight percent aqueous potassium hydroxide solutions at approximately $473°K$ ($392°F$), and saturated carbonate-bicarbonate aqueous solutions at temperatures of approximately $333°K$ ($140°F$). Although alkaline hydrogen-oxygen (H_2-O_2) fuel cells have been investigated by many companies, principally

Union Carbide, Shell (U.K.), Allis Chalmers, United Technologies, Exxon, Alsthom (France), Varta (G.F.R.) and Siemens (G.F.R.), only a small fraction of the overall effort has been devoted to the use of carbonaceous fuel gases. This neglect is principally due to problems associated with carbonation of the electrolyte by carbon dioxide in air or the fuel gas, or from the oxidation of carbon monoxide from the dissolution of the carbon dioxide formed in the alkaline electrolyte.

Various techniques (6) have been employed to overcome this problem in bench-scale systems. The electrolyte is circulated in the Allis-Chalmers methanol-oxygen system and is regenerated externally. The Shell (U.K.) methanol-air system employs a silver-palladium tube to separate the hydrogen from the gas mixture after external reformation, so that carbon-containing gases do not have access to the electrolyte. A similar scheme has been employed by United Technologies to provide hydrogen, formed by steam reformation of carbonaceous fuels, to the fuel electrodes of Bacon cells. This technique is impractical from technical and economic considerations in fuel cell power plants of the type considered in this study.

Commercial acid gas-scrubbing systems—for example, Lurgi Rectisol and Benfield—are available, which will permit reduction of the carbon dioxide content of feed streams to less than 10 ppm. There is, however, a serious question as to whether this level is sufficiently low to prevent electrolyte carbonation and, worse still, solid potassium carbonate formation in the pores of the gas diffusion electrodes, thereby leading to relatively rapid performance degradation. The problem of electrolyte carbonation may be dealt with by electrolyte circulation, followed ultimately by replacement with fresh caustic solution, or by cyclic decarbonation of the electrolyte in an external electrolytic cell, as proposed for the Exxon-Alsthom methanol-air fuel cell power system (7).

Scrubbing carbon dioxide from the air feed to the alkaline fuel cells may also be accomplished by the Lurgi and Benfield processes. This would probably be as expensive as the fuel gas scrubbing described above, despite the much lower carbon dioxide content of air, as the capital costs of the processes are a strong function of the total number of mols of throughput gas. Another approach involves simple caustic scrubbing of the incoming air. This could cost as little as $5/kW (8). Rather than dispose of the approximately one ton of potassium carbonate formed each day in a 25 MW fuel cell power plant, it is probably more desirable to regenerate the caustic by the Exxon-Alsthom technique alluded to above.

Despite the problems associated with scrubbing carbon dioxide from the fuel gas and air feeds, alkaline fuel cell subsystems are very attractive for large-scale power generation from the standpoints of cost and useful life. For example, long-lived air cathodes in alkaline solution operate with significantly better polarization characteristics than they do in acidic solutions. This is all the more remarkable since these cathodes do not require noble metals. Silver may be employed as the perhydroxide ion elimination catalyst with a concomitant substantial lowering of catalyst cost. A further advantage of alkaline fuel cell power systems is that nickel at approximately $4.41/kg ($2/lb) is stable in the cell environment. In contrast, acid systems are restricted to graphite and other carbon products because of the prohibitive expense of tantalum at $117/kg ($53/lb) and niobium at $33/kg ($15/lb).

Estimated overall system efficiencies for alkaline fuel cell power systems, using carbonaceous gas as a fuel, are similar to those quoted for acid systems, about 35 to 40%. The efficiency advantage of the alkaline fuel cell, resulting from lower cathodic polarization, is almost totally negated by the efficiency penalty associated with elimination of carbon dioxide from the air and fuel gas streams (9).

An alkaline fuel cell, operating in 30 weight percent potassium hydroxide solution at approximately 343°K (158°F) has been selected to represent this class of fuel cells in the rest of this study. It is preferred over the Bacon fuel cell, which operates typically in 75 weight percent potassium hydroxide solutions at temperatures in excess of 473°K (392°F), because of the well-known severe corrosion problems of this latter system. In contrast, the Exxon-Alsthom carbonate-bicarbonate fuel cell is disqualified from further consideration because of problems relating to low cathodic activity due mainly to excessive concentration polarization (8).

Molten Carbonate Fuel Cells

No large molten carbonate fuel cell batteries have been produced. Systems studies have been made for 15 to 22 kW units by Texas Instruments (10) and IGT (11)(12); and United Technologies has a development program on which few details have been published (13)(22). Broers (14)(15) did much of the earlier research on the devices upon which current technology is based, but did little work on multiple-cell devices.

Texas Instruments designed a 1 kW test unit which was delivered to the U.S. Army Mobility Equipment R&D Center at Fort Belvoir, Virginia. IGT has operated fuel cell batteries in excess of 2 kW, but the results of their recent work are not published. These units use a separate reformer to supply a suitable mixture of hydrogen and carbon monoxide to the cells.

An economic assessment of molten carbonate fuel cells for large-scale power production with propane feed to an integrated reformer-molten carbonate fuel cell system was published by Hart and Womack (16) in 1967. The calculations they presented were based in part on the unpublished work at the Central Electricity Generating Board Research Laboratories and Marchwood Engineering Laboratories. They used a fuel cell model representing their projection of the best fuel cell performance to be expected in the near future plus an estimation of the cost of individual parts of the model cell. The fuel cell operating at maximum power was assumed to use its waste heat to generate steam. This steam supplied a steam turbine generator for additional power generation.

For this system a plant efficiency of 46.5% was calculated based on propane feed as the fuel and assuming that a propane reformer was integrated with the fuel cell system. They concluded that initial capital costs for the system would be at least 25% higher than for a coal-fired steam turbine generator system of similar life. Since the fuel cell life was expected to be short [optimistically 157.7 Ms (5 years) compared to perhaps 315.4 to 630.7 Ms (10 to 20 years) for a gas turbine and 630.7 to 946.1 Ms (20 to 30 years) for a steam turbine], they concluded that such a fuel cell plant was not economically attractive. The economics of a fuel cell for domestic power are different, and von Fredersdorff (IGT) published an analysis of the molten carbonate fuel cell for this use in

1963 (17), concluding that the system could be economical if the fuel cell investment cost was no more than $300 over a 315.4 Ms (10 year) operation.

All costs—fuel costs in particular—are much higher now than they were at the time of the earlier studies. The molten carbonate fuel cell power capability has been improved significantly; and there are stricter regulations on emissions from power plants, substantially increasing capital cost and reducing efficiency.

The practical efficiencies which may be attained for large fuel cell batteries of this type, based on small battery performance, are still in dispute, but 45% efficiency based on the higher heating value (HHV) of natural gas is probably feasible (18).

The cells with highest performance reported recently use a paste electrolyte of alkali aluminates suggested by Broers (15), a nickel anode and lithiated nickel oxide cathode. The cells give useful power densities at 923°K (1202°F) of up to about 161 mW/cm² (150 W/ft²) using air as oxidant, depending on the fuel gas. Oxygen allows greater power densities. Cell temperatures of 873° to 1023°K (1112° to 1382°F) have been used, and corrosion problems increase at the higher temperatures.

In large systems it is anticipated that increased efficiency can be attained by recovering waste heat and using it to generate steam for use in a steam turbine or as process steam, as suggested by Hart and Womack (16). The highest efficiency plants would be large ones to minimize the turbogenerator plant costs.

If methanated (high-Btu) gas or methanol is used as a fuel, a reforming step is necessary. A higher efficiency is attainable if this can be accomplished on the fuel cell electrodes or nearby surfaces within the cell, since the reforming process absorbs heat which would be supplied in situ by the heat produced at the electrodes of the molten carbonate cell. Although a commercial reformer operaates at a higher temperature than does the molten carbonate cell, the large electrode area within the cell and the probable slower throughput per unit area makes internal reforming a reasonable possibility.

Experiments on internal reforming or reforming on a catalyst at the same temperature as the fuel cell have been successful (19)(20), but the feasibility of internal reforming at high power densities in multicell systems is still to be demonstrated. The fuel cell plant would have a clean exhaust, since most of the sulfur is removed from the fuel gases initially, and oxides of nitrogen are not formed in significant quantities at these low temperatures. Unused fuel gases can be burned with excess oxygen in the cathode exit gas stream.

In addition to the fuel cell plant cost analysis reported by Hart and Womack for large-scale power production by a molten carbonate system, Bockris and Srinivasan (21) present a list, from different sources, of some 1969 and projected fuel cell costs. A projected cost for the molten carbonate cell system attributed to Broers is $600/kW for large-scale production using natural gas and a reformer. More recently, United Technologies reports (13)(22) an estimate of ~ $225/kW capital cost (EPRI RP114 Program) for the molten carbonate system with a lower heating value (LHV) efficiency of 47%. They also report multicell life tests, in conjunction with the Electric Power Research Institute, of 36 Ms (10,000 hours) between overhauls, with good performance stability during subscale tests, some of which lasted more than 18 Ms (5,000 hours).

Broers and IGT have also reported up to 36 Ms (10,000 hours) fuel cell life, but there is no consensus on the eventual maximum which may be possible.

Variations of the Broers type of molten carbonate fuel cell have been suggested, such as the use of a liquid lead electrode catalyst for oxidation of solid forms of carbon or coal (23), but none has received an extended study effort or been developed as yet.

In summary, on the basis of available information, the eventual useful fuel cell life, cost, type of reforming necessary and efficiency are still very questionable.

Stabilized Zirconia Fuel Cells

High-temperature solid electrolyte fuel cells have many advantages over other types of fuel cells. As Markin describes (24), these are:

(a) There are no liquids involved, so problems associated with pore flooding and maintenance of a stable three-phase interface are totally avoided;

(b) The electrolyte composition is invariant and does not depend on the composition of the fuel and oxidant streams; and

(c) Activation polarization losses are negligible.

Further, a power system, based on the use of stabilized zirconia at operating temperatures of approximately 1273°K (1832°F), has a unique advantage when coal is employed as a fuel in that the waste heat, generated because of the thermodynamic and electrochemical inefficiencies of the fuel cells, may be used directly in the gasification of coal, thus providing fuel gas for the fuel cells. This thermal coupling leads to high overall efficiencies for the power system, and practical efficiencies of greater than 60% are considered possible (25). This efficiency is all the more remarkable because this type of power system employs coal, rather than natural gas, methanol or naphtha, as a fuel.

The disadvantages of this fuel cell are related to:

(a) The relatively high electrolyte resistivity;

(b) The need for an effective, low-cost method of interconnection of cells to form a battery; and

(c) Problems involving battery component interactions and adequate sealing techniques.

Many companies and other research organizations [e.g., Westinghouse, General Electric, C.G.E. (France), Brown Boveri (G.F.R.), Battelle (Geneva, Switzerland) and AERE (Harwell, England)] have explored possible solutions to these problems in their efforts at component development and device fabrication. The largest device demonstrated to date was based on a bell and spigot geometry. It was constructed and operated by the Westinghouse Electric Corporation and delivered 100 W of electrical power (26).

The Westinghouse thin-film concept provides an economical and effective method for the series-connection of individual cells in a solid-electrolyte battery. The largest of these devices demonstrated to date delivered 8 W and lived for about

360 ks (100 hours)—the first 108 ks (30 hours) saw a 5% voltage degradation at a given current output (28). The life-limiting problem lay in the interconnection-electrolyte junction where inadequate sealing led to fuel gas leakage, resulting in air electrode reduction and, ultimately, performance degradation. This design has been selected for this study.

DESCRIPTION OF PARAMETRIC POINTS

To facilitate a comprehensive comparison of the four fuel cell power-generation systems, the parameter values listed in Table 4.1 below were fixed for the base cases of all systems.

TABLE 4.1: BASE CASE VALUES COMMON TO ALL SYSTEMS

Parameter	Value
Size of power plant	25 MW dc
Type of fuel	High-Btu gas
Type of oxidizer	Air
Fuel cell useful life	10,000 hr
	(5% efficiency degradation)

The smallest system to be considered in this study, 25 MW, was chosen for the base cases, as no economies of scale are expected for the fuel cell subsystem. Further, no fuel cell power system of even two orders of magnitude lower in rating has been constructed or operated, so that even in this small a power plant the system problems can only be addressed in the most general fashion, as will be evident from the schematics provided in the following subsections.

A 25-MW fuel cell power plant is too small to justify the expense of a dedicated coal gasification reactor. Thus, it will be fueled with hydrogen, high-Btu gas, or methanol, all of which may be derived from coal. Because of the flexibility afforded by the existing extensive network of natural gas lines, high-Btu gas was selected as the fuel for the base cases.

A value for the useful lives of all of the fuel cell subsystems was arbitrarily fixed for the base cases at 36 Ms (10,000 hours) at a constant power output, with a 5% degradation in terminal voltage. The degradation specification is important in the determination of the power conditioning costs. The useful life value specified is considered a reasonable estimate of the state of the art for molten carbonate fuel cells, an overestimate by a factor of approximately 100 for the solid electrolyte fuel cell, and an underestimate by a factor of approximately 2 for aqueous acid and alkaline fuel cells, as described in the previous section.

In general, the parametric assessment involved the variation of one parameter, with the others retaining the value used in the base case. In certain cases, however, a change in one parameter caused the variation of other parametric values. The change from air to oxygen as the oxidant, for example, was thought to cause an increase in the current density in the acid, alkaline and molten carbonate fuel cells; and an increase in the cell voltage in the solid electrolyte fuel cells.

Point 1 of the parametric point list for each fuel cell type is the base case. The abbreviations AC, AL, MC and SE were chosen for the acid, alkaline, molten carbonate, and solid electrolyte fuel cell power systems, respectively. Thus, AC1 represents the base case in the phosphoric acid system. The first eight parametric value changes common to all four systems are shown in Table 4.2. Points 2 through 4 explored the economics of scale realized by increasing the power plant rating. Because of their modular nature, no economy of scale was assumed for the fuel cell subsystems. Any economic benefits realized, therefore, come from the other subsystems. The effect of the replacement of air with oxygen from a dedicated oxygen plant was tested for all systems in Point 5. Three more parametric points per system (6 through 8) were expended to investigate the impact of fuel cell subsystem useful life on the electricity costs. The effect of a terminal voltage decrease of 15% (at constant power output) after 360 Ms (100,000 hours) of plant operation was explored in Point 9.

TABLE 4.2: PARAMETRIC CHANGES COMMON TO ALL FOUR SYSTEMS

Point No.	Parameter	Value	Base Case
2	Size of power plant	100 MW	25 MW
3	Size of power plant	250 MW	25 MW
4	Size of power plant	900 MW	25 MW
	Fuel type	Medium-Btu gas	High-Btu gas
5	Oxidant type	Oxygen	Air
6	Fuel cell useful life	30,000 hr (5% Voltage degradation)	10,000 hr
7	Fuel cell useful life	50,000 hr (5%)	10,000 hr
8	Fuel cell useful life	100,000 hr (5%)	10,000 hr
9	Fuel cell useful life	100,000 hr (15%)	10,000 hr

A cross-comparison of the efficiencies and electricity costs is possible for all systems in Points 4 through 9, and for the acid, alkaline and molten carbonate systems in Points 2 and 3. The solid electrolyte power system is not available for comparison purposes in Points 2 and 3 because medium-Btu gas was employed as a fuel instead of the high-Btu gas used in the other three systems. Further details of Points 2 through 8, and a description of the parametric points specific to each system, are provided in the following subsections.

Phosphoric Acid Fuel Cell Power System

As indicated previously, the base case in this power system, AC1, involves a fuel cell subsystem that has a 25-MW dc rating, a useful life of 36 Ms (10,000 hours) of operation, and a 5% voltage degradation at constant power output. The fuel is a high-Btu gas; air is the oxidant. In the fuel cell subsystem, the fuel cell is considered to have a bipolar design similar to that described by

Baker et al (29). The anode and cathode, each fabricated from carbon as described by Kordesch and Scarr (30), are catalyzed by a platinum addition to each electrode of 1 mg/cm^2 (0.002 lb/ft^2). The addition of other electrocatalysts for the reduction of performance sensitivity to carbon monoxide in the fuel stream (31) are considered desirable but were not included in the economic analysis. The electrolyte is 85 weight percent aqueous phosphoric acid immobilized in a zirconium pyrophosphate matrix with an effective resistivity of 2 Ω-cm. The electrolyte thickness is 0.5 mm (0.020 inch). The fuel cell subsystem operates at 463°K (375°F). Based on a conservative estimate of anticipated advances in the state of the art, beyond that reported by Schiller and Meyer (32) in 1971, values of 0.7 V and 200 mA/cm^2 (186 A/ft^2) were selected for the cell voltage and current density, respectively.

A schematic of the complete power system is shown in Figure 4.1. High-Btu gas from a 0.689 MPa (100 psi) abs line is assumed to be available, and is fed after preheating to a steam-methane reformer operating at a pressure of 0.689 MPa (100 psi) abs and a temperature of 1144°K (1600°F). The reformer effluent, consisting mainly of carbon monoxide, hydrogen, carbon dioxide and steam, is cooled and fed to a shift converter operating at 0.483 MPa (70 psi) abs and 700°K (800°F). The shift converter is operated at as low a temperature as possible in order to minimize the carbon monoxide concentration in the exit gas. The hydrogen-rich fuel gas is further cooled to approximately 422°K (375°F) and is fed to the ten fuel cell modules. Air is supplied to the modules by means of blowers, as shown in Figure 4.1.

Steam, required for the steam reformation of methane (the principal constituent of high-Btu gas) is raised in the cooling of the fuel gas between the reformer and the shift converter, in the shift converter, and between the shift converter and fuel cell subsystem. The water required for the steam generators is reclaimed from the fuel-gas exhaust by the knock-out process shown in Figure 4.1. The water-vapor depleted exhaust gases, containing approximately 10% of the hydrogen fed to the fuel cell modules, and the unused carbon monoxide, is mixed with a portion of the incoming high-Btu gas, and the mixture is burned to supply the heat required to cover the endothermic processes in the reformer. Further details of the fuel processing subsystem are provided in Appendix 1.

The fuel cell subsystem, rated at 25 MW dc, consists of ten 2.5 MW modules. Each module is wired into a dedicated power conditioning unit consisting of a force-commutated dc-to-ac inverter, a transformer and filters. A more detailed description of the power conditioning subsystem is provided in Appendix 2.

As indicated previously, the changes from the base case, characterized by Points 2 through 9 and shown in Table 4.2, are common to all systems. AC2 and AC3 differ from AC1 only in that the fuel cell subsystems are rated at 100 MW and 250 MW dc, respectively. In AC4, the fuel cell subsystem is rated at 900 MW dc. The fuel used in the 900 MW plant is medium-Btu gas, instead of the high-Btu gas employed in AC1, AC2 and AC3. Thus, the steam-methane reformer, shown in Figure 4.1, is unnecessary in this power plant.

AC5 involves the use of oxygen as the fuel cell oxidant instead of air. Because of the reduction of concentration polarization at the fuel cell cathode, the apparent current density was considered to have doubled, i.e., from 200 mA/cm^2

(186 A/ft^2) to 400 mA/cm^2 (372 A/ft^2), despite the concomitant increase in
the fuel electrode polarization and the cell ohmic polarization losses.

FIGURE 4.1:　SCHEMATIC OF A PHOSPHORIC ACID FUEL CELL POWER SYSTEM WHICH EMPLOYS HIGH-Btu GAS AS FUEL

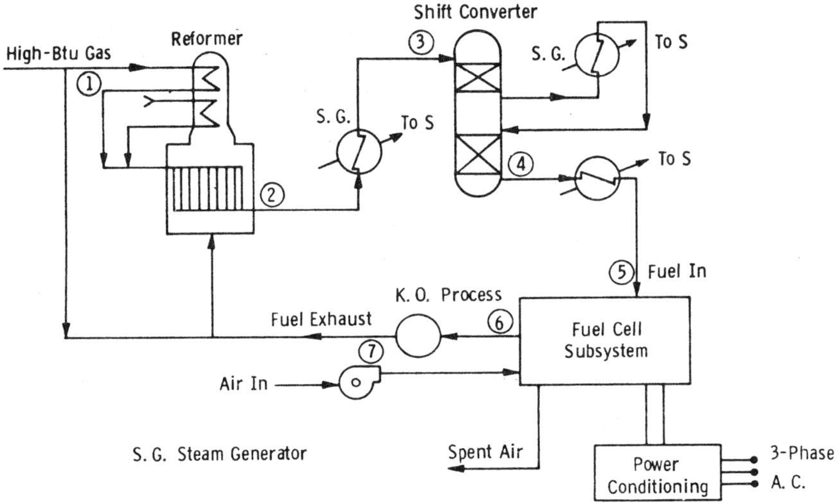

Point No	Temperature, °F	Pressure, psia	Point No	Temperature, °F	Pressure, psia
1	60	100	5	300	25
2	1600	85	6	375	20
3	800	70	7	120	18
4	800	40			

In parametric points AC6, AC7 and AC8, the fuel cell subsystem useful life in
a 25 MW dc plant is increased to 108 Ms (30,000 hours), 180 Ms (50,000 hours)
and 360 Ms (100,000 hours), respectively. In all three points, as in the base
case, AC1, a 5% efficiency or voltage degradation, by comparison with the initial performance, is assumed at the end of useful life. Thus, the initial cell voltage, 0.7 V, will have fallen to 0.665 V at end of life. In AC9, a voltage degradation of 15% (0.105 V) is assumed at the end of a 360 Ms (100,000 hour) useful
life.

The changes from the base case, represented in Points AC10 through AC16, are
specific to the phosphoric acid fuel cell power system. These are shown in
Table 4.3.

AC10 and AC11 were included to explore the effect of fuel cell subsystems cost
reductions, resulting from increases in the apparent current density of the base
case, 200 mA/cm^2 (186 A/ft^2), to 300 mA/cm^2 (279 A/ft^2) and 400 mA/cm^2
(372 A/ft^2). Similarly, AC12 and AC13 represent advances in the state of the

art which result in the lowering of platinum loadings in both cathode and anode from 1 mg/cm^2 (0.002 lb/ft^2) in the base case to 0.3 mg/cm^2 (6 x 10^{-4} lb/ft^2) and 0.1 mg/cm^2 (2 x 10^{-4} lb/ft^2) respectively.

TABLE 4.3: PARAMETRIC POINTS—ACID FUEL CELL

Point No.	Parameter	Value/Type	Base Case Value
10	Current density	300 mA/cm^2	200 mA/cm^2
11	Current density	400 mA/cm^2	200 mA/cm^2
12	Catalyst loading	0.3 mg Pt/cm^2	1 mg Pt/cm^2
13	Catalyst loading	0.1 mg Pt/cm^2	1 mg Pt/cm^2
14	Type of fuel	Methanol	High-Btu gas
15	Electrolyte thickness	0.25 mm	0.5 mm
16	Oxidant type	Oxygen	Air
	Size of power plant	250 MW	25 MW

High-Btu gas is replaced by methanol as the fuel in AC14. This will involve storage of methanol in tanks at the power plant. The electrolyte thickness of 0.5 mm (0.020 inch) in the base case is reduced to 0.25 mm (0.010 inch) in AC15. Because of the resultant reduction in ohmic losses in the electrolyte, the cell voltage at 200 mA/cm^2 (180 A/ft^2) is considered to have increased from 0.70 V to 0.71 V.

The effects of replacement of air by oxygen in a 250 MW dc power plant are explored in AC16. Once again, as in AC5, the assumed apparent current density of the cell was doubled to 400 mA/cm^2 (372 A/ft^2). This point was included to demonstrate, by comparison with AC5, the effect of scale in lowering the oxygen cost.

Alkaline Fuel Cell Power System

A schematic of the alkaline fuel cell power system base case, AL1, is shown in Figure 4.2. The temperatures and pressures at various system locations are also presented in this figure. For the purposes of this study, it has been assumed that the alkaline power system is similar to its acid counterpart, described in the previous subsection, but with the following differences:

(A) Carbon dioxide scrubbers—To prevent carbonation of the electrolytes, the Lurgi Rectisol process is employed to scrub both the air and the fuel gas inlet streams, as shown in Figure 4.2. Further details of this process, which employs refrigerated methanol to scrub acid gases, are found in Appendix 1.

(B) Fuel cell subsystem—A bipolar configuration, identical with that described for the acid system, is assumed also for the alkaline fuel cell battery. However, the apparent current density in the base case for this system, 100 mA/cm^2 (93 A/ft^2), is only half that in the acid fuel cell. The cell voltage, 0.8 V, is higher because of the generally lower cathodic polarizations in the alkaline fuel cell. These values represent a conservative estimate of advances in the state of the art since 1969 (33) for this fuel cell. The anode and cathode are catalyzed with 1 mg of platinum/cm^2 (0.002 lb/ft^2) and 5 mg of silver/cm^2 (0.1 lb/ft^2), respectively. The electrolyte thickness is 0.5 mm (0.020 inch) with an assumed effective resistivity of 2 Ω-cm (5 ohm-inch). The fuel cell operates at a temperature of 343°K (158°F), not only for the base case but also for all parametric points described below.

FIGURE 4.2: SCHEMATIC OF AN ALKALINE FUEL CELL POWER SYSTEM WHICH EMPLOYS HIGH-Btu GAS AS FUEL

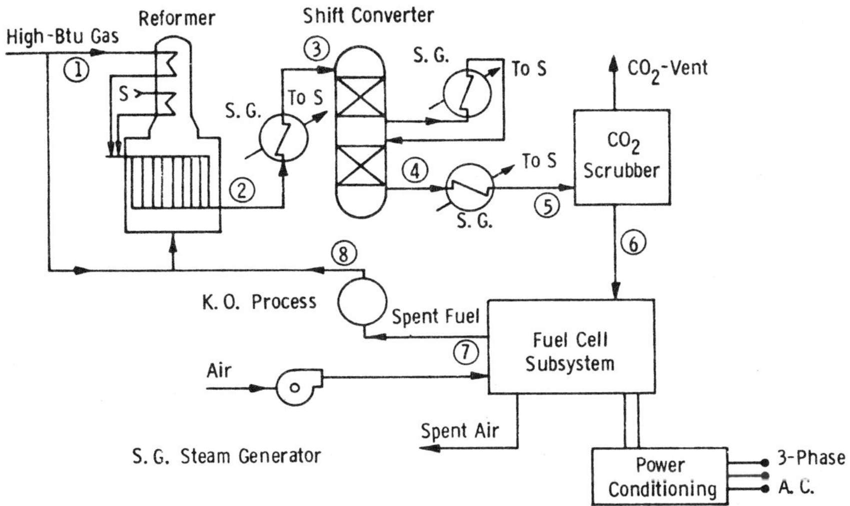

Point No	Temperature, °F	Pressure, psia	Point No	Temperature, °F	Pressure, psia
1	60	100	5	400	25
2	1600	85	6	100	25
3	800	70	7	160	20
4	800	40	8	100	18

Points AL2 through AL9 involve the same changes in parameter values as those for AC1 through AC9 discussed in the previous subsection. Once again, replacement of air as oxidant by oxygen from a dedicated oxygen plant (AL5) is considered to result in a doubling of the base case apparent current density, i.e., from 100 mA/cm^2 (93 A/ft^2) to 200 mA/cm^2 (186 A/ft^2).

The changes from the base case, represented by parametric points AL10 through AL16, are specific to the alkaline fuel cell system, and are shown in Table 4.4.

TABLE 4.4: PARAMETRIC POINTS—ALKALINE FUEL CELL

Point No.	Parameter	Value/Type	Base Case Value
AL10	Current density	175 mA/cm^2	100 mA/cm^2
AL11	Current density	250 mA/cm^2	100 mA/cm^2
AL12	Anode catalyst loading	0.1 mg Pt/cm^2	1 mg/cm^2
AL13	Anode catalyst loading	0.01 mg Pt/cm^2	1 mg/cm^2
AL14	Anode and cathode materials	Raney Nickel	Pt/C
AL15	Cathode catalyst loading	1 mg Ag/cm^2	5 mg/cm^2
AL16	Electrolyte thickness	0.25 mm (0.010 in)	0.5 mm (0.020 in)

The effects of increases in the fuel cell apparent current density from 100 mA/cm^2 (93 A/ft^2) to 175 mA/cm^2 (163 A/ft^2), brought about by improvements in the state of the art, are tested in parametric points AL10 and AL11, respectively. Points AL12 and AL13 explore the effects of reductions of platinum loadings in the anode from 1 mg/cm^2 (0.002 lb/ft^2). Replacement of the carbon gas-diffusion electrodes by Raney nickel electrodes is examined in parametric point AL14. Point AL15 tests the effect of the reduction of the silver loading in the cathode from 5 mg/cm^2 (0.01 lb/ft^2) to 1 mg/cm^2 (0.002 lb/ft^2). In Point AL16, the electrolyte thickness, or electrode separation, is 0.025 cm (0.01 inch), i.e., half of that in the base case. This results in a cell voltage increase of 5 mV, so that the cell voltage at 100 mA/cm^2 (93 A/ft^2) is 0.805 V in AL16.

Molten Carbonate Fuel Cell Power System

The effect of ten plant and operating variables on the cost of electricity was investigated. These are the effect of power plant size, fuel cell life, cell output degradation, current density at fuel cell electrodes, electrolyte thickness, temperature of the fuel cell, replacement of air by oxygen, oxygen plant size, fuel type and recovery of waste heat from the fuel cell by a steam turbine generator system.

MC1, the reference or base case, has a 25-MW dc rating with a filter press design similar to that described by IGT (22), with a porous nickel anode and lithiated nickel oxide cathode. The electrolyte is the Broers type (15), consisting of a semisolid paste of alkali aluminate powder and molten alkali carbonates with the ternary eutectic composition (Li$_2$CO$_3$—43.5 mol percent, Na$_2$CO$_3$—31.5 mol percent, K$_2$CO$_3$—25.0 mol percent, MP 670°K. Electrolyte resistance (23) is assumed to be 1.5 times that of the free electrolyte of the same thickness. This value would actually vary with the paste structure and chemical composition. High-Btu gas is used as fuel, and it is assumed that it can be reformed on internal cell surfaces. A schematic of the plant configuration for MC1 is shown in

Figure 4.3. It consists of a split series of fuel cell modules consisting of ten separate banks, each with its own dc-to-ac inverter, transformer and filters. It operates at 923°K, as do all other plants except MC12 and MC13, which operate at 973° and 1023°K, respectively.

FIGURE 4.3: SCHEMATIC OF MOLTEN CARBONATE FUEL CELL POWER SYSTEMS (MC1-3, MC6-14 AND MC16)

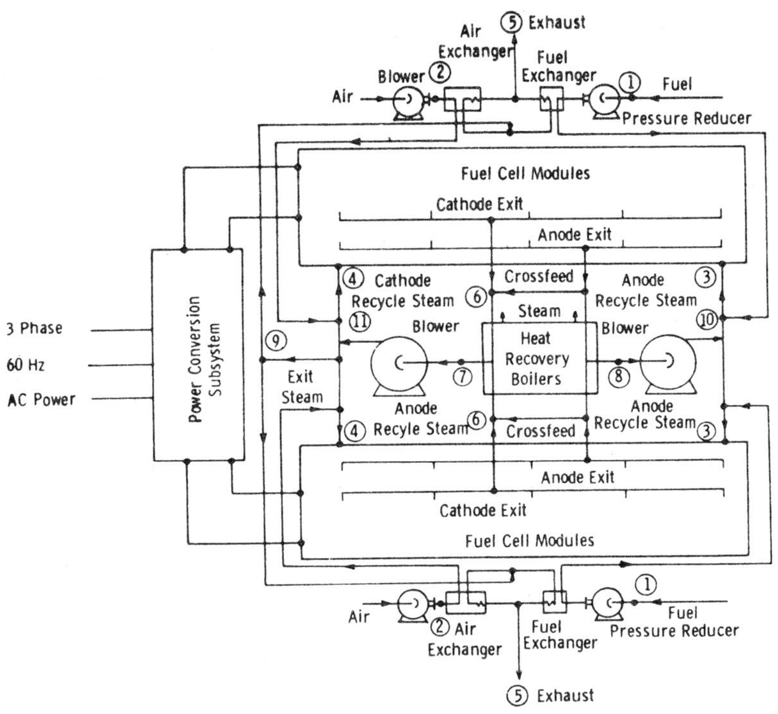

Point No.	Temperature, °F	Pressure, psia	Point No.	Temperature, °F	Pressure, psia
1	ambient	100	7	752	15
2	ambient	35	8	752	15
3	752	30	9	752	30
4	752	30	10	752	30
5	302	15	11	752	30
6	1202	25			

Fuel and oxidant streams enter the fuel cell at **3** and **4** and react electrochemically at the anode and cathode, respectively, at the fuel cell temperature of 923°K. After leaving the fuel cell, part of the anode (fuel) gas stream is added to the cathode gas stream at **6** to replace carbon dioxide lost in the cathode electrochemical reactions. The two gas streams, thus modified, enter a heat recovery steam generator and exit at a temperature between 673° and 823°K, depending on the recycle flow rate necessary to minimize concentration polarization at the fuel cell electrodes and to remove excess heat from the fuel cell. Part of the cathode gas stream is then diverted at **9** through a counterflow heat exchanger, leaving as stack exhaust at **5** at 423°K, and in the process preheating incoming air and desulfurized fuel from ambient temperature to 573° to 723°K. The exhaust gas stream contains carbon dioxide, water vapor, nitrogen, a small amount of oxygen, but no combustible gases.

Preheated air and desulfurized fuel, entering at **2** and **1**, are then combined with the cathode and anode gas streams at **11** and **10** and enter the fuel cell again at **4** and **3**. This completes the cycle. The steam produced can be sold as process steam, used to operate a steam turbine to produce more electricity as in MC4, or operate a turbine drive for an oxygen plant as in Points MC5 and MC17. MC1, MC2 and MC3 differ only in plant size, being 25-MW dc, 100-MW dc and 250-MW dc, respectively. MC4 differs in size (900-MW dc) and also in utilizing a steam turbine generator to produce additional electricity from waste heat, as shown in Figure 4.4. The plant configuration differs from MC1 not only in size but also in allowing space for the steam turbine generator system with its ac transformer.

FIGURE 4.4: SCHEMATIC OF A MOLTEN CARBONATE FUEL CELL POWER PLANT WITH HEAT RECOVERY BY A STEAM TURBINE GENERATOR (MC4)

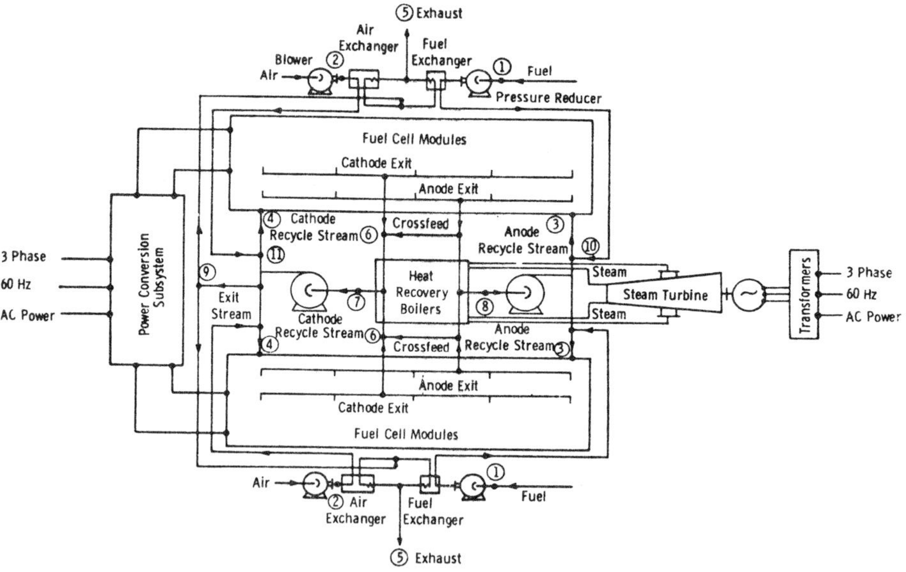

Point No.	Temperature, °F	Pressure, psia	Point No.	Temperature, °F	Pressure, psia
1	ambient	100	7	752	15
2	ambient	35	8	752	15
3	752	30	9	752	30
4	752	30	10	752	30
5	302	15	11	752	30
6	1202	25			

MC5 is similar to MC1 except for inclusion of an oxygen plant capable of supplying about 3.461 to 3.623 kg/s (330 to 345 tons/day) of oxygen to be used in place of air. The schematic is shown in Figure 4.5.

FIGURE 4.5: SCHEMATIC OF A MOLTEN CARBONATE FUEL CELL SYSTEM USING OXYGEN (MC5 AND MC17)

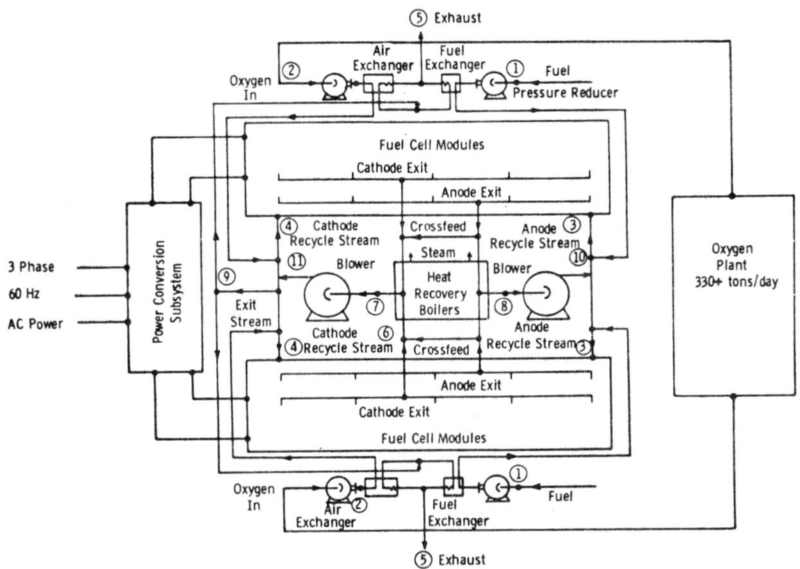

Point No.	Temperature, °F	Pressure, psia	Point No.	Temperature, °F	Pressure, psia
1	ambient	100	7	752	15
2	ambient	35	8	752	15
3	752	30	9	752	30
4	752	30	10	752	30
5	302	15	11	752	30
6	1202	25			

MC6, MC7 and MC8 differ from MC1 only in fuel cell lifetimes of 108, 180 and 360 Ms (30,000, 50,000 and 100,000 hours), respectively; and MC9 is similar to MC8 except for the assumption of 15 rather than 5% voltage degradation. MC10 and MC11 assume a current density of 150 and 250 mA/cm^2 (139 A/ft^2 and 232 A/ft^2) as compared to 200 mA/cm^2 (186 A/ft^2) for MC1.

MC12 and MC13 assume an operating temperature of 973° and 1023°K (1292° and 1382°F), respectively, as compared to 923°K (1202°F) for MC1. Electrolyte resistance (23) is a little lower at the higher temperatures, but corrosion is worse.

MC14 uses medium-Btu gas as fuel rather than the high-Btu gas used by MC1. MC15 uses methanol as fuel, which requires a storage tank as shown in Figure 4.6. Otherwise, it is similar to MC1.

MC16 differs from MC1 in using an electrolyte thickness of 0.5 mm (0.020 inch) rather than 1 mm (0.040 inch) used for MC1. MC17 is similar to MC3 (250-MW dc) but uses oxygen rather than air as oxidant. This is included as a comparison to MC5 (25-MW dc) to show the effect of the lower oxygen cost with larger plant size.

FIGURE 4.6: SCHEMATIC OF A MOLTEN CARBONATE FUEL CELL SYSTEM USING METHANOL AS FUEL (MC15)

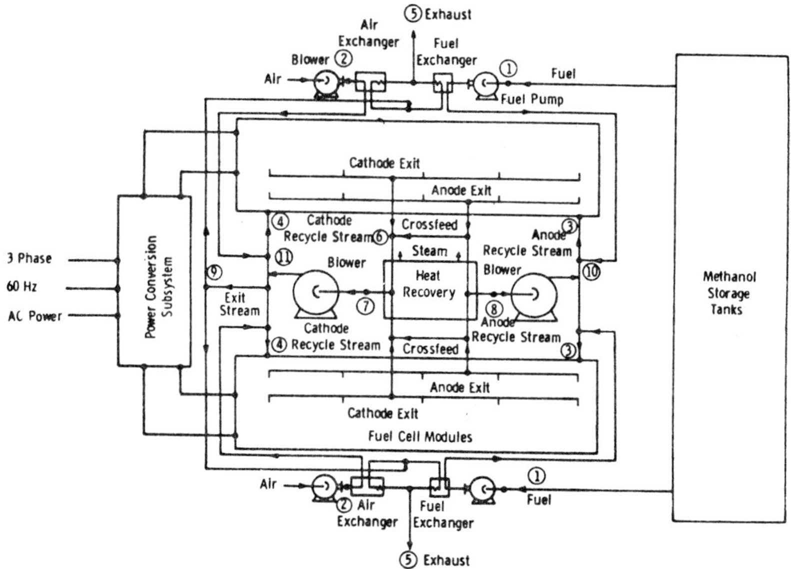

Point No.	Temperature, °F	Pressure, psia	Point No.	Temperature, °F	Pressure, psia
1	ambient	15	7	752	15
2	ambient	35	8	752	15
3	752	30	9	752	30
4	752	30	10	752	30
5	302	15	11	752	30
6	1202	25			

Solid Electrolyte Fuel Cell Power System

A schematic of the power plant corresponding to the base case, SE1, is shown in Figure 4.7. The basic plant layout is relatively simple in that it consists of the fuel cell generator, the power conditioning subsystem, recuperative heat exchangers (to allow the fuel cell exit gases to heat up the incoming fuel and air streams), and an air blower. The temperatures and pressures at various locations in the power system are also tabulated in Figure 4.7.

All parametric points, including SE1, employ the Westinghouse thin-film solid-electrolyte fuel cell battery in the fuel cell subsystem. This device, built on porous tubes of stabilized zirconia, is described in greater detail (25). The electrolyte film of yttria-stabilized zirconia (ZrO$_2$—10% Y$_2$O$_3$) is gas-impervious and 40 μm (1.6 mils) in thickness. The air electrode is a porous layer of tin oxide-doped indium sesquioxide activated with praseodymium cobaltite. A porous layer of a nickel-stabilized zirconia cermet serves as the fuel electrode. The interconnection layer, which serves to series-connect adjacent cells, consists of a gas-impervious layer of chromium sesquioxide, 20 μm (0.8 mil) in thickness. The device is assumed to operate at 1273°K (1832°F) and with high-Btu gas as fuel. The current density in the electrolyte region of the unit cell is taken as 400 mA/cm^2 (372 A/ft^2).

FIGURE 4.7: SCHEMATIC OF A SOLID-ELECTROLYTE FUEL CELL POWER PLANT EMPLOYING HIGH- OR MEDIUM-Btu GAS AS A FUEL (SE1-3, SE6-17, SE20)

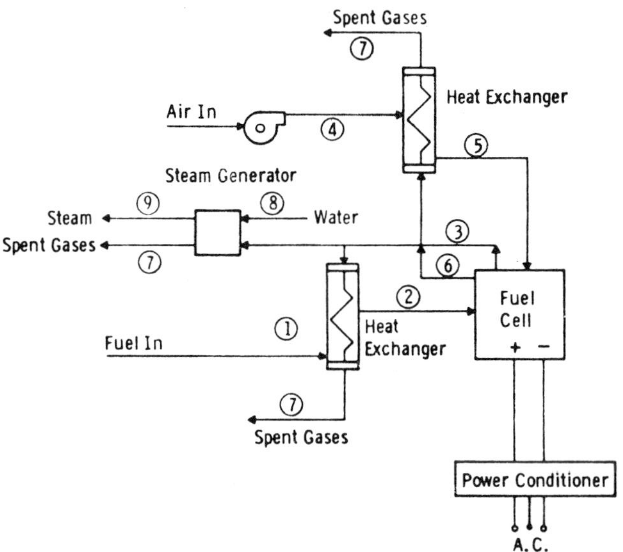

Point No.	Temperature (°F)	Pressure (psia)	Point No.	Temperature (°F)	Pressure (psia)
1	80	60	6	1300	30
2	1100	45	7	200	15
3	1300	30	8	70	650
4	80	60	9	1000	600
5	1100	45			

Parametric points SE2, SE3 and SE4 explore the economies of scale achievable in power plants. The fuel cell subsystems are rated at 100-MW dc, 250-MW dc and 900-MW dc, respectively. These three power plants are fueled with medium-Btu gas, as are all plants, except in the base case (high-Btu) and in parametric point SE19, in which low-Btu gas from an integrated gasifier system is employed. A steam bottoming plant is included in the 900-MW power plant, SE4, similar to that described for MC4 in the previous subsection. With medium-Btu gas, the average cell voltage at 400 mA/cm^2 (372 A/ft^2) is 0.66 V, corresponding to a voltage efficiency of 80%. The plant layout corresponding to SE4 is shown schematically in Figure 4.8.

FIGURE 4.8: SCHEMATIC OF A SOLID-ELECTROLYTE FUEL CELL POWER PLANT WHICH EMPLOYS A STEAM BOTTOMING PLANT FOR WASTE HEAT RECOVERY (SE4)

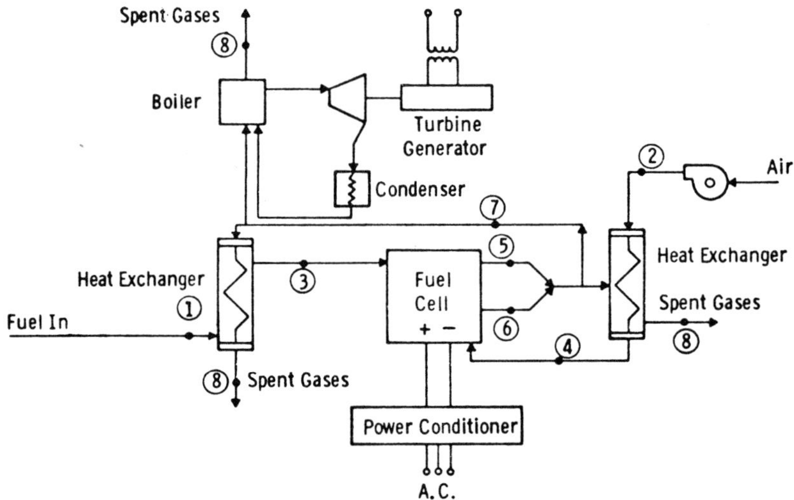

Point No.	Temperature (°F)	Pressure (psia)	Point No.	Temperature (°F)	Pressure (psia)
1	80	60	5	1300	30
2	80	60	6	1300	30
3	1100	45	7	1300	30
4	1100	45	8	200	15

Point SE5 explores the effect of the replacement of air as the oxidant by oxygen from a dedicated oxygen plant (see Figure 4.9). The average cell voltage increases to 0.76 V, as most of the concentration polarization in the fuel cell battery is associated with the cathode. Points SE6 through SE8 explore the effects of increases in the useful life of the fuel cell batteries. The effect of increased power conditioning costs associated with an increase in the permissible voltage degradation and described in the previous subsections is investigated in Point SE9.

FIGURE 4.9: SCHEMATIC OF A SOLID-ELECTROLYTE FUEL CELL POWER PLANT EMPLOYING OXYGEN AS AN OXIDANT (SE5)

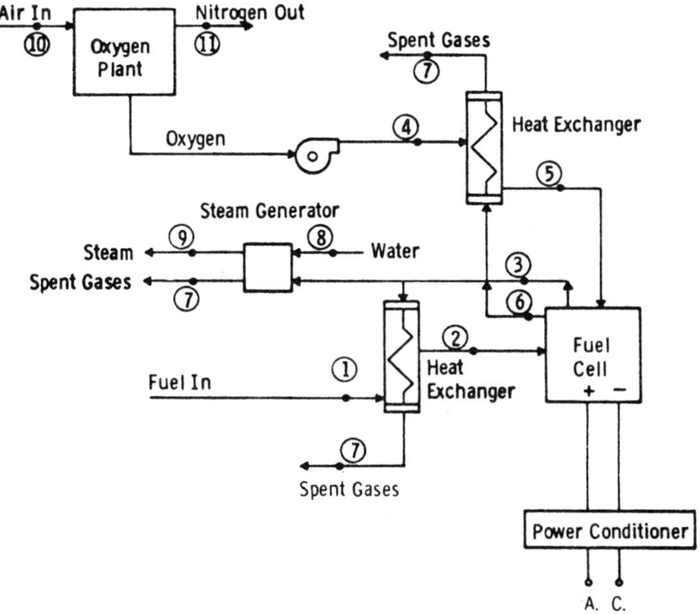

Point No.	Temperature (°F)	Pressure (psia)	Point No.	Temperature (°F)	Pressure (psia)
1	80	60	7	200	15
2	1100	45	8	70	650
3	1300	30	9	1000	600
4	80	60	10	70	15
5	1100	45	11	70	15
6	1300	30			

The changes represented by Points SE10 through SE20 are specific to the solid electrolyte power system and are presented in Table 4.5. Points SE10 and SE11 investigate the effect of battery operation at higher power densities. At current densities of 600 and 800 mA/cm² (557 and 743 A/ft²), the cell voltages were 0.59 and 0.51 V, respectively, because of increased ohmic and concentration polarization losses. A reduction of electrolyte thickness in SE12 from 40 to 20 μm (1.6 to 0.8 mil) leads to a cell voltage increase from 0.66 to 0.67 V. The effect of the replacements of chromium sesquioxide by manganese-doped cobalt chromite, and of tin-doped indium oxide by antimony-doped tin oxide, are explored in SE13 and SE15. Substitution of yttria-stabilized zirconia by calcia-stabilized zirconia in SE14 leads to a reduction of the cell voltage from 0.66 to 0.64 V because of the higher resistivity of the better electrolyte. If the fuel cell subsystem operates at 1173°K (1652°F), resistive losses in the electrolyte and interconnection regions lead to a reduction in cell voltage from 0.66

to 0.55 V. Conversely, an increase in the temperature of operation to $1373°K$ $(2012°F)$, as in SE17, results in an increase in cell voltage to 0.77 V.

TABLE 4.5: PARAMETRIC POINTS—SOLID-ELECTROLYTE FUEL CELL

Point No.	Parameter	Value/Type
10	Current density	$600 \ mA/cm^2$ ($557 \ A/ft^2$)
11	Current density	$800 \ mA/cm^2$ ($743 \ A/ft^2$)
12	Electrolyte thickness	$20 \ \mu m$ (0.8 mil)
13	Interconnection material	Mn-doped $CoCr_2O_4$
14	Electrolyte material	Calcia-stabilized zirconia
15	Air electrode material	Sb-doped SnO_2
16	Temperature	$1173°K$ ($1652°F$)
17	Temperature	$1373°K$ ($2012°F$)
18	Type of fuel	Coal
	Current density	$800 \ mA/cm^2$ ($743 \ A/ft^2$)
	Power plant size	250 MW dc
	Electrolyte thickness	$20 \ \mu m$ (0.8 mil)
19	Type of fuel	Low-Btu gas
	Power plant size	900 MW dc
20	Current density	$800 \ mA/cm^2$ ($743 \ A/ft^2$)
	Electrolyte thickness	$20 \ \mu m$ (0.8 mil)
	Type of fuel	High-Btu gas

The Westinghouse–OCR fuel cell power generation system is represented by Point SE18. In a 250-MW dc plant, shown schematically in Figure 4.10, a coal gasification reactor and the fuel cell subsystem are thermally coupled.

The heat released because of thermodynamic and electrochemical inefficiencies in the fuel cell batteries is employed to gasify coal. The fuel gas thus generated is then fed to the fuel electrodes of the fuel cell subsystem.

All the oxygen reaching the gasifier as carbon dioxide and water vapor enters through the fuel cell electrolyte. An electrolyte thickness of 20 μm (0.8 mil), an electrolyte current density of 800 mA/cm^2 (743 A/ft^2), and an average cell voltage of 0.68 V are assumed.

Point SE19 explores the advantages and disadvantages of coupling a 900-MW dc fuel cell subsystem with a low-Btu coal gasifier, as shown schematically in

Figure 4.11. Because of the nitrogen diluent in the fuel gas, the fuel electrode concentration polarization is relatively higher. Consequently, the cell voltage is lowered to 0.56 V.

Point SE20 investigates the effect of a higher current density, 800 mA/cm^2 (743 A/ft^2), and a reduced electrolyte thickness, 20 μm (0.8 mil). Even with high-Btu gas as fuel, the average cell voltage is 0.69 V compared with 0.84 V in the base case.

FIGURE 4.10: SCHEMATIC OF THE WESTINGHOUSE FUEL CELL POWER SYSTEM INVOLVING THERMAL COUPLING OF THE FUEL CELL SUBSYSTEM AND THE COAL GASIFIER (SE18)

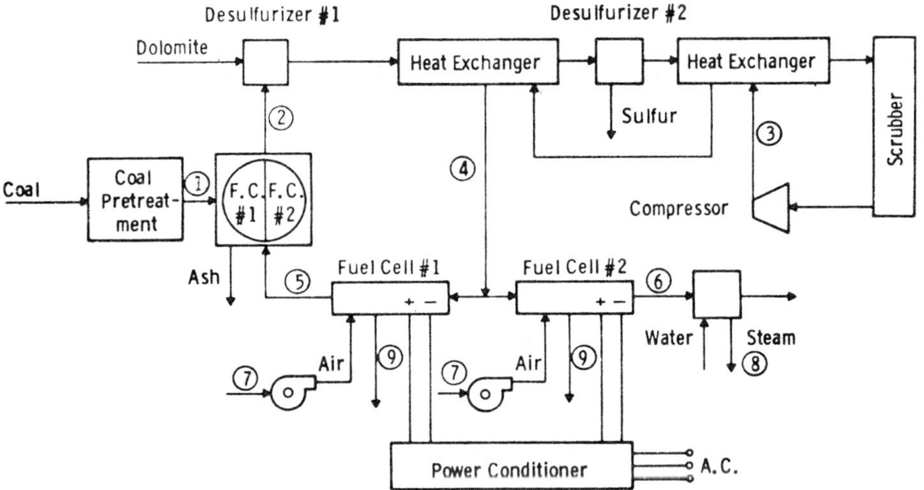

Point No.	Temperature (°F)	Pressure (psia)	Point No.	Temperature (°F)	Pressure (psia)
1	70	−	6	1870	23.5
2	1750	19	7	70	15
3	200	35	8	1000	600
4	1550	35	9	200	17
5	1650	35			

FIGURE 4.11: SCHEMATIC OF A SOLID-ELECTROLYTE FUEL CELL
POWER SYSTEM EMPLOYING LOW-BTU GAS AS A FUEL. A STEAM
BOTTOMING PLANT IS EMPLOYED FOR WASTE HEAT RECOVERY (SE19)

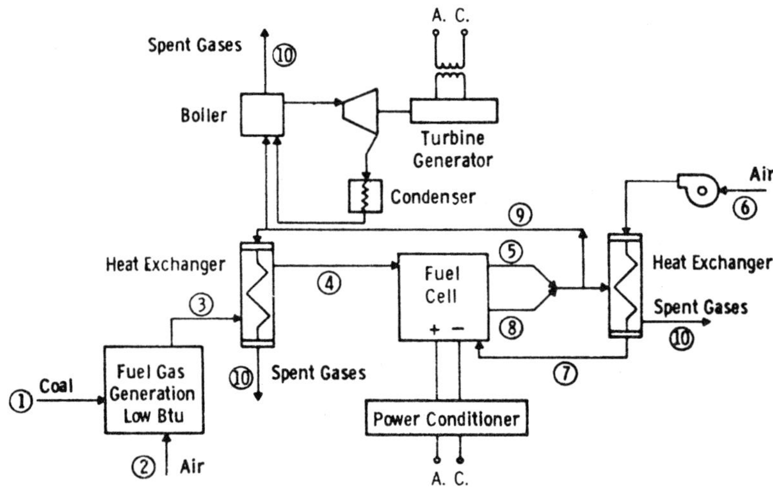

Point No.	Temperature (°F)	Pressure (psia)	Point No.	Temperature (°F)	Pressure (psia)
1	70	—	6	70	15
2	70	75	7	1100	60
3	100	60	8	1300	30
4	1100	45	9	1300	45
5	1300	30	10	250	15

APPROACH TO EFFICIENCY CALCULATIONS

Efficiency calculations without the detailed mass and energy balances provided
by a comprehensive conceptual design study are, of necessity, approximate.
Thus, these calculations have been performed with a number of assumptions,
some with less justification than desirable. These assumptions, presumably,
will undergo further exploration and clarification as part of a further study.

The approaches for the aqueous acid and alkaline power systems were similar;
those for the high-temperature systems were specific to the system. Where possible, the approach was consistent, e.g., in the efficiency assumptions for the
power conditioning and fuel processing subsystem.

Phosphoric Acid Fuel Cell Power System

The power plant efficiency is defined as the quotient of the power output at
grid voltages and the higher heating value (HHV) of the fuel (e.g., high-Btu gas,

etc.) fed to the power system. Similarly, the overall energy efficiency is the quotient of the power output and the HHV of the coal required to produce the fuel.

High-Btu gas, medium-Btu gas and methanol were fuels employed in the acid fuel cell power plants. These are discussed in the first subsection. In the subsequent subsection, the electrical losses associated with the various parasitic plant subsystems are detailed. The methods of calculating the power plant and overall energy efficiencies are discussed in the final subsection.

Quantities of Fuel Required: All parametric points involve the use of high-Btu gas, with the exceptions of AC4 (medium-Btu gas) and AC14 (methanol). High-Btu gas must be reformed with steam to accomplish the following reaction:

$$CH_4 + H_2O \longrightarrow CO + 3H_2$$

This reaction is endothermic, having an enthalpy of 226 kJ/g-mol (97,380 Btu/lb-mol) at 1144°K (1600°F). The exit stream from the reformer cannot be fed directly to the fuel cell subsystem because the fuel electrode may be deactivated by the presence of carbon monoxide, as discussed earlier. Consequently, it is passed through a shift converter in which the mildly exothermic reaction

$$CO + H_2O \longrightarrow CO_2 + H_2$$

is caused to occur at approximately 700°K (800°F). The carbon dioxide/carbon monoxide ratio in the shift converter exit stream is considered to be approximately 50 to 1. The fuel processing subsystem is discussed further in Appendix 1.

The molar compositions of the high-Btu gas fed to the reformer (34) and to the fuel cell subsystem (after reformation and shift conversion) are given in Table 4.6.

TABLE 4.6: FUEL GAS MOLAR COMPOSITIONS

Component	Initial Composition	After Shift Conversion
H_2	2.49	377.55
CH_4	94.23	–
CO_2	0.39	92.84
CO	0.08	1.86
N_2	2.81	2.81
S	–	–
HHV	959.2 Btu/scf	–
LHV	864 Btu/scf	–
Molecular weight	16.15	–

Thus, 1 std m³ of high-Btu gas yields 3.776 std m³ of hydrogen. Hydrogen comprises 79.5% of the exit stream from the shift converter, and the concentration of carbon monoxide is 0.39%.

The hydrogen requirements of the fuel cell subsystems are a function of the dc ratings of the subsystems, the average cell voltage and the hydrogen utilizations. The four ratings are 25-MW dc (all parametric points with the exception of AC2, AC3, AC4 and AC16), 100-MW dc (AC2), 250-MW dc (AC3 and AC16) and

900-MW dc (AC4). To deliver one kilowatt dc from a fuel cell operating at a terminal voltage of 0.7 V requires a cell current of 1429 A. Thus, 1429/96,489 equivalents or 14.9 mg/s (3.29×10^{-5} lb/s) of hydrogen are required if the utilization of hydrogen in the fuel gas is 100%. Published data (35) for phosphoric acid fuel cells, however, indicated that hydrogen utilization of 75% represented the state of the art in 1966–1967.

For simplicity, this treatment assumes a hydrogen utilization of 90% based on the assumptions of a better shift-conversion capability so that the minimum hydrogen-carbon monoxide ratio constraint (i.e., at near-exhaust compositions) can be met at this utilization, and an advance in fuel electrodes technology leading to more porous, thinner electrodes with lower concentration polarizations. Thus, the actual hydrogen requirement is 59.60 g/kWh (0.1314 lb/kWh), corresponding to a high-Btu gas requirement of 51.64 std cm^3/kJ (6.56 scf/kWh).

As shown in Figure 4.1, the high-Btu gas stream, fed to the steam-methane reformer, is split in two parts. One part is converted to hydrogen, as described above, while the other is mixed with air and used to provide the heat required for the endothermic reformation process. It can be shown that, under ideal conditions, the minimum volume of the high-Btu gas necessary for firing comprises 22% of the total feedstock. Practical systems in 1963 (36) used 41% of the feedstock in this manner. In this study, a value of 30% has been assumed on the basis of

   (a) Firing the hydrogen not utilized in the fuel cell subsystem to
        provide some of the reformer heat requirement, and

   (b) Advances in reformer technology.

Thus, high-Btu gas is fed to the fuel processing subsystem at the rate of 73.49 std cm^3/kJ (9.34 scf/kWh) of dc output from the fuel cell subsystem.

The medium-Btu gas, used as a feedstock in the 900-MW dc power plant, does not require reformation, but it must be shift-converted to yield a fuel gas suitable for use in the fuel cell subsystem. The compositions before (37) and after shift conversion are shown in Table 4.7.

TABLE 4.7: MEDIUM-BTU FUEL GAS MOLAR COMPOSITIONS

Component	Initial Composition	After Shift Conversion
H_2	0.3276	0.8618
CO_2	0.0573	0.5915
CO	0.5460	0.0118
N_2	0.0043	0.0043
H_2O	0.0647	–
HHV	281.5 Btu/scf	–
LHV	265.0 Btu/scf	–
Molecular weight	19.76	–

Because of the higher concentration of carbon monoxide in the exit stream of the shift converter, a hydrogen utilization of 80% has been assumed in this case.

Thus, the overall medium-Btu gas requirement is 253.6 std cm^3/kJ (32.24 scf/kWh) of dc output from the fuel cell subsystem.

The methanol required in parametric point AC14 may be shown to be 0.1183 g/kJ [426 g/kWh (0.939 lb/kWh)] with the assumptions employed in the above discussion of high-Btu gas reformation and shift conversion.

AC Power Outputs: An efficiency of 95.5% was assumed for the force-commutated inverter power-conditioning subsystem of all 25-MW dc power systems. The rationale for the selection of this type of power-conditioning equipment is outlined in Appendix 2. In addition, in the acid power system a further 2% of the ac output of the power-conditioning equipment was allocated for air blowers and sundry control systems. Thus, the net ac power output from the acid power system, with a dc rating of 25 MW, is (25 MW)(0.955)(0.98), or 23.4 MWe.

In the cases of the 100-MW (AC2), 250-MW (AC3 and AC16), and 900-MW dc (AC4) power plants, a line-commutated inverter was employed in the power-conditioning subsystem (see Appendix 2). An efficiency of 95% was assumed. Double transformation was considered necessary in the 100-MW and 250-MW dc systems, since the contract statement of work specified that all power plants of greater than 50 MWe should deliver ac power to the grid at 500 kV. It was assumed that the added transformer would operate at an efficiency of 99.5%. The 900-MW dc system was considered not to need this double transformation. As in the case of the 25-MW system, 2% of the ac output was assigned to the air blowers, etc. Thus, the net ac outputs were 92.6 MWe, 231.5 MWe and 838 MWe, respectively.

When a dedicated oxygen plant is employed, as in parametric points AC5 (25-MW dc) and AC16 (250-MW dc), the ac output to the grid must be reduced further. The power required for oxygen plant operation may be calculated from information given in Appendix 3. The power plant in AC5 requires approximately 3.15 kg/s (300 tons/day) of oxygen. The energy required is 0.0975 kJ/Mg (318.4 kWh/ton) or 95.52 MWh corresponding to a power usage of 3.98 MW. With a 4% allowance for oxygen vented to the atmosphere, the power required is 4.14 MW. The net ac power output of a 25-W dc power plant employing oxygen as the oxidant (AC5) is 19.3 MW. Similarly, the net ac power output from the power plant of parametric point AC16 is 190.3 MW.

Calculation of Efficiencies: The heat rates of the power plants may be calculated as follows:

$$\text{Heat Rate} = \frac{(\text{Fuel Rate})(\text{HHV of Fuel})}{(\text{Net ac Power Output})/(\text{Nominal dc Rating})}$$

where the required fuel rate and the fuel HHV are given in units of scf/kWh and Btu/scf, or lb/kWh and Btu/lb, respectively. For example, for AC1:

$$\text{Heat Rate} = \frac{(9.34 \text{ scf/kWh})(959.2 \text{ Btu/scf})}{(23.4 \text{ MW})/(25 \text{ MW})} = 9,570 \text{ Btu/kWh}$$

Alkaline Fuel Cell Power System

The approach to the calculation of heat rates in the alkaline power system is very similar to that employed for the acid system. The higher cell voltages in alkaline cells (0.8 V vs 0.7 V for acid fuel cells) result in a lower hydrogen requirement, 0.0145 g/kJ [52.2 g/kWh (0.1151 lb/kWh)], and a lower rate at which fuel is fed to the power system. The assumptions of the efficiencies of the power-conditioning subsystems are identical to those given in the previous section and in Appendix 2.

The Lurgi Rectisol process for the removal of carbon dioxide from both fuel and air streams before they are fed to the fuel cell subsystems has been arbitrarily assumed to consume an additional 6% of the ac power output from the power-conditioning subsystem when high-Btu fuel is used as feedstock. When added to the 2% consumed by the air blowers and control systems, the inefficiencies amount to 8% of the total ac output. In parametric point AL4, the use of the medium-Btu feedstock results in a doubling of the carbon dioxide partial pressure in the shift converter exit stream. The inefficiencies are now assumed to total 10% of the ac output of the power-conditioning subsystem.

The oxygen requirements in the power plants corresponding to parametric point AL5, in which oxygen from a dedicated oxygen plant is employed as the fuel cell oxidant, are lower than in the similar acid plant (AC5) because of the higher cell voltages. The power requirement is 3.6 MW, calculable from the value given for AC5 by multiplication by the reciprocal of the ratio of the cell voltages, 0.875.

Thus, all the heat rates for the parametric points of the alkaline fuel cell power system are calculable from the corresponding values of the acid system. For example, in the base case, AL1,

$$\text{Heat Rate} = \frac{(9{,}570 \text{ Btu/kWh})(0.7 \text{ V}/0.8 \text{ V})}{(92\%/98\%)} = 8.920 \text{ Btu/kWh}$$

Molten Carbonate Fuel Cell Power System

The plant efficiency is calculated on the basis of the higher heating value of the fuel and of steady-state operation at the voltages and current densities described earlier.

All cells produce an excess of heat during operation at rated power over that necessary to heat the fuel and oxidant streams to the fuel cell temperature. This excess heat is partially converted to steam but does not contribute to the power plant efficiency for MC1 through MC3 and MC6 through MC16. For MC4, this excess heat is used to operate a steam turbine generator to produce additional ac electricity and thereby increase the overall plant electrical efficiency. MC5 and MC17 also use the excess heat, but to operate a steam turbine drive for the compressor in the oxygen plant. The steam turbine drive was assumed to convert the available energy to shaft work with an efficiency of 40%.

The efficiency of plants using pure oxygen was not reduced by 2% since the electrical output is not derated to provide power for the electrical motor drive compressor that is commonly used for an oxygen plant. The nominal dc output of the fuel cell is further derated by the dc-to-ac power conversion subsystem as described in Appendix 2. It is assumed that 90% of the fuel is consumed by the fuel cell. The remaining 10% is burned in the exit gases by excess oxygen from the cathode exit gas and appears as waste heat.

It is further assumed that various electrical auxiliaries (such as recirculating fans) consume electrical energy equivalent to 3% of the HHV of the fuel. This may be high.

For simplicity, reformed high-Btu gas or methanol are assumed to have only carbon monoxide and hydrogen as fuel species after reforming. Some CH_4 and other fuel species, however, will, in practice, be present.

Equation 1 is used to calculate all molten carbonate fuel cell efficiencies except for MC4.

$$(1) \quad \text{Efficiency} = \frac{0.9 \ [(\Sigma N)(n)(F)(\epsilon)] \ PCE - 0.03 \ HHV}{HHV} = \frac{(A)(PCE) - B}{HHV}$$

where   0.9 is the fraction of the fuel utilized by the fuel cell,

ΣN is the sum mol fraction of carbon monoxide and hydrogen present after reforming 1 mol of the original fuel gas,

n is 2 (the number of electrons involved in electrochemical oxidation of 1 molecule of carbon monoxide or hydrogen with ½ molecule oxygen),

PCE is the power-conditioning efficiency (see Appendix 2),

HHV is the higher heating value per mol of original fuel gas,

F is the Faraday constant, 23.062 kcal/eV, and

ϵ is the cell voltage.

For MC4 waste heat is utilized to increase the plant efficiency. It is assumed that heat is exchanged between exhaust gas and incoming fuel and air, with the exhaust leaving the stack at 423°K (302°F), and that other heat loss to the surroundings is 2% of the HHV per mol of fuel gas. It is further assumed that a steam turbine generator with 40% efficiency is used to convert the waste heat to ac electricity. The overall efficiency of MC4 is calculated by Equation 2.

$(2) \quad$ Overall Efficiency $MC_4 -$

$$\frac{(A)(PCE) - B + 0.4(\Delta H^c_{650°} - A) - 0.02 \ HHV - [\Delta H_{650°} - \Delta H_{25°}]_{i.g.} + [\Delta H_{650°} - \Delta H_{150°}]_{e.g.}}{HHV}$$

where   0.4 is the turbine-generator efficiency,

$\Delta H^c_{650°}$ is the heat of combustion of the fuel gases at 650°C,

$[\Delta H_{650°} - \Delta H_{25°}]_{i.g.}$ is the sensible heat difference in the given °C temperature interval for the input gases to the fuel cell, and

$[\Delta H_{650°} - \Delta H_{150°}]_{e.g.}$ is the sensible heat difference in the given °C temperature interval for the exhaust gases from the plant.

Other symbols are as in Equation 1. ΔH values were taken from the JANAF Thermochemical Tables (38).

High-Temperature Solid Electrolyte Fuel Cell

Thermodynamic Efficiency: The quotient of average cell voltage and the voltage corresponding to the higher heating value of the fuel is defined as the thermodynamic efficiency n_{th}.

$$n_{th} = \frac{E_{cell}}{E_{HHV}}$$

E_{cell} was calculated in the cases SE2–SE18 on the basis of actual measurements on cells and batteries.

$$E_{cell} = E_O - E_R - E_P$$

where E_O is the open-cell voltage, average over the total range of fuel/combustion product ratios,

E_R is the voltage losses due to ohmic resistance in battery components, and

E_P is the polarization loss due to diffusion problems of fuel and combustion products.

The voltage which corresponds to the higher heating value of the fuel, E_{HHV}, is calculated from the following relation:

$$E_{HHV} = \frac{HHV}{n\,F}$$

where HHV is the higher heating value of fuel in kcal/mol,

n is the number of electrons transferred per mol of fuel gas reacted, and

F is the Faraday constant, 23.06 kcal/eV.

For SE1 and SE20 an average open-cell voltage was calculated from the free energy change of the oxidation of methane at 1300°K (1880°F). In this instance, a high average open-cell voltage of 1.04 V is achieved. The reason for this is the thermodynamic instability of methane at 1300°K (1880°F). Kinetically, however, it is possible to burn methane electrochemically at this temperature without carbon deposition if, except for the fuel electrode, the cell contains no metal surfaces. An in situ reformation and oxidation of the methane takes place at the anode. The question, however, remains whether this concept is practical, as experimental data are limited.

Power Output of the Plant: Since the ac power output is included in the heat rate calculations, it is necessary to explain how these figures were derived. The power output of the fuel cell subsystem is reduced by the inefficiencies of the power-conditioning subsystem, as described in Appendix 2. As in the case of molten carbonate fuel cells, it is assumed that an amount of energy equivalent to 3% of the HHV of the fuel is consumed to provide plant auxiliary power. The power output P_{net} was calculated according to the following equation:

$$(3) \qquad P_{net} = (MW)(PCE) - (0.03)(MW/n_{th}) + P_{turb}$$

where MW is the nominal power output,

PCE is the power conditioning efficiency,

P_{turb} is the ac power output from steam turbine generator, and

n_{th} is the thermodynamic efficiency.

Heat Rate: Heat rate (HR) calculations for all parametric points except SE18 were performed in the following sequence:

(1) Establish dc power rating in kilowatts,

(2) Calculate thermal equivalent from (1) in Btu/hr,

(3) Calculate fuel rate from (2) by considering thermodynamic inefficiencies in Btu/hr, and

(4) Calculate heat rate in Btu/kWh in dividing (3) by the actual ac power output of the power plant.

For example, for a Point SE1

$$HR = \frac{MW}{n_{th}\ P_{net}} = \frac{(25 \times 10^6)(3,413)}{(0.76)(22.9 \times 10^6)} = 4,900 \text{ Btu/kWh}$$

The heat rate for SE18 was established earlier by Westinghouse under the sponsorship of the Office of Coal Research. Based on a detailed mass and energy balance for a 100-kW power plant (25) a heat balance of 6,370 Btu/kWh was calculated. This heat rate was used in the present calculations taking no credit for possible improvements as the plant size was increased from 100 kW to 250 MW.

Power Plant Efficiency: The power plant efficiency, n_{PP}, is the quotient of the theoretical and calculated heat rates for all parametric points.

$$n_{PP} = \frac{3,413}{HR}$$

In the cases of SE4 and SE19, additional power is produced by utilizing the waste heat of the fuel cell generator. This, of course, is only possible in large installations and was not considered for smaller plants. The waste heat was used via a steam turbine generator with an assumed 40% efficiency. This is why the power plant efficiency is higher than the thermodynamic efficiency of the electrochemical generator. This is not a contradiction, because the fuel cell for which the thermodynamic efficiency is calculated delivers only a part of the electrical energy of these plants which could be considered combined-cycle plants. The heat rates of SE4 and SE19 must be viewed this way also.

CAPITAL, SITE-LABOR, AND OPERATION AND MAINTENANCE COSTS

This section is devoted mainly to a description of the approaches taken in calculating the capital costs of the fuel cell subsystems. The power-conditioning costs for the power plants corresponding to all 69 parametric points are given in Appendix 2. Similarly, oxygen plant costs for Point 5 in all fuel cell power systems and for AC16 and MC17 are discussed in Appendix 3. The fuel-processing subsystem, required for the low-temperature aqueous acid and alkaline power systems, is costed as described in Appendix 1. This appendix also includes a brief description of costing the air blowers for the low-temperature fuel cell power systems.

It must be emphasized here that without a conceptual design of the fuel cell modules, the costing techniques are at best approximate. The assumptions employed as to design and materials of construction are stated, even if extensive justification is not provided. Refinement of these assumptions should be made in the future.

The site labor costs for installing the four types of fuel cell module were assumed arbitrarily to be \$5/kW for the acid and alkaline systems, \$8/kW for the molten carbonate system, and \$10/kW for the solid electrolyte system. These relatively low site labor charges reflect the modularity of the assumed fuel cell subsystem, which leads to relatively straightforward and simple installation procedures at the power plant location. The higher charges for the high-temperature system reflect the slightly more complex procedures of subsystems interconnection, because of increased insulation requirements.

The operation and maintenance (O&M) costs for the fuel cell power plants have, in general, three component charges related to: (a) plant labor; (b) material and component replacement in subsystems other than the fuel cell subsystem; and (c) full replacement of the fuel cell subsystem. These costs are described in greater detail below.

(a) An hourly labor cost is calculable, as described previously, by the formula:

$$\text{O\&M (labor)} = Z \times \frac{\text{ac output}}{0.6 + 0.004 \text{ (ac output)}} \times \frac{\$15,000}{8,740 \text{ hr (capacity factor)}}$$

where Z, a factor related to the complexity of O&M procedures, is taken as 1.0 for steam plants. For all plants with a dc rating of 25 MW, it has been assumed that, because of their probable substation locations, remote operation will be possible with control exercised from a central station. Also, because of this probable remote control, the value of Z has been arbitrarily assumed to be 0.2 for these plants. For all power plants of 250 MW or larger, however, a factor of 0.4 has been chosen, reflecting the need for full-time personnel at the plant location. This Z value is substantially less than that for steam plants and may be justified on the basis of the relative cleanliness of the fuels employed in the fuel cell plants, and the ease of maintenance of the power-conditioning subsystem.

(b) To allow for material and component replacement in the fuel-processing and power-conditioning subsystems, and other major components, an O&M charge, amounting to 5% of the total capital and site costs of the major components (with the exception of the fuel cell subsystem), is allotted for each power plant.

(c) The fuel cell subsystem useful life is only 36 Ms (10,000 hours) (exceptions: Points 6 through 9). This is much less than the 946 Ms (30 year) life assumed for the other plant components. Accordingly, a special charge, calculated by dividing the replacement cost of the fuel cell subsystem by the product of the ac output and the life, is included in the O&M costs for every parametric point. Because of uncertainties, this calculation omits price escalation, learning, and sinking fund factors.

Phosphoric Acid Fuel Cell Power System

A bipolar design has been assumed for the acid fuel cell battery stack. The basic element is a corrugated or embossed graphite plate which serves as the anode current collector of one cell and the cathode current collector of an adjacent cell. The anode and cathode are assumed to be thin carbon electrodes, similar to those described by Kordesch and Scarr (30). These contain active carbon catalyzed by platinum. The electrolyte consists of a matrix-immobilized 85 weight percent phosphoric acid. The fuel gas stream flows in the corrugations between the thin carbon anode and the graphite piece. Air flows similarly between the cathode and the corrugated graphite plate.

The cost of the corrugated graphite plate, which may be fabricated by extrusion, is estimated (39) to be approximately $10.76/m^2 ($1/ft^2). If battery stack end requirements are ignored, only one of these plates is required per cell. The active carbon electrodes, without catalyst, are similarly estimated (39) to cost approximately $10.76/m^2 ($1/ft^2) each. For an interelectrode separation (i.e., electrolyte thickness) of 0.5 mm (19.7 mils), the quantity of electrolyte required, assuming that it forms 80% by volume of the matrix, is approximately 100 g (0.22 lb). At $0.349/kg ($15.85/100 lb) (40), the electrolyte cost is only $0.377/m^2 ($0.035/ft^2). The electrolyte matrix component has been assumed to be cost-determining, so that the cost of the immobilized phosphoric acid is taken as $5.38/m^2 ($0.50/ft^2). Thus, the cost of the bipolar battery stack materials, unassembled and without catalyst, is approximately $37.67/m^2 ($3.50/ft^2).

Doubling of this materials cost results in a manufactured cost for the battery stack of $75.35/m^2 ($7/ft^2). A further allowance of $5.38/m^2 ($0.50/ft^2) for gasketing and insulating the battery stack yields a manufactured cost of $80.73/m^2 ($7.50/ft^2) for the fuel cell module. Adding a profit of 20% results in a selling price of $96.88/m^2 ($9/ft^2) for a module with uncatalyzed electrodes.

Most parametric points (except for AC12 and AC13) involve platinum loadings in the anode and cathode of 1 mg/cm^2 (0.00205 lb/ft^2). If chloroplatinic acid (H_2PtCl_6) is employed in the catalyst-addition process, the cost of platinum per unit cell area is $117.33/m^2 ($10.90/ft^2), for a chloroplatinic acid cost of $5.8675/g of contained platinum ($182.48/oz t platinum) (40). Thus, for most parametric points, the fuel cell module cost is $214.20/m^2 ($19.90/ft^2). In Point AC12, where the platinum loadings in the anode and cathode are decreased to 0.3 mg/cm^2 (6.144 x 10^{-4} lb/ft^2), the catalyst cost is reduced to $35.20/m^2 ($3.27/ft^2), resulting in a fabricated module cost of $132.07/m^2 ($12.27/ft^2). The catalyst and module costs for AC13 [0.1 mg Pt/cm^2 (2.05 x 10^{-4} lb/ft^2)] are calculated in the same manner.

The cost per kW of the fuel cell module is assumed to be inversely proportional to the power density. For AC1, the power density is 140 mW/cm^2 (130.1 W/ft^2). This corresponds to a cell area requirement of 0.7143 m^2/kW (7.689 ft^2/kW), and a module cost of $153/kW, based on the costs given above. This leads to a vendor selling price of $3.8 million for the 25-MW dc fuel cell subsystem. The costs of all fuel cell subsystems are calculated in the same manner, based on the cell power densities, the dc ratings and the catalyst loadings.

The replacement cost of the acid fuel cell subsystem is calculated under the following assumptions:

(a) The cost of recovering the platinum at the end of battery life and reprocessing it into chloroplatinic acid is $1.00/g ($31.10/oz t).

(b) The rest of the fuel cell module has zero salvage value.

(c) The installation cost remains at $5/kW.

Thus, the catalyst cost in the replacement modules is only $20/m² ($1.86/ft²), resulting in a fabricated module cost of $116.88/m² ($10.86/ft²). In parametric point AC1, the replacement cost of the 25-MW dc fuel cell subsystem is $2,087,000 plus $125,000 for installation. The corresponding O&M charge is:

$$\frac{\$2,087,000 + \$125,000}{(23,400 \text{ kW})(10,000 \text{ hr})} = 9.45 \text{ mills/kWh}$$

or a charge of $221/operational hour.

Alkaline Fuel Cell Power System

A bipolar design, identical to that selected for the acid system, has been assumed also for the alkaline fuel cell modules. Thus, the selling price of $96.88/m² ($9/ft²) for a module with uncatalyzed electrodes given for the acid fuel cell system is applicable here also. The major differences between the acid and alkaline modules lie in the areas of cathodic catalyst type and power density.

The carbon anode is catalyzed by platinum with a loading of 1 mg/cm² (0.00205 lb/ft²), and the cathode is catalyzed by silver with a loading of 5 mg/cm² (0.0103 lb/ft²). Thus, the platinum loading is half of that in the acid system and adds $58.68/m² ($5.45/ft²) to the uncatalyzed cost quoted above. It is assumed that the cathodes are silver catalyzed with silver nitrate, which costs $0.094/g of contained silver ($2.69/oz av Ag) (40). This adds a further $4.745/m² ($0.4408/ft²) to the uncatalyzed module cost. The catalyzed cell cost is, therefore, $160.30/m² ($14.89/ft²).

The power density for most parametric points is 80 MW/cm² (74.32 W/ft²) leading to a cell area requirement of 1.25 m²/kW (13.46 ft²/kW). The module cost is, therefore, $200/kW, or $5 million for a 25-MW fuel cell subsystem. Subsystem costs for other power density values are calculated on the basis of the inverse relationship assumed between the cost and the power density. The replacement cost is calculated, as in the acid system, with recovery and reprocessing costs for platinum and silver of $1/g ($31.10/oz t) and $0.03/g ($0.93/oz av), respectively.

Molten Carbonate Fuel Cell Power System

This section describes cost estimates for the fuel cell, gas recirculation, and heat recovery systems. Plant costs for siting, construction and electrical controls are described in an appendix available only in the original report. Oxygen plant costs are described in Appendix 3, and power-conditioning costs are given in Appendix 2.

The same cost for materials, fabrication and assembly per unit area of electrode has been assumed for all molten carbonate fuel cells. Since there have been no cell assemblies of more than a few kilowatts, and constructional details of current test units (22) are not available, only very rough estimates can be made. As a first approach to costing the fuel cell, assume use of a filter press design;

estimate a cost based on the somewhat analogous filter press technology, and approximate cell material costs suggested by IGT (11), with 0.715 m^2 (7.7 ft^2) of electrode area per kilowatt. From Perry's Handbook (41), the lowest cost (1970) of a filter press with filters uninstalled was about $161.50/m^2 ($15/ft^2) for iron or wood materials. Update to July 1974 prices by the factor 1.4; subtract a material cost of about $21.50/m^2($2/ft^2) for the wood or iron materials; add a cell material cost suggested by IGT in 1966 (11) ($20 to $40/kW escalated to July 1974 prices of about $31 to $62/kW or an average of $46/kW), and add a further 20% for assembly and leakproofing; using these figures a cost of

$$($15/ft^2)(7.7 \text{ ft}^2)(1.4) - ($2/ft^2)(7.7 \text{ ft}^2) + (31 \text{ to } 62), \text{ i.e., } $177 \text{ to } $208/kW$$

may be calculated. If 20% is added to allow for vendor profit, the cost would be $213 to $250/kW, uninstalled.

A second approach to costing the cell is to take the values estimated by Hart and Womack (16). The list price was £27.5/kW dc. Adjusting this for an exchange rate of $2.4/£ (in 1966), for the ratio of power densities assumed in that and the present study of 0.2 to 0.3 (in terms of kW/ft^2), a commodity escalation price of $1.58 and a profit factor of 1.2 for fabrication gives

$$(27.5)(2.4)(0.2/0.13)(1.58)(1.2)$$

or $192.51/kW uninstalled.

For this study an uninstalled capital cost of $190/kW has been arbitrarily selected as being about the lower limit of what might reasonably be expected. An installation cost of $8/kW is assumed.

Without a conceptual design of the fuel cell system, costing of the waste heat recovery system is also arbitrary. Refinement of the estimates on the basis of a detailed design was anticipated as a part of a further study. For example, normal waste heat boilers have a once-through passage of the hot gases, but in the present system the fuel cell exit gases are recycled between the fuel cell and boiler temperatures. Such a system would have to be specifically designed for this application. The installed cost of the heat recovery steam generator is assumed to be $60/kW. Installation cost is estimated as 33% of the total, which is about the same as for the simple steam turbine power plant boiler but much more than the ~5 to 10% for a simple once-through waste heat boiler. The cost of the associated turbine-generator combination in MC4 is taken from a Westinghouse price list (42) adjusted to July 1974 prices.

Heat exchangers are arbitrarily assumed to cost $12.50/kW dc installed. Of this cost about 45% is for the heat exchangers and 55% involves the installation labor, piping, etc.

Blower costs will depend on the rate of recirculation necessary both to remove heat and to prevent cell concentration polarization. About $1.15/kW plus 10% installation is assumed for this item for all plants except MC4; $2.55/kW is assumed for MC4, the 900-MW plant which may require higher temperature blowers.

The sum totals of the blower, heat exchanger and heat recovery steam generator costs for MC1, the base case, represent only about 3.85% of the total plant capital costs. Substantial errors in these estimates, therefore, will not be significant in view of the large uncertainty in estimating the ultimate lifetimes and cost of the fuel cell itself, which represents about 38.5% of the total capital costs.

Capital costs are much greater where an oxygen plant (see Appendix 3) is included, even though the fuel cell is significantly reduced in size due to the higher power output per cell that has been assumed. Due to economics of size the oxygen plant represents about 33% of the total capital cost for MC5, a 25-MW plant, but only 20.5% for MC17, a 250-MW plant. This is more than the corresponding fuel cell capital costs of roughly 15 and 19.6% of the total, respectively. The high capital cost in addition to large power requirement appears to preclude the use of oxygen rather than air in the molten carbonate plants.

Solid Electrolyte Fuel Cell Power System

The costing of the fuel cell subsystem is based on a cost analysis for the Westinghouse thin-film battery, discussed in detail in (43). For a power output of 0.5 W/cm (1.27 W/inch) of a tubular battery having an outside diameter of 1.27 cm (0.5 inch), the cost of the battery raw material was $21/kW in mid-1970. The following set of assumptions have been employed in calculating the cost per kilowatt dc output of a fuel cell module in mid-1974:

(a) Active electrolyte and active interconnection region lengths are equal to each other and to 2 mm (0.079 inch).

(b) Fuel electrode and air electrode gap lengths are equal 0.5 mm (20 mils).

(c) A fabrication-cost/materials-cost ratio of 7 to 3 is based on Westinghouse manufacturing experience in thick-film device technology, which is comparable to that required for thin-film battery fabrication.

(d) A factor of 1.34, corresponding to the increase in the Marshall and Stevens index from mid-1970 to mid-1974, is used to estimate the mid-1974 costs of the fabricated batteries.

(e) An additional $18/kW is allowed for materials and assembly labor necessary to manifold and sheath the fuel cell batteries in the fabrication of the fuel cell module.

(f) 3% of the cost of the fabricated fuel cell module is taken as the cost of insulation (44).

(g) A profit of 20% is assumed in the estimation of a vendor price.

Thus, for SE2 through SE4, SE6 through SE9, SE13 and SE15, the selling price in dollars per kilowatt of a fuel cell module is calculated as follows:

$$P = \left(\frac{\$21}{kW} \times \frac{0.5 \text{ W/cm}}{0.4224 \text{ W/cm}} \times \frac{100}{30} \times 1.34 + \frac{\$18}{kW} \right) 1.03 \times 1.2 = \$160/kW$$

For all other points the selling price of the fuel cell module was calculated using the above value and an assumed inverse relationship between cost and power density.

No known heat exchanger technology is available for application in high-temperature fuel cells operating up to 1373°K (2012°F). It was assumed, therefore, that heat exchange above 873°K (1112°F) must be accomplished in the fuel cell generator itself. At this temperature, heat shock problems are largely reduced. Based on this assumption, heat exchanger calculations were performed where nearly a 50% reduction of heat exchange surface is achieved and where the maximum metal temperature in the heat exchanger is reduced to 873°K (1112°F). The calculations took into account a 100°K (180°F) mean temperature difference and a base cost of $215/m² ($20/ft²) of heat exchange surface. The installed cost was assumed to be 230% of this base cost, or $495/m² ($46/ft²). These costs were broken down into material and site labor costs, following Guthrie (44), who recommends values of 73% (materials) and 27% (site labor) of the total installed cost.

The cost of the integrated coal gasification reactor in SE18, the Westinghouse Fuel Cell Power System, was estimated on the basis of an economic evaluation of a 200-MW coal-burning fuel cell power plant, performed by IGT in early 1969 (45). The updated installed cost is broken down according to the above Guthrie recommendations for shell-and-tube heat exchangers—73% for materials and 27% for site labor—because fully 75% of the installed cost of the reactor is attributable to the Incoloy 800 pipe required for encapsulation of the fuel cell modules and for adequate heat exchange between the fuel cell batteries and the coal undergoing gasification.

Estimates of the cost of gas compressors, venturi scrubbers and waste heat boilers were based on information provided (44). The waste-heat recovery system (i.e., the steam bottoming plant) of SE4 was costed as described earlier.

RESULTS OF PARAMETRIC ASSESSMENT

The power plant efficiency is defined as the quotient of the power output at grid voltages and the higher heating value (HHV) of the fuel fed to the power system. As described earlier, a heat rate was calculated for the power plant corresponding to each of the 69 points in the parametric assessment. These values were input to a computer program in which they were converted to fractional efficiencies, after a minor allowance was made for power plant electrical requirements.

The program was also employed to calculate an overall energy efficiency, defined as the quotient of the power plant ac output and the higher heating value of the coal required in the production of the fuel. The factors employed to convert the power plant efficiencies to overall energy efficiencies are described in another report (not republished here) for high-Btu gas, medium-Btu gas and methanol, respectively.

All of these fuels were considered to have been derived from Illinois No. 6 coal. Point SE18, the Westinghouse Fuel Cell Power System, involves the use of a coal (Illinois No. 6) as fuel so the power plant and overall energy efficiencies for this system are identical. North Dakota Lignite was selected for the low-Btu gasification reactor of SE19 because of its higher gasification efficiency.

The cost of electricity (COE) and its three component parts, ascribable to capital, fuel and O&M charges, were also calculated by means of the computer program from the cost input described earlier. These costs were based on

NASA-mandated values for labor rate, contingency charge, escalation rate, interest during construction, fixed charge rate, fuel cost and capacity factor. The values specified are listed in Table 4.8.

TABLE 4.8: VALUES SELECTED FOR VARIABLES IN PLANT CONSTRUCTION AND OPERATION

Labor rate	$10.60/hr
Contingency charge	4.5%
Escalation rate	6.5%
Interest during construction	10%
Fixed charge rate	18%
Capacity factor	65%
Fuel costs	
High-Btu gas	$2.60/$10^6$ Btu
Medium-Btu gas	$2.00/$10^6$ Btu
Methanol	$2.70/$10^6$ Btu
Coal	$0.85/$10^6$ Btu

The effect of changes in all of these base case values on the COE were explored for every point in the parametric assessment.

Earlier the efficiency and COE results were shown and discussed for each fuel cell power system. A comprehensive comparison of the four types of fuel cell power systems is given later in justification of the conclusions and recommendations of this study.

Phosphoric Acid Fuel Cell Power Systems

The power plant and overall energy efficiencies, the capital cost and COE, and the estimated time of construction for each of the plants corresponding to the sixteen parametric points of the acid fuel cell system are shown in Table 4.9. All relevant information pertaining to the operation of each of the fuel cell subsystems in these plants is provided also in this table.

When air is employed as the oxidant, the plant efficiency lies in the range of 35 to 36%, for all points. The overall energy efficiency with high-Btu gas is approximately 24%. For AC4 the overall efficiency is better than 29%, reflecting the use of medium-Btu gas in this power plant. There is no comparable advantage to the use of methanol as a fuel, as shown by the value of 25% for the overall energy efficiency of the power plant corresponding to Point AC14.

Points AC5 and AC16, corresponding to 25-MW dc and 250-MW dc power plants which use oxygen instead of air as the oxidant, display power plant and overall energy efficiencies of 30 and 20%, respectively. The efficiency reduction, amounting to a sixth of the total, is attributable to the power required by the dedicated oxygen plants.

The capital cost and the COE broken into its three components—capital, fuel and O&M—are shown in Table 4.9 for the values of the construction and operation variables of Table 4.8.

TABLE 4.9: VALUES OF ALL RELEVANT PARAMETERS FOR THE PARAMETRIC POINTS OF THE AQUEOUS-ACID FUEL CELL POWER SYSTEM

	Parametric Point, AC Number															
	1	2	3	4	5	6	7	8	9	10	11	12	13	14	15	16
Power output, MWe	23.4	92.6	231.5	838	19.3	23.4	23.4	23.4	23.4	23.4	23.4	23.4	23.4	23.4	23.4	190.3
Fuel cell rating, MW	25	100	250	900	25	25	25	25	25	25	25	25	25	25	25	250
Fuel																
High Btu gas	x	x	x	-	x	x	x	x	x	x	x	x	x	-	x	x
Medium Btu gas	-	-	-	x	-	-	-	-	-	-	-	-	-	-	-	-
Methanol	-	-	-	-	-	-	-	-	-	-	-	-	-	x	-	-
Oxidant																
Air	x	x	x	x	-	x	x	x	x	x	x	x	x	x	x	-
Oxygen	-	-	-	-	x	-	-	-	-	-	-	-	-	-	-	x
Fuel cell life, 10^3 hr	10	10	10	10	10	30	50	100	100	10	10	10	10	10	10	10
Voltage degradation, %	5	5	5	5	5	5	5	5	15	5	5	5	5	5	5	5
Temperature, °C	190	190	190	190	190	190	190	190	190	190	190	190	190	190	190	190
Electrolyte																
Type, 85 w/o H_3PO_4	x	x	x	x	x	x	x	x	x	x	x	x	x	x	x	x
Thickness, cm	0.05	0.05	0.05	0.05	0.05	0.05	0.05	0.05	0.05	0.05	0.05	0.05	0.05	0.05	0.025	0.05
Anode																
Type, Pt/C	x	x	x	x	x	x	x	x	x	x	x	x	x	x	x	x
Catalyst loading, mg Pt/cm^2	1.0	1.0	1.0	1.0	1.0	1.0	1.0	1.0	1.0	1.0	1.0	0.3	0.1	1.0	1.0	1.0
Cathode																
Type, Pt/C	x	x	x	x	x	x	x	x	x	x	x	x	x	x	x	x
Catalyst loading, mg Pt/cm^2	1.0	1.0	1.0	1.0	1.0	1.0	1.0	1.0	1.0	1.0	1.0	0.3	0.1	1.0	1.0	1.0
Current density, mA/cm^2	200	200	200	200	400	200	200	200	200	300	400	200	200	200	200	400
Average cell voltage, V	0.7	0.7	0.7	0.7	0.7	0.7	0.7	0.7	0.7	0.7	0.7	0.7	0.7	0.7	0.71	0.7
Power plant efficiency, %	35.5	35.1	35.1	34.8	29.6	35.5	35.5	35.5	35.5	35.5	35.5	35.5	35.5	35.8	36.0	29.9
Overall efficiency, %	23.9	23.6	23.6	29.3	19.9	23.9	23.9	23.9	23.9	23.9	23.9	23.9	23.9	24.9	24.2	20.1
Total capital cost × 10^{-6}, $	10.4	39.5	97.6	367	16.0	10.4	10.4	10.4	10.8	8.8	7.9	8.6	8.1	10.9	10.2	117
Capital costs, $/kWe	448	429	424	440	833	448	448	448	463	377	339	371	346	468	440	595
Cost of electricity, mills/kWh																
Capital	14.2	13.6	13.4	13.9	26.3	14.2	14.2	14.2	14.6	11.9	10.7	11.7	10.9	14.8	13.9	18.8
Fuel*	25.0	25.3	25.3	19.6	29.9	25.0	25.0	25.0	25.0	25.0	25.0	25.0	25.0	25.8	24.7	29.7
Operation and maintenance	10.9	10.9	10.8	10.5	9.9	4.6	3.3	2.3	2.4	8.0	6.5	9.9	9.6	11.0	10.8	9.2
Total	50.1	49.7	49.5	44.0	66.2	43.8	42.5	41.5	42.1	44.9	42.2	46.6	45.5	51.5	49.4	57.7
Estimated time of construction, yr	1.5	2.0	2.5	4.0	1.5	1.5	1.5	1.5	1.5	1.5	1.5	1.5	1.5	1.5	1.5	2.5
Estimated availability date	1985	1985	1990	1990	1990	1990	1990	1990	1990	1990	1990	1990	1990	1990	1990	1990

*Use base delivered fuel cost.

The COE for AC1, 13.9 mills/MJ (50.1 mills/kWh) has as its major component the fuel charge, which comprises 50% of the total cost. This is understandable in terms of the low efficiency of this power plant and the costliness of the fuel gas at \$2.60/10^6 Btu. The importance of the cost of high-Btu gas and its effect on the cost of electricity are shown in Figure 4.12. At \$1.50/10^6 Btu, the electricity cost is 11.0 mills/MJ (39.5 mills/kWh), and at \$4.00/10^6 Btu, the cost is 17.7 mills/MJ (63.5 mills/kWh).

FIGURE 4.12: DEPENDENCE OF THE COST OF ELECTRICITY ON THE FUEL COST AND THE USEFUL LIFE OF A 25-MW dc PHOSPHORIC ACID FUEL CELL POWER SYSTEM

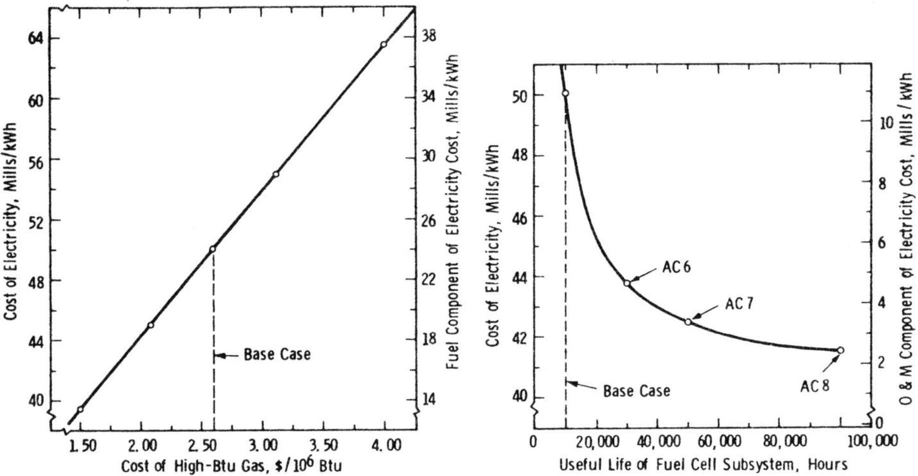

A graphic display of the capitalization breakdown is shown as Figure 4.16 page 131. The costs of the installed fuel cell, power-conditioning and fuel processing subsystems are \$169, \$62 and \$38/kWe of ac output, respectively. The balance of plant cost is \$67/kWe, and the indirect costs amount to \$112/kWe. The total capitalization is \$448/kWe.

In the standard case, defined by the values of Table 4.8, that portion of the COE ascribable to capital is 3.94 mills/MJ (14.2 mills/kWh). Because of the short time (1½ years) required for plant construction, changes in the escalation rate and the interest rate during construction have very little effect on the cost of electricity. Similarly, because of the modular nature of the fuel cell and power-conditioning subsystems, which minimizes the site labor required, even a doubling of the labor rate results in a relatively small increase (5%) in the electricity cost. Changes in the fixed charge rate from the base value of 18%, however, have a more pronounced effect, e.g., a lowering of the rate to 10% results in a reduction in the COE of 1.75 mills/MJ (6.30 mills/kWh).

Of the 3.03 mills/MJ (10.9 mills/kWh) O&M charge calculated for the power plant corresponding to Point AC1, only 0.388 mill/MJ (1.4 mills/kWh) is attributable to labor and to material and component replacement in all parts of the plant, excluding the fuel cell subsystem. The balance, amounting to 87% of the total, is due to the need for fuel cell replacement after 36 Ms (10,000 hours) of operation. The hyperbolic relationship between the fuel cell useful life and the COE (and its O&M component) is evident from Figure 4.12, which is plotted from data generated for Points AC6, AC7 and AC8, as well as AC1. Lifetimes of 144 Ms (40,000 hours) and 360 Ms (100,000 hours) lead to reductions in the base case COE of 1.97 mills/MJ (7.1 mills/kWh) and 2.36 mills/MJ (8.5 mills/kWh), respectively.

The effect of reducing the platinum loadings in both cell electrodes from 1.0 mg/cm^2 (0.002 lb/ft^2) in the base case to 0.3 mg/cm^2 (6.1 x 10^{-4} lb/ft^2) and 0.1 mg/cm^2 (2 x 10^{-4} lb/ft^2) is explored in Points AC12 and AC13. These results are also included in Table 4.9 and are shown graphically in Figure 4.13. The order of magnitude reduction in electrode platinum loadings caused a reduction of only 1.28 mills/MJ (4.6 mills/kWh) in the cost of electricity. This result is surprising in view of the widely accepted tenet that it is desirable, from an economic standpoint, to avoid platinum as an electrocatalyst.

Figure 4.13 indicates that, provided a 0.56 mill/MJ (2 mills/kWh) penalty can be absorbed and an adequate supply of platinum is available, there is little point to efforts to reduce the electrode platinum loadings much below 0.4 mg/cm^2 (8.2 x 10^{-4} lb/ft^2), corresponding to a platinum usage of approximately 6 g/kW (13.2 lb/MW). This conclusion is supported by Abens, Baker, DiPasquale and Michelko (29), who stated in a recent paper that "much obfuscation of cell costs has been caused by belaboring the advances made in catalyst cost reductions."

The power density, the product of the electrode current density and the cell voltage, has a marked effect on the COE, as shown by the data for AC10 and AC11 in Table 4.9. These results are also presented in Figure 4.13. A doubling of the base-case power density results in a 2.2 mills/MJ (7.9 mills/kWh) reduction in the COE. This graph may be used also to compute the effect of reductions in the selling price of the fuel cell subsystem because of the inverse relationship between the selling price and the power density.

Because of the modular nature of the fuel cell and power-conditioning subsystems, there is little, if any, economy of scale in fuel cell power plants. This is illustrated by the data presented in Table 4.9 for 100-MW (AC2) and 250-MW dc (AC3) power plants. Medium-Btu gas is used in the 900-MW dc power plant (corresponding to Point AC4), and because of its relatively lower cost (Table 4.8), the fuel component of the COE is reduced to 5.44 mills/MJ (19.6 mills/kWh).

In contrast, substitution of methanol for high-Btu gas, as in Point AC14, results in an increase in the COE despite a minimal increase in the power plant efficiency. This was caused by the cost of facilities for storing methanol at the power plant, and by the slightly higher charge for this fuel (Table 4.8). The COE is only minimally affected by an electrolyte-thickness reduction (Point AC15). An increase in the versatility of the power-conditioning subsystem to permit handling the lower current or voltage input expected as a result of fuel cell performance deterioration with time (Point AC9) has only a minor effect on the COE.

Oxygen is substituted for air as the oxidant in the fuel cell subsystems of the 25-MW and 250-MW dc power plants corresponding to Points AC5 and AC16, respectively. A comparison of the COE for AC1 and AC5 reveals a 4.5 mill/MJ (16 mill/kWh) penalty for the plant employing oxygen. A further comparison of the data for 250-MW dc power plants (Points AC3 and AC16) indicates that economy of scale in oxygen plants reduces this differential to 2.3 mills/MJ (8.3 mills/kWh). These cost penalties and the efficiency reduction noted above combine to make very unattractive the replacement of air by oxygen from a dedicated oxygen plant.

FIGURE 4.13: DEPENDENCE OF THE COST OF ELECTRICITY FOR PHOS-
PHORIC ACID FUEL CELL POWER SYSTEMS ON THE PLATINUM LOAD-
INGS IN THE ANODE AND CATHODE, AND ON THE POWER DENSITY
PER UNIT CELL AREA

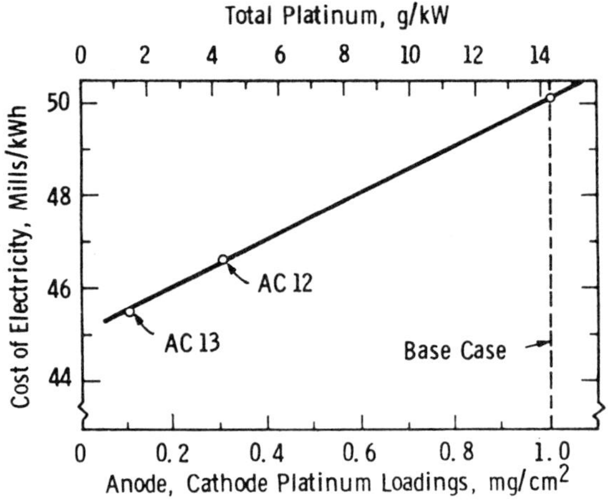

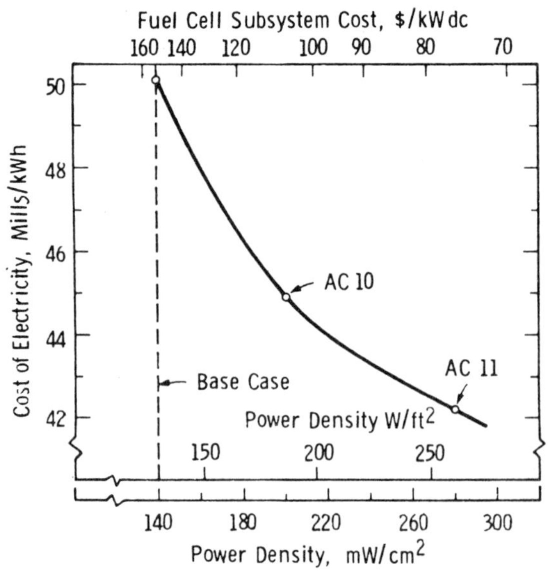

Alkaline Fuel Cell Power System

The results of the parametric assessment for the alkaline fuel cell power system are shown in Table 4.10. In general, the power plant and overall energy efficiencies are higher than those for the acid power system. The margin is not as great as expected with the higher cell voltages—0.8 V vs 0.7 V—in this system. The lower margin is due to the parasitic losses associated with operation of the Rectisol units for scrubbing the carbon dioxide from the fuel gas and air streams fed to the power system.

The capitalization required for alkaline power plants is greater because of the lower power density and, hence, higher costs of the fuel cell subsystem, and because of the costs of the Rectisol units, which are not required in the acid system. Similarly, the O&M charges are higher because of the greater replacement costs of the fuel cell subsystem. The net effect is that the COE for the alkaline fuel cell power systems is approximately 20% higher than that for the phosphoric acid system.

With high-Btu gas as fuel and air as oxidant, the power plant and overall energy efficiencies lie in general at 38% and 25 to 26%, respectively. When medium-Btu gas is employed as the fuel (AL4), the overall energy efficiency increases to 31%, while the power plant efficiency, 37%, is slightly lower. As discussed earlier for the acid system, the substitution of oxygen for air as the oxidant (AL5) results in a substantial lowering of both efficiency values. The power plant and overall energy efficiencies move sharply downward to 32 and 21%, respectively.

A simplified direct costs breakdown is shown later in Figure 4.16. It should be noted that the Rectisol unit for air scrubbing is included in the fuel processing cost in this diagram. The cost of electricity breakdown for the values of the construction and operation variables listed in Table 4.8 is as follows: 5.44 mills/MJ (19.6 mills/kWh) for capital, 6.48 mills/MJ (23.3 mills/kWh) for fuel and 4.95 mills/MJ (17.8 mills/kWh) for O&M. The total COE is 16.9 mills/MJ (60.8 mills/kWh).

As for the acid power system, the high O&M charge is mostly due to the high replacement cost of the fuel cell subsystem. Of the total, 4.4 mills/MJ (16 mills/kWh) are ascribable to the need for total replacement after 36 Ms (10,000 hours) of operation. The data for AL6, AL7 and AL8, shown in Table 4.10, indicate the substantial effect of fuel cell useful life on the COE. Increasing the useful life to 180 Ms (50,000 hours) reduces the COE by 4.0 mills/MJ (14.4 mills/kWh).

An equally profound effect is shown by comparing the data for AL10 and AL11 with those for the base case AL1. A power density increase of 150% results in lowering the COE by 4.05 mills/MJ (14.6 mills/kWh). A lowering of catalyst loadings, however, as in AC12, AC13 and AC15, leads to less significant reductions of the COE, just as in the case of the acid system.

No economy of scale is observed. The COE for the 100-MW dc (AL2) and 250-MW dc (AL3) power systems differs very little from those for the 25-MW dc base case. The full benefit of the use of the cheaper medium-Btu gas is not realized in the 900-MW dc plant (AL4) because of the additional Rectisol process

TABLE 4.10: VALUES OF ALL RELEVANT PARAMETERS FOR THE PARAMETRIC POINTS OF THE AQUEOUS-ALKALINE FUEL CELL POWER SYSTEM

	Parametric Point, AC Number															
	1	2	3	4	5	6	7	8	9	10	11	12	13	14	15	16
Power output, MWe	22	87	218	770	18.8	22	22	22	22	22	22	22	22	22	22	22
Fuel cell rating, MW	25	100	250	900	25	25	25	25	25	25	25	25	25	25	25	25
Fuel																
High Btu gas	x	x	x	–	x	x	x	x	x	x	x	x	x	x	x	x
Medium Btu gas	–	–	–	x	–	–	–	–	–	–	–	–	–	–	–	–
Oxidant																
Air	x	x	x	x	–	x	x	x	x	x	x	x	x	x	x	x
Oxygen	–	–	–	–	x	–	–	–	–	–	–	–	–	–	–	–
Fuel cell life, 10^3 hr	10	10	10	10	10	30	50	100	100	10	10	10	10	10	10	10
Voltage degradation, %	5	5	5	5	5	5	5	5	15	5	5	5	5	5	5	5
Temperature, °C	70	70	70	70	70	70	70	70	70	70	70	70	70	70	70	70
Electrolyte																
Type, 30 w/o KOH	x	x	x	x	x	x	x	x	x	x	x	x	x	x	x	x
Thickness, cm	0.05	0.05	0.05	0.05	0.05	0.05	0.05	0.05	0.05	0.05	0.05	0.05	0.05	0.05	0.05	0.025
Anode																
Type, Pt/C	x	x	x	x	x	x	x	x	x	x	x	x	x	–	x	x
Type, Raney Ni	–	–	–	–	–	–	–	–	–	–	–	–	–	x	–	–
Catalyst loading, mg Pt/cm^2	1.0	1.0	1.0	1.0	1.0	1.0	1.0	1.0	1.0	1.0	1.0	0.1	0.01	0.01	1.0	1.0
Cathode																
Type, Ag/C	x	x	x	x	x	x	x	x	x	x	x	x	x	–	x	x
Type, Raney Ni	–	–	–	–	–	–	–	–	–	–	–	–	–	x	–	–
Catalyst loading, mg Ag/cm^2	5.0	5.0	5.0	5.0	5.0	5.0	5.0	5.0	5.0	5.0	5.0	5.0	5.0	5.0	1.0	5.0
Current density, mA/cm^2	100	100	100	100	200	100	100	100	100	175	250	100	100	100	100	100
Average cell voltage, V	0.8	0.8	0.8	0.8	0.8	0.8	0.8	0.8	0.8	0.8	0.8	0.8	0.8	0.8	0.8	0.8
Power plant efficiency, %	38.1	37.7	37.7	36.6	31.8	38.1	38.1	38.1	38.1	38.1	38.1	38.1	38.1	38.1	38.1	38.3
Overall efficiency, %	25.6	25.4	25.4	30.7	21.4	25.6	25.6	25.6	25.6	25.6	25.6	25.6	25.6	25.6	25.6	25.8
Total capital cost $\times 10^{-6}$, $	13.6	51.9	130	53.6	16.6	13.6	13.6	13.6	13.9	10.9	9.8	11.6	11.3	13.6	13.5	13.6
Capital costs, $/kWe	620	599	599	700	890	620	620	620	636	500	448	529	517	620	615	620
Cost of electricity, mills/kWh																
Capital	19.6	18.9	18.9	22.1	28.1	19.6	19.6	19.6	20.1	15.8	14.2	16.7	16.3	19.6	19.4	19.6
Fuel*	23.3	23.6	23.6	18.7	27.9	23.3	23.3	23.3	23.3	23.3	23.3	23.3	23.3	23.3	23.3	23.1
Operation and maintenance	17.8	17.9	17.9	18.1	11.2	7.2	5.0	3.4	3.6	11.4	8.7	16.5	16.5	17.8	17.4	17.8
Total	60.8	60.4	60.4	58.9	67.2	50.1	48.0	46.4	47.0	50.6	46.2	56.5	56.1	60.8	60.1	60.6
Estimated time of construction, yr	1.5	2.0	2.5	4.0	1.5	1.5	1.5	1.5	1.5	1.5	1.5	1.5	1.5	1.5	1.5	1.5
Estimated availability date	1985	1985	1990	1990	1985	1985	1985	1985	1985	1985	1985	1985	1985	1985	1985	1985

*Used base delivered fuel cost.

costs associated with the need to eliminate additional carbon dioxide, which now forms 40% of the shift converter effluent instead of the 20% when high-Btu gas is employed as the fuel.

As for the corresponding case in the acid fuel cell parametric assessment, there is COE penalty associated with the replacement of air by oxygen (AL5) from a dedicated liquid-air distillation plant. The penalty of the alkaline system is less than half that for the acid system, however, because acid gas scrubbing of the oxidant stream is no longer necessary. The effects of electrolyte-thickness reduction (AL16) and increasing the flexibility of the power conditioning subsystem (AL9) are small.

Molten Carbonate Fuel Cell Power System

Table 4.11 is a summary of initial plant and overall efficiencies for the various parametric points. The efficiencies do not represent optimized systems, but allow a comparison of the relative effect of different parameters. The efficiency of the fuel cell for any one fuel is assumed to be the same, with current density being changed by changes in operating variables such as the use of oxygen instead of air or alteration of the electrolyte thickness. Consequently, all systems using high-Btu gas have nearly the same plant and overall efficiencies of 48 to 49% and 32 to 33%, respectively. The small differences are due to different efficiencies of the power-conditioning system for different plant size, as described in Appendix 2.

There will be an efficiency decrease with time from these initial values, depending on the amount of voltage degradation of the cell at constant power.

The fuel type has an important effect on the efficiency. Thus, plant efficiencies of about 36.5 and 45% are obtained for medium-Btu and methanol fuels, respectively, compared to about 49% for high-Btu gas. On the other hand, the overall efficiency of about 30.5% for medium-Btu gas is not much less than the 33 and 31% efficiencies for high-Btu and methanol fuels. This is due to the higher gasification efficiency of the medium-Btu gas.

The use of medium-Btu gas involves much larger heat losses in the cell due to entropy factors in the electrochemical reactions. If this heat is recovered and used to produce additional ac power via a steam turbine generator as in MC4, the power plant efficiency is increased beyond that of the other systems which do not use the waste heat in this way. If the analogous combined fuel cell/turbine generator system is used with high-Btu or methanol fuels, a plant efficiency would be obtained of about 59%, but one of a lower overall efficiency (40 to 41%) than the 45.5% for MC4 (medium-Btu fuel). This is due to the lower gasification efficiency for the higher-grade fuels.

MC15 uses methanol as fuel. The ~45% power plant efficiency calculated for this system is only 4% less than the value for the corresponding system using high-Btu gas (MC1). Internal reforming of methanol, however, will probably be easier than for high-Btu gas. This might reduce or eliminate efficiency differences between the high-Btu (MC1) and methanol (MC15) systems.

The systems using oxygen, MC5 and MC17, are arbitrarily operated at the same efficiency as MC1 but at higher current densities to reduce fuel cell capital and

TABLE 4.11: VALUES OF ALL RELEVANT PARAMETERS FOR THE PARAMETRIC POINTS OF THE MOLTEN-CARBONATE FUEL CELL POWER SYSTEM

	Parametric Point, MC Number.																
	1	2	3	4	5	6	7	8	9	10	11	12	13	14	15	16	17
Power output, MWe	22.6'	88.8	22.2	1,190	22.5	22.6	22.6	22.6	22.6	22.6	22.6	22.6	22.6	22.2	22.5	22.6	223.8
Fuel cell rating, MW dc	25	100	250	900	25	25	25	25	25	25	25	25	25	25	25	25	250
Fuel																	
High-Btu gas	x	x	x	–	x	x	x	x	x	x	x	x	x	–	–	x	x
Medium-Btu gas	–	–	–	x	–	–	–	–	–	–	–	–	–	x	–	–	–
Methanol	–	–	–	–	–	–	–	–	–	–	–	–	–	–	x	–	–
Oxidant																	
Air	x	x	x	x	–	x	x	x	x	x	x	x	x	x	x	x	–
Oxygen	–	–	–	–	x	–	–	–	–	–	–	–	–	–	–	–	x
Fuel cell life, 10^3 hr	10	10	10	10	10	30	50	100	100	10	10	10	10	10	10	10	10
Voltage degradation, %	5	5	5	5	5	5	5	5	15	5	5	5	5	5	5	5	5
Temperature, °C	650	650	650	650	650	650	650	650	650	650	650	700	750	650	650	650	650
Electrolyte type																	
Paste of Li, Na, K, carbonates and alkali aluminates	x	x	x	x	x	x	x	x	x	x	x	x	x	x	x	x	x
Thickness, cm	0.1	0.1	0.1	0.1	0.1	0.1	0.1	0.1	0.1	0.1	0.1	0.1	0.1	0.1	0.1	0.05	0.1
Anode type																	
Ni	x	x	x	x	x	x	x	x	x	x	x	x	x	x	x	x	x
Cathode type																	
Lithiated NiO	x	x	x	x	x	x	x	x	x	x	x	x	x	x	x	x	x
Current density, MA/cm^2	200	200	200	200	300	200	200	200	200	150	250	201	202	200	200	203	300
Average cell voltage, V	0.7	0.7	0.7	0.7	0.7	0.7	0.7	0.7	0.7	0.7	0.7	0.7	0.7	0.7	0.7	0.7	0.7
Power plant efficiency, %	48.8	48.0	48.0	54.4	48.8	48.8	48.8	48.8	48.8	48.8	48.8	48.8	48.8	36.4	44.8	48.8	48.0
Overall efficiency, %	32.9	32.3	32.3	45.7	32.9	32.9	32.9	32.9	32.9	32.9	32.9	32.9	32.9	30.6	31.2	32.9	32.3
Total capital cost x 10^{-6}, $	11.5	45.0	114	570	19.9	11.5	11.5	11.5	11.9	14.6	10.7	11.5	11.5	11.9	11.8	11.4	150
Capital costs, $/kWe	514	505	511	482	888	514	514	514	530	653	477	513	511	540	528	510	674
Cost of electricity, mills/kWh																	
Capital	16.3	16.0	16.2	15.2	28.1	16.3	16.3	16.3	16.7	20.7	15.1	16.2	16.2	17.1	16.7	16.1	21.3
Fuel*	18.2	18.5	18.5	12.6	18.2	18.2	18.2	18.2	18.2	18.2	18.2	18.2	18.2	18.8	20.6	18.2	18.5
Operation and maintenance	23.4	23.5	23.6	16.1	20.8	8.6	5.7	3.5	3.6	30.7	19.0	23.2	23.1	23.9	23.5	23.1	18.9
Total	57.8	57.9	58.2	43.9	67.0	43.1	40.1	37.9	38.5	69.6	52.2	57.7	57.4	59.7	60.8	57.4	58.7
Estimated time of construction, yr	1.5	2.0	2.5	5.0	1.5	1.5	1.5	1.5	1.5	1.5	1.5	1.5	1.5	1.5	1.5	1.5	2.5
Estimated availability date	1990+	1990+	1990+	1990+	1990+	1990+	1990+	1990+	1990+	1990+	1990+	1990+	1990+	1990+	1990+	1990+	1990+

*Used base delivered fuel cost.

O&M costs, as explained below. Note that for these systems, power for the air compressor of the oxygen plant is derived from a steam turbine operating with free excess heat from the fuel cell. If an electric drive compressor were used, it would severely reduce the plant efficiency.

In summary, the best overall efficiency of about 45% for the molten carbonate system investigated is attained for a combined system with the fuel cell using medium-Btu fuel and a steam turbine generator to convert excess heat from the fuel cell to electrical energy.

The highest plant efficiencies of about 59% can be obtained for the analogous combined systems using high-Btu or methanol with internal reforming, but their overall efficiencies are only 40 to 41%, due to the inefficiency of the gasification process.

For the fuel cell system only, plant efficiencies of ~49, ~45 and ~36% and overall efficiencies of ~33, ~31 and ~30.5% are obtained with the high-Btu, methanol and medium-Btu fuels, with little to choose between high-Btu and methanol since internal reforming may be easier with the latter fuel.

The cost estimates for COE from the various unoptimized molten carbonate fuel cell plants selected for analysis vary from about 10.8 to 19.4 mills/MJ (39 to 70 mills/kWh), as shown in Table 4.11. If a ten-year fuel cell lifetime with excess heat recovery were possible, a COE value close to 8.3 mills/MJ (30 mills/kWh) could perhaps be realized.

Of the parameters investigated, power plant size has only a small effect of about 0.8% reduction of COE for a factor of ten increase in plant size, going from MC1 (25-MW) to MC3 (250-MW). If oxygen rather than air is used as oxidant, however, plant size is important, since the cost of oxygen varies substantially with the amount of oxygen required. Thus, MC5, the 25-MW plant, has a COE of about 18.6 mills/MJ (67 mills/kWh) compared to 16.3 mills/MJ (58.7 mills/kWh) for the corresponding 250-MW plant (MC17) using oxygen. The latter figure, however, is no improvement over the 25-MW plant using air and assumes the availability of free waste heat from the fuel cell to operate the turbine-drive compressor of the oxygen plant.

The effect of fuel type is surprisingly small, with medium-Btu gas and methanol for the 25-MW plants (MC14 and MC15) having a COE only about 3 to 5% greater, respectively, than for the corresponding plant using high-Btu gas (MC1) and no conversion of waste heat. If the waste heat from the fuel cell, however, is converted to ac electrical energy via a steam turbine generator, as in MC4, the COE is better for medium-Btu than for either high-Btu or methanol fuels, since the overall efficiency is better and capital and O&M costs are about the same.

If the power density can be increased without decreasing the efficiency by any means, this has a strong effect on the COE. This is illustrated by MC10 and MC11, in which the power density is reduced by 25% and increased by 25%, respectively, from that of the base case, MC1. This results in about a 20% increase and about a 9% decrease, respectively, in the COE.

MC16 shows that the effect of reducing the electrolyte thickness by a factor of 2 from an already thin 1 mm (39 mils) is small, only about 0.7% less than the COE for MC1.

MC6, MC7 and MC8 show that the effect of cell lifetime is the most important factor in reducing the COE. A change from a 36 Ms (10,000 hour) cell lifetime to 108, 180 and 360 Ms (30,000, 50,000 and 100,000 hour) gave corresponding reductions of the COE of about 25, 30 and 34%, respectively.

In summary, fuel cell life longer than the presently possible 36 Ms (10,000 hour) production of additional ac energy from excess fuel cell heat, or a substantial increase in power output over that assumed, are the principal factors capable of appreciably reducing the COE below about 16.7 mills/MJ (60 mills/kWh) estimated from available state-of-the-art developments to date for the molten carbonate fuel cell.

Fuel cell size and fuel type (except for combined fuel cell and steam turbine generator systems), electrolyte thickness and moderate voltage degradation have relatively small effects on the COE.

Oxygen is not competitive with air as the oxidant except possibly in the extreme case of a large (e.g., 250-MW) fuel cell plant, with waste heat being used to run a turbine which operates the air compressor of the oxygen plant.

Solid Electrolyte Fuel Cell Power System

The power plant and overall energy efficiencies for all twenty points of the parametric assessment of the solid electrolyte power system are listed in Table 4.12. This table also provides information pertinent to the operation of the fuel cell subsystem in the power plant corresponding to each of the twenty points.

In the base case SE1 high-Btu gas is employed as the fuel. Because of the high average cell voltages, predicted on the direct electrochemical oxidation of methane at the fuel electrode, the power plant and overall energy efficiencies are high, lying at 69.7 and 46.9%, respectively. These values are unlikely to be realized in practice because of the metastability of methane at this temperature, resulting in either reformation on any available metallic surface, including that of the metallic component of the fuel electrode, in the presence of an adequate supply of water vapor, or carbon deposition (46).

For most points in the parametric assessment, medium-Btu gas is the fuel, so that for comparison purposes, Point SE2, corresponding to the 100-MW dc power plant, is much more useful than Point SE1. With medium-Btu gas as fuel, the power plant efficiency lies in the range 40 to 42% (SE2, SE3, SE5 through SE9 and SE12 through SE15), while the overall energy efficiency moves between 33 and 36%. Because of the lowered cell voltages in the fuel cell subsystems, corresponding to Points SE10 and SE11, increases in the power density cause a reduction in the plant and overall efficiencies.

The heat which must be rejected by the fuel cell subsystem results from thermodynamic and electrochemical inefficiencies. This waste heat may be employed in a steam-bottoming plant (SE4) or to supply the heat required by the endothermic processes occurring in a coal gasification reactor, which then meets the

TABLE 4.12: VALUES OF ALL RELEVANT PARAMETERS FOR THE PARAMETRIC POINTS OF THE SOLID-ELECTROLYTE FUEL CELL POWER SYSTEM

Parametric Point, SE #	1	2	3	4	5	6	7	8	9	10	11	12	13	14	15	16	17	18	19	20
Power Output, MWe	22.9	87.6	219	1154	18.3	22.3	22.3	22.3	22.3	22.3	22.3	22.3	22.3	22.3	22.3	22.3	22.3	219	1064	22.7
Fuel Cell Rating, MW	25	100	250	900	25	25	25	25	25	25	25	25	25	25	25	25	25	250	900	25
Fuel																				
High-Btu Gas	X																			X
Medium-Btu Gas		X	X	X	X	X	X	X	X	X	X	X	X	X	X	X	X	X		
Low Btu Gas																			X	
Oxidant																				
Air	X	X	X	X		X	X	X	X	X	X	X	X	X	X	X	X	X	X	X
Oxygen					X															
Fuel Cell Life, 10^3 hr	10	10	10	10	10	30	50	100	100	10	10	10	10	10	10	10	10	10	10	10
Voltage Degradation, %	5	5	5	5	5	5	5	5	15	5	5	5	5	5	5	5	5	5	5	5
Temperature, °C	1000	1000	1000	1000	1000	1000	1000	1000	1000	1000	1000	1000	1000	1000	1000	900	1100	1000	1000	1000
Electrolyte																				
$(ZrO_2)_{1-x}(CaO)_x$														X				X		
$(ZrO_2)_{1-x}(Y_2O_3)_x$	X	X	X	X	X	X	X	X	X	X	X	X	X		X	X	X		X	X
Thickness, cm 10^{-3}	4	4	4	4	4	4	4	4	4	4	4	2	4	4	4	4	4	2	4	2
Anode																				
Ni-ZrO_2 — Cermet	X	X	X	X	X	X	X	X	X	X	X	X	X	X	X	X	X	X	X	X
Cathode																				
$In_2O_3/PrCoO_3$-x	X	X	X	X	X	X	X	X	X	X	X	X	X	X		X	X	X	X	X
SnO_2/Sb-Doped															X					
Interconnection																				
$CoCr_2O_2$ (Mn-Doped)													X							
Cr_2O_3	X	X	X	X	X	X	X	X	X	X	X	X		X	X	X	X	X	X	X
Thickness, cm 10^{-3}	2	2	2	2	2	2	2	2	2	2	2	2	2	2	2	2	2	2	2	2
Current Density, mA/cm^2	400	400	400	400	400	400	400	400	400	600	800	400	400	400	400	400	400	800	400	800
Average Cell Voltage, Volts	0.84	0.66	0.66	0.66	0.76	0.66	0.66	0.66	0.66	0.59	0.51	0.67	0.66	0.64	0.66	0.55	0.77	0.68	0.56	0.69
Thermodynamic Eff, % (1)																				
Powerplant Eff, %	69.7	40.9	40.9	60.2	39.7	41.6	41.6	41.6	41.6	37.1	32.0	42.3	41.6	40.3	41.6	34.7	48.5	53.0	47.8	56.7
Overall Eff, %	46.9	34.3	34.3	50.6	33.3	35.0	35.0	35.0	35.0	31.2	26.9	35.5	35.0	33.9	35.0	29.2	40.8	53.0	47.8	38.2
Total Capital Cost × 10^{-6}, $	9.67	42.0	106	539	17.7	10.7	10.7	10.7	11.0	9.63	9.28	10.5	10.7	10.7	9.04	11.9	9.80	205	893	7.67
Capital Costs, $/kWe	424	482	488	466	972	481	481	481	496	434	418	474	481	483	407	535	442	948	859	340
Cost of Elect, Mills/kWh																				
Capital	13.4	15.2	15.4	14.7	30.7	15.2	15.2	15.2	15.7	13.7	13.2	15.0	15.2	15.3	12.9	16.9	14.0	30.0	27.1	10.7
Fuel (2)	12.7	16.7	16.7	11.3	17.2	16.4	16.4	16.4	16.4	18.4	21.3	16.1	16.4	16.9	16.4	19.7	14.1	5.5	6.3	15.7
Oper. & Maint.	16.0	20.8	21.1	14.2	24.2	7.9	5.3	3.4	3.5	16.2	14.5	20.2	20.6	21.1	20.6	24.3	17.9	12.2	18.4	10.9
Total	42.1	52.7	53.2	40.2	72.2	39.5	36.9	35.0	35.6	48.3	49.0	51.3	52.2	53.3	49.9	60.9	45.9	47.7	51.9	37.3
Est. Time of Construction, yr	1.5	2.0	2.5	5.0	1.5	1.5	1.5	1.5	1.5	1.5	1.5	1.5	1.5	1.5	1.5	1.5	1.5	3.0	5.5	1.5
Est. Availability Date	2000+	2000+	2000+	2000+	2000+	2000+	2000+	2000+	2000+	2000+	2000+	2000+	2000+	2000+	2000+	2000+	2000+	2000+	2000+	2000+

Notes:
(1) Where Applicable
(2) Use Base Delivered Fuel Cost

fuel requirements of the fuel cell subsystem (SE18 and SE19). The net effect of waste heat recovery is to substantially raise the efficiency of the power system. For Point SE4, the power plant and overall energy efficiencies are 60.2 and 50.6%, respectively. As the fuel in the power plant corresponding to Point SE18, the Westinghouse Fuel Cell Power System, is coal; the power plant and overall energy efficiencies are identical at 53%.

For point SE19, which involves the use of a low-Btu gasification reactor, employing coal, air and steam as input, the power plant efficiency is 47.8% because of the use of a steam-bottoming plant. This is lower than for Point SE18, in which medium-Btu is generated by recycling fuel gas which has been partially oxidized in the fuel cell subsystem. The overall efficiency for Point SE19 is identical with the power plant efficiency because the low-Btu gasifier is considered to be fully integrated with the power system. Thus, coal and air are the inputs.

The COE and their breakdowns into capital, fuel and O&M charges, for the values of plant construction and operation shown in Table 4.8, are given in Table 4.12.

As discussed above, the results for Point SE2 provide a better basis for discussion of the parametric assessment of the solid electrolyte power system because of the general use of medium-Btu gas for most points, and also because of the technical uncertainty surrounding the direct use of high-Btu gas as a fuel. The COE for the power plant corresponding to SE2 is 14.6 mills/MJ (52.7 mills/kWh). The portion of the COE ascribable to capital is 4.23 mills/MJ (15.2 mills/kWh), to fuel 4.64 mills/MJ (16.7 mills/kWH) and to O&M 5.77 mills/MJ (20.8 mills/kWh).

The O&M charge is high because of the high replacement cost—unlike the platinum-laden acid fuel cell modules, the solid electrolyte modules are assumed to have a scrap value of zero. Of the total charge, 5.35 mills/MJ (19.3 mills/kWh) is ascribable to the need to replace the modules after 36 Ms (10,000 hours) of useful life. Points SE6 through SE8 explore the effect of increasing the useful life. The electricity costs are 11.0 mills/MJ (39.5 mills/kWh) for 108 Ms (30,000 hours) and 9.73 mills/MJ (35.0 mills/kWh) for 300 Ms (100,000 hours).

Comparison of the COE and efficiency data for SE2, SE10 and SE11, reveal the effect of variations in the power density. An increase of the active cell power density from 264 to 354 mW/cm^2 (245 to 329 W/ft^2) results in a COE reduction from 14.6 to 13.4 mills/MJ (53.7 to 48.3 mills/kWh). A further increase in the power density to 480 mW/cm^2 (379 W/ft^2), however, caused a small increase in the COE, as the increase in the cost component ascribable to fuel (because of the lower cell voltage and thus plant efficiency) more than outweighs the sum of the reductions in the costs ascribable to capital and O&M.

The power density changes, also, when the temperature of operation is changed from 1273°K (1832°F) as in SE2 to 1373°K (2012°F) and 1173°K (1652°F) in SE17 and SE16, respectively. The decline in electrolyte and interconnection resistivities with increasing temperature is assumed to result in a parallel increase in cell voltages. Thus, an increase in the temperature of 100°K (180°F) causes a 20% increase in cell voltage and power plant efficiency resulting in a COE decrease of approximately 1.9 mills/MJ (6.8 mills/kWh). A lowering of the fuel cell temperature by 100°K (180°F) results in COE penalty of 2.27 mills/MJ (8.2 mills/kWh).

Substitution of oxygen for air as the oxidant, as in Point SE5, results in a substantial COE penalty. The penalty is greater here than for Point MC5 because the power required for the oxygen plant is taken from the ac output of the power-conditioning subsystem instead of being generated by a turbine powered by waste heat from the fuel cell subsystem.

Replacement of tin-doped indium oxide by antimony-doped tin oxide, as in SE15, causes a small decrease in the electricity cost. Even smaller decreases are registered by reducing the electrolyte thickness (SE12) and by substituting manganese-doped cobalt chromite for chromium sesquioxide as the interconnection material (SE13). Despite the lower cost of calcia as a stabilizing agent for zirconia (relative to yttria), the use of calcia-stabilized zirconia as an electrolyte (SE14) results in a small electricity cost penalty. This is caused by the higher resistivity of this electrolyte, which, in turn, causes a cell voltage reduction and, consequently, a loss in plant efficiency.

The use of a steam-bottoming plant for waste-heat recovery in the power plant corresponding to Point SE4 results in a dramatic reduction in the electricity cost from 14.6 mills/MJ (52.7 mills/kWh) to 11.2 mills/MJ (40.2 mills/kWh). Similarly, a reduction of 1.4 mills/MJ (5 mills/kWh) is observed for the Westinghouse Fuel Cell Power System (SE18). The results for SE19 indicate that partial thermal coupling of the fuel cell subsystem with a low-Btu gasifier offers little advantage from a COE standpoint, despite the use of a steam-bottoming plant for waste heat recovery.

CONCLUSIONS AND RECOMMENDATIONS

Considerable caution must be exercised when a comparison between fuel cell power systems is attempted on the basis of data provided for individual systems. The calculated power plant efficiencies may be relied upon to within a few percent either way, as they compare favorably with estimates available from other fuel cell work. They are, of course, dependent on the correctness of the assumptions of fuel cell subsystem performance, cited earlier.

As will be evident from the approaches taken in the costing of the fuel cell subsystems, there is a much greater possibility of error in the estimation of COE. The costing procedures employed represent an unbiased effort to estimate the possible costs of the fuel cell subsystems on the basis of a realization of the performance targets, as discussed earlier. Because they are founded on so many arbitrary assumptions, the comparison of the different fuel cell power systems based on the COE derived in this study must be approached with care.

The parametric assessment of the four cell power systems was based on a matrix of 69 points—16 points each for the phosphoric acid and alkaline systems, 17 for molten carbonate and 20 for solid electrolyte. The parameters of the power systems, which were varied, are listed in Table 4.13.

TABLE 4.13: PARAMETERS VARIED IN FUEL CELL ASSESSMENT

Fuel cell useful life	Fuel cell plant rating
Power density	Electrolyte thickness
Fuel type	Voltage degradation
Oxidant type	Waste heat recovery system**
Catalyst loading*	Temperature of operation**

*Applicable for acid and alkaline systems.
**Applicable for molten carbonate and solid electrolyte
systems.

For each of the 69 points, power plant and overall energy efficiencies, and COE
(broken down into capital, fuel and O&M components) were calculated. Analysis of these results indicated that four of the parameters listed in Table 4.13
were of particular importance in improving efficiency and reducing COE. The
four parameters and their areas of impact are shown in Table 4.14.

TABLE 4.14: IMPORTANT PARAMETERS AND AREAS OF IMPACT

Fuel cell useful life	O&M costs
Fuel cell power density	Capital and O&M costs
Waste heat recovery system	Plant and overall efficiencies
Fuel type	Overall efficiency

The importance of fuel cell subsystems useful life is seen in Figure 4.14. The
decrease in COE with increasing life is most pronounced for the molten carbonate and solid electrolyte systems. The effect is least for the phosphoric acid
power system because, at $152/kW dc, the acid fuel cell subsystem not only is
the cheapest, but also has a sizeable salvage value due to its platinum content.
Although the alkaline system has a similar salvage value, the lower power density of the cell results in a higher replacement cost for the final cell subsystem
and, consequently, a greater dependency of the COE on the useful fuel cell life.

Figure 4.15 shows the marked effect of power density, i.e., power output per
unit electrode area, on the COE for the acid, alkaline and molten carbonate
systems. Increasing power density at constant efficiency implies advances in
the state of the art of cathode and anode fabrication technology. The more
conventional technique of power density variation is to increase the current
density, accepting a cell voltage reduction and, therefore, an efficiency penalty.
This results in an increased fuel charge which serves to offset the reductions in
the capital and O&M components of the COE (as discussed for Points SE10
and SE11 earlier).

A further complication of operation at a higher power density was discussed by
Kordesch (47) for alkaline fuel cells. The useful life is inversely proportional
to the current density, so that operation at a higher power density would result
in more frequent replacement of the fuel cell subsystem and, thus, in a higher
O&M charge.

The coupling of a steam-bottoming plant to the 900-MW solid electrolyte and
molten carbonate fuel cell subsystems of the power plants, corresponding to
Points SE4 and MC4, raises the ac outputs to 1,164 and 1,170 MWe, respectively,

and the overall energy efficiencies to 50.6 and 45.7%, respectively.

Thermal coupling of the fuel cell subsystem with a coal gasifier (another form of waste heat recovery) results in an overall energy efficiency of 53.0%, the highest derived in this study (Point SE18).

Because of the greater efficiency of the production of medium-Btu gas, relative to high-Btu gas, a 25% gain in overall energy efficiency may be registered by the use of this fuel.

The lower cost of medium-Btu gas (Table 4.8) results also in a lowering of the fuel component of the COE by greater than 20%, as shown by comparing the data for Points AC1 and AC4.

FIGURE 4.14:   THE EFFECT OF THE USEFUL LIFE OF THE FUEL CELL SUBSYSTEM ON THE COST OF ELECTRICITY

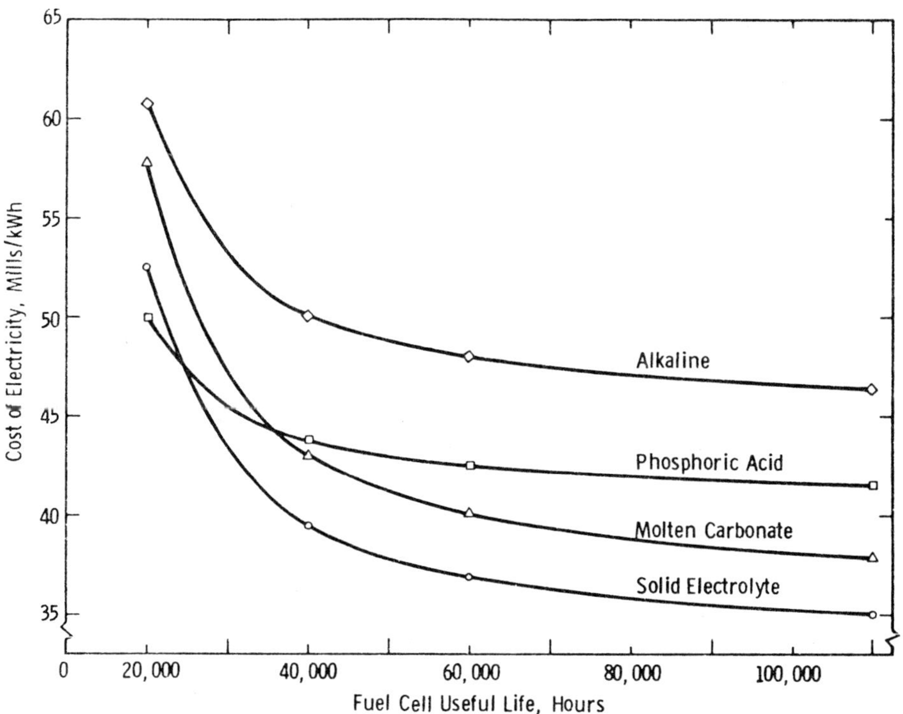

The fuel gas for the acid, alkaline and molten carbonate systems is high-Btu gas, while the solid electrolyte system costs were calculated on the basis of the use of medium-Btu gas.

FIGURE 4.15: DEPENDENCE OF THE ELECTRICITY COST ON THE POWER DENSITY AT CONSTANT EFFICIENCY

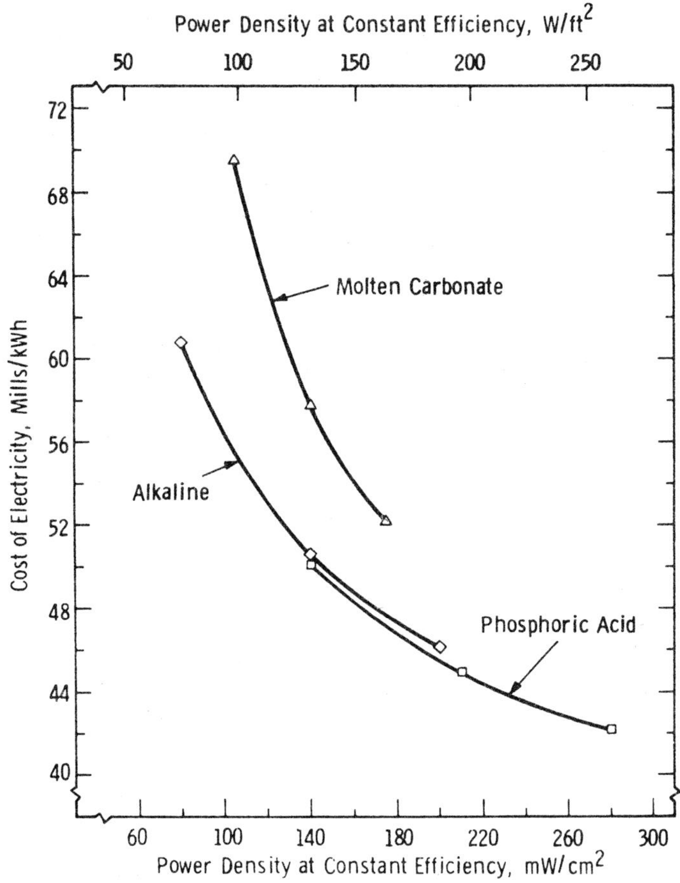

A comparison of the capital cost breakdowns for all four base cases is provided in Figure 4.16. The balance of plant was calculated by subtracting the sum of the material and site labor costs for the major components—e.g., for the phosphoric acid system, the fuel cell, power-conditioning and fuel processing subsystems—from the total direct costs of the power plant. The indirect costs, which include the interest during construction, escalation and contingency charges, and the profit and owner costs, were calculated similarly and represent the difference between the total capitalization and total direct costs.

The dominance of the fuel cell subsystem cost, lying in the range of 35 to 42% of the total capitalization, is apparent for every power system. The indirect costs, averaging 25% of the total, are also of considerable importance. Although the fuel processing cost for the phosphoric acid system is small, approximately $38/kWe (8.6%), this is not the case for the alkaline system, in which scrubbing of the carbon dioxide from the fuel gas and air before they enter the fuel cell subsystem is necessary. The cost of both scrubbers is included in the total of

$87/kWe shown for the alkaline fuel processing. The power-conditioning costs are similar, and lie in the range of $62 to $66/kWe. The differences arise from the assumptions of dissimilar parasitic losses for each system. The balance of plant costs are slightly higher for the two high-temperature power systems because they include charges for recuperative heat exchangers necessary for the heating of the input air and fuel gas streams.

FIGURE 4.16: BREAKDOWN OF CAPITAL COSTS

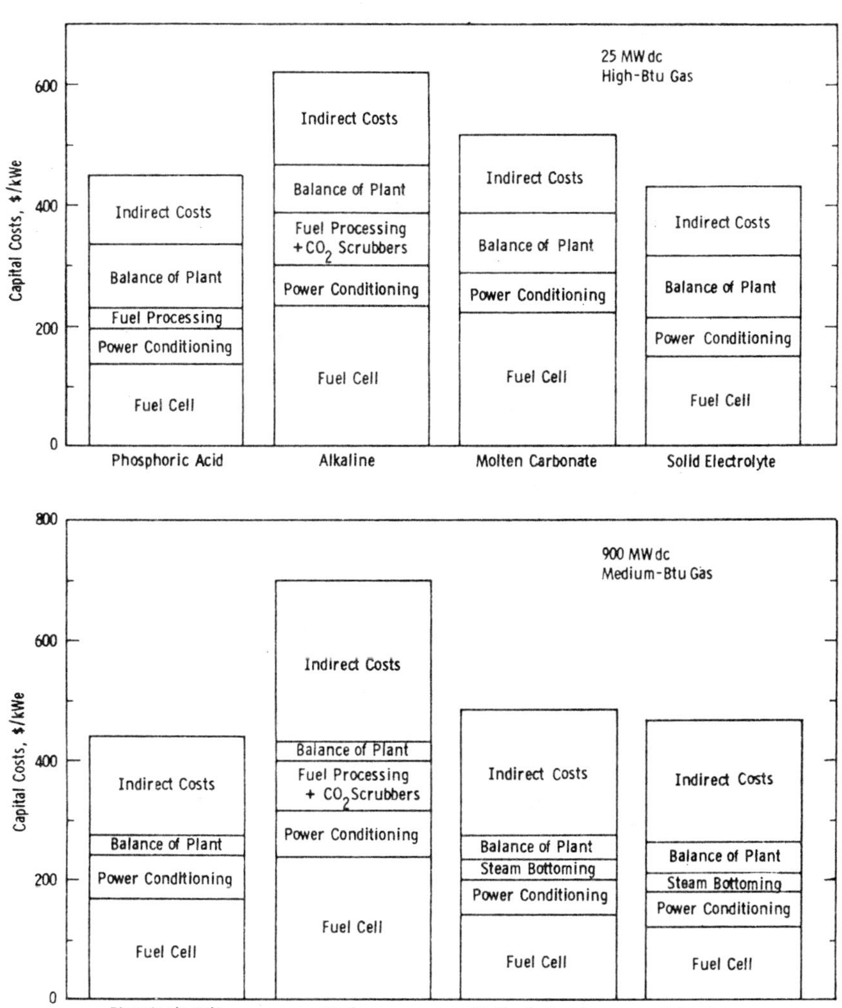

The base cases correspond to parametric points AC1, AL1, MC1 and SE1 and the 900-MW fuel cell power plants AC4, AL4, MC4 and SE4.

From a total capitalization standpoint, the alkaline power system, at $620/kWe, is the most expensive, due to the low power density assumed for the fuel cell subsystem and the need for carbon dioxide scrubbing. The phosphoric acid and solid electrolyte systems, at $448 and $424/kWe, respectively, require the least investment, and the molten carbonate system requires an intermediate capitalization of $514/kWe. The relative importance of the capitalization, at a fixed charge rate of 18%, is shown in Figure 4.17, which presents a breakdown of the COE for the base cases of all four power systems. The power plant and overall energy efficiencies for each case are also shown. The fuel gas is high-Btu gas costing $2.46/GJ ($2.60/10^6 Btu); the useful life of all fuel cell subsystems was assumed to be 36 Ms (10,000 hours). The fuel charges are less for the high-temperature systems because of their greater efficiencies. Their O&M charges are greater, however, because of their higher fuel cell subsystem replacement costs.

FIGURE 4.17: BREAKDOWN OF ELECTRICITY COSTS FOR THE BASE
CASE FUEL CELL POWER PLANTS

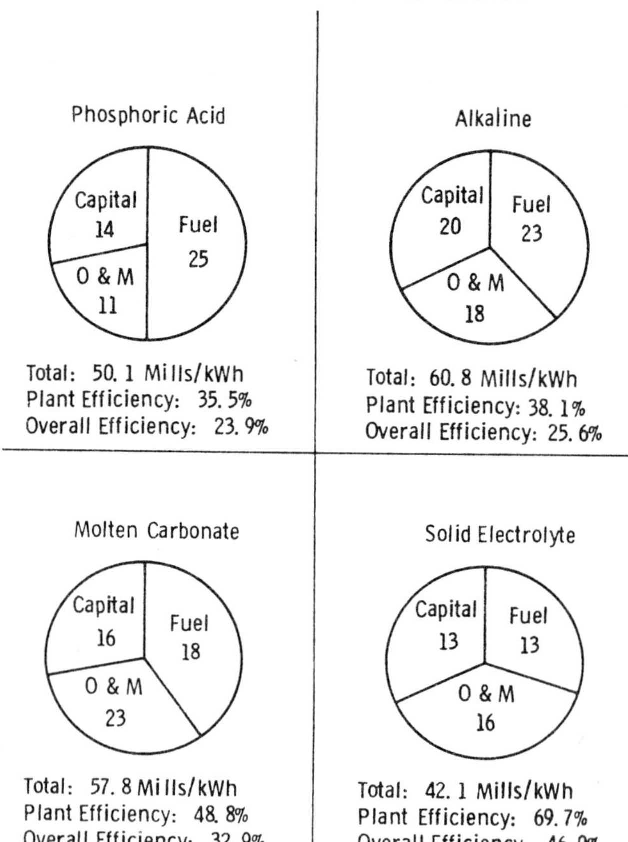

Total: 50. 1 Mills/kWh
Plant Efficiency: 35. 5%
Overall Efficiency: 23. 9%

Total: 60. 8 Mills/kWh
Plant Efficiency: 38. 1%
Overall Efficiency: 25. 6%

Total: 57. 8 Mills/kWh
Plant Efficiency: 48. 8%
Overall Efficiency: 32. 9%

Total: 42. 1 Mills/kWh
Plant Efficiency: 69. 7%
Overall Efficiency: 46. 9%

25-MW dc; useful life of 10,000 hours. The fuel gas is high-Btu gas (OTF).

A better basis for comparison is afforded by the data for Point 4 of every system. In every case, the 900-MW dc fuel cell subsystem uses medium-Btu gas as a fuel and is assumed to have a useful life of 36 Ms (10,000 hours). In addition, each of the high-temperature systems includes a steam-bottoming plant consisting of steam generators, a steam turbine and a heat rejection facility. This raises the ac output of the molten carbonate and solid electrolyte plants to 1,170 and 1,164 MW, respectively.

A breakdown of the capital costs for each system is shown in Figure 4.16. The total capitalizations required for the phosphoric acid, molten carbonate and solid electrolyte systems are seen to be virtually identical, lying in the range of $440 to $480/kWe. These costs would be even closer but for the assumption of a 157.7 Ms (5 year) construction period for the high-temperature systems, as against a 126.1 Ms (4 year) period for the acid system.

Accordingly, the indirect costs amount to approximately 37.5% of the total for the acid system versus approximately 44% for the high-temperature systems. The alkaline system, with additional problems posed by the use of medium-Btu gas, necessitating the removal of carbon dioxide at twice the rate of that when high-Btu gas is employed as the fuel, is now noncompetitive, lying at $700/kWe. For convenience of presentation, the fuel processing costs for the acid system have been included in the balance of plant costs in Figure 4.16.

The COE breakdowns and both efficiencies for each of the 900-MW dc systems are shown in Figure 4.18. The overall energy efficiencies for the high-temperature systems are, as expected, much higher than for either of the low-temperature systems. The higher fuel cost for the acid system, however, is offset by the lower O&M charge due to the lower fuel cell subsystems replacement cost, so that the total COE is essentially the same for the molten carbonate and acid systems at 12.2 mills/MJ (44 mills/kWh). The COE for the solid electrolyte system is lower still at 11.2 mills/MJ (40.2 mills/kWh). The alkaline system displays a COE of 16.4 mills/MJ (58.9 mills/kWh), which is substantially higher than for any of the other systems.

Projections as to the lowest COE possible for each system may be made on the basis of the data shown in Figure 4.18. These projections are highly tentative and are based on the multitude of assumptions presented and discussed earlier. If medium-Btu gas, costing $2.46/GJ ($2.60/10^6 Btu), is employed as the fuel, air is the oxidant, and the fuel cell subsystem life is at least 144 Ms (40,000 hours), then the COE will be as shown in Table 4.15.

TABLE 4.15: PROJECTIONS OF POSSIBLE EFFICIENCIES AND ELECTRICITY COSTS OF FUEL CELL POWER SYSTEMS

System Type	Overall Energy Efficiency, %	Possible COE, mills/kWh
Phosphoric acid	~30	High 30s
Alkaline	~30	Low 40s
Molten carbonate	~45	Low 30s
Solid electrolyte	~50	High 20s

## FIGURE 4.18:	BREAKDOWN OF ELECTRICITY COSTS FOR ALL 900-MW DC FUEL CELL POWER PLANTS

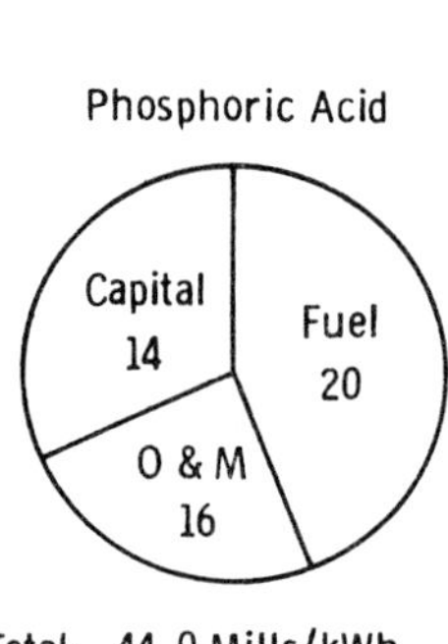

Total: 44. 0 Mills/kWh
Plant Efficiency: 34. 8%
Overall Efficiency: 29. 3%

Total: 58. 9 Mills/kWh
Plant Efficiency: 36. 8%
Overall Efficiency: 30. 7%

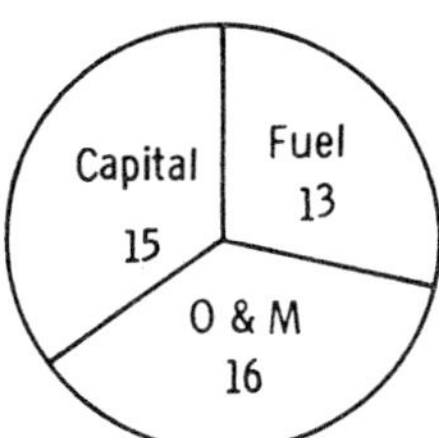

Total: 43. 9 Mills/kWh
Plant Efficiency: 54. 4%
Overall Efficiency: 45. 7%

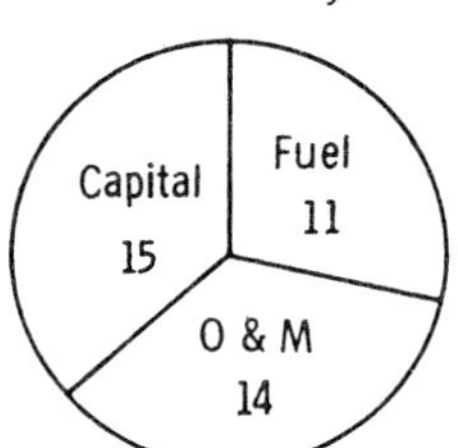

Total: 40. 2 Mills/kWh
Plant Efficiency: 60. 2%
Overall Efficiency: 50. 6%

The fuel gas is medium-Btu gas (OTF), and the useful life of all fuel cell subsystems is 10,000 hours.

The costs and efficiencies for the high-temperature systems are predicated on the use of a waste-heat recovery system. This probably limits the minimum size of a fuel cell subsystem to approximately 200-MW dc in order to allow economical and efficient recovery of the rejected heat.

The selection of fuel cell power systems for inclusion in further studies (Task II, Conceptual Design Preparation, and Task III, Implementation assessment) was based on the criterion of an overall energy efficiency significantly in excess of 35%. This eliminates all of the low-temperature fuel cell plants, and most of the high-temperature plants, which do not incorporate a waste-heat recovery system. Because of their high overall efficiencies, a solid electrolyte and molten carbonate power plant, as typified by Points SE4 and MC4, is recommended for the further refinement of efficiency and electricity cost estimates specified for Task II.

The Westinghouse Solid Electrolyte Fuel Cell Power System is recommended also for inclusion in Tasks II and III. Inspection of Figure 4.19, which presents all of the data pertinent to the recommended cases, reveals why.

FIGURE 4.19: BREAKDOWN OF ELECTRICITY AND CAPITAL COSTS

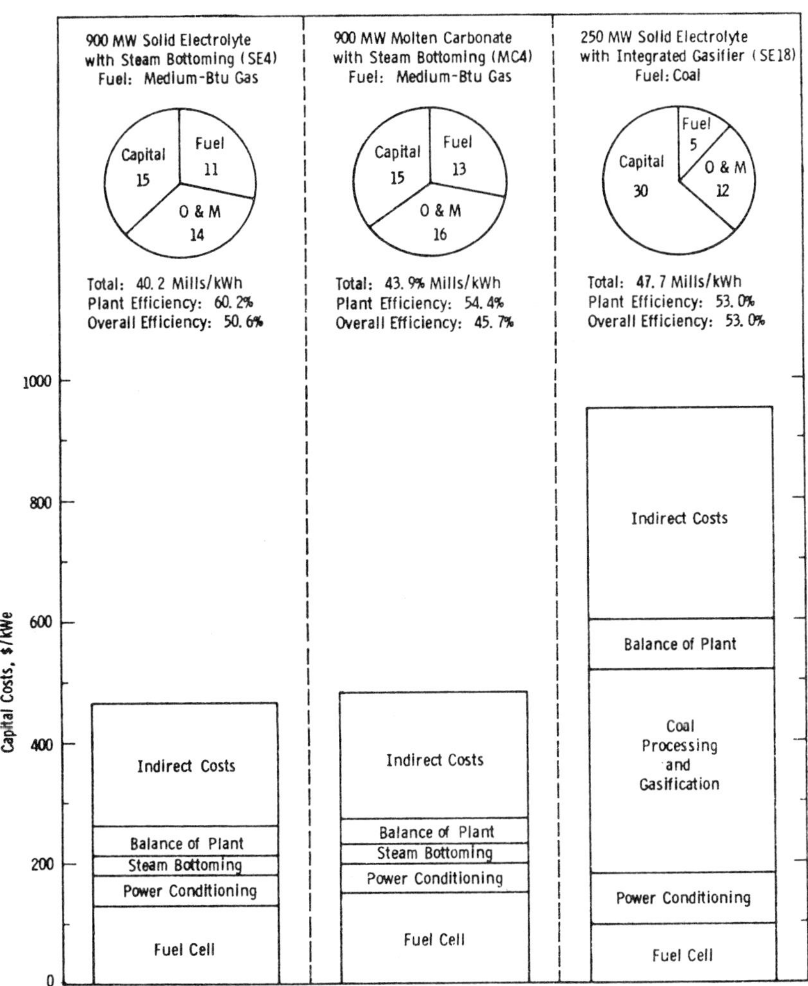

This breakdown is for the three power plants (corresponding to parametric points SE4, MC4 and SE18) recommended for further investigation in Tasks II and III. Note that the assumed useful life of all fuel cell subsystems is 10,000 hours.

The overall efficiency has been estimated very conservatively in this study. The estimated value of 53%, lower than previously published values of 60 and 57.5% (25) is, nevertheless, the highest determined in this study.

The COE of 13.3 mills/MJ (47.7 mills/kWh) is inflated by the capitalization associated with coal gasification. This estimate of $335/kWe is based on an evaluation performed in late 1968 (45), in which approximately 75% of the installed cost of the special fluidized bed coal gasification reactor was attributable to the cost of Incoloy 800 sheathing, considered necessary for efficient heat transfer from the fuel cell modules. In Task II, alternative materials and methods for efficient and economical thermal coupling of the gasifier and fuel cell subsystem should be explored. The potential for a reduction in the COE, and specifically its capital component, is obvious from Figure 4.19.

APPENDIXES

Appendix 1—Fuel Processing for Low-Temperature Fuel Cell Power Plants

The fuels employed in the parametric assessment of the low-temperature fuel cell power systems are medium-Btu gas (AC4 and AL4), methanol (AC14) and high-Btu gas (all other points). In order for these fuels to be usable at the anodes of the acid or alkaline fuel cell modules, they must be converted to a fuel gas consisting principally of hydrogen.

For high-Btu gas, consisting principally of methane, steam reformation coupled with shift conversion is the most economical method of producing this fuel gas. Carbon dioxide removal from the fuel gas stream is necessary in the alkaline case to prevent conversion of the potassium hydroxide electrolyte to potassium carbonate. This step is not required in the acid fuel cell system, as phosphoric acid does not react with carbon dioxide. Auxiliary equipment necessary in both power systems includes steam generators to supply the steam requirements of the reformer. Thus, an acid fuel cell power plant, fueled with high-Btu gas, requires a steam reformer, a shift converter and a steam generator. In addition to these components, the alkaline fuel cell system must include a carbon dioxide removal subsystem.

When medium-Btu gas, which is principally comprised of carbon monoxide (approximately 55% by volume) and hydrogen (approximately 33% by volume), is the fuel, there is obviously no need for steam reformation. Shift conversion and steam generation, to meet the steam requirements of the shift converter, are still required.

As stated earlier in this chapter, the carbon dioxide content of the medium-Btu gas stream after shift conversion is approximately double that in the shift-converter effluent for high-Btu gas. Thus, for alkaline power plants using medium-Btu gas as the fuel, carbon dioxide must be scrubbed at twice the rate necessary in plants operating on high-Btu gas. Although methanol may be cracked directly to form hydrogen and carbon monoxide (5), this study assumed that methanol, used in the power plant corresponding to Point AC14, is fed to a steam reformer, just as in the case of high-Btu gas.

To deliver 1 kW of electrical power from a phosphoric acid fuel cell operating at a terminal voltage of 0.7 V, a cell current of 1429 A is required. Thus, a minimum of 1429/96,489 equivalents or 14.9 mg/s (3.28×10^{-5} lb/s) of hydrogen must be delivered to the anode. The hydrogen requirement for an alkaline fuel cell is lower because of the higher cell voltage and may be calculated by the use of a multiplier, 0.7 V/0.8 V.

The optimum level of hydrogen utilization in a fuel cell is a complex function of the fuel cell performance and the relative costs of the fuel cell subsystem, the fuel processing subsystem, and the fuel, as the unused hydrogen may be employed to provide the thermal requirement of the steam reformer, as described below. The preparation of a detailed conceptual design of the complete power system, coupled with the knowledge of fuel cell performance as a function of hydrogen utilization, would permit accurate estimation of this optimum level. For the purposes of this preliminary study, however, utilization rates of 90 and 80% were assumed for feedstocks of high- and medium-Btu gas, respectively, for the reasons outlined earlier in the chapter. These assumptions lead to the hydrogen requirements for the four ratings of fuel cell subsystems shown in Table 4.16. The 900-MW power plants operate with medium-Btu gas as fuel; all others employ high-Btu gas as a feedstock.

TABLE 4.16: HYDROGEN REQUIREMENTS OF LOW-TEMPERATURE FUEL CELL POWER PLANTS

Fuel Cell	Hydrogen Requirements, tons/d	
Rating, MW	Acid	Alkaline
25	39.5	34.5
100	158	138
250	395	345
900	1,600	1,400

In the following subsections, the procedures employed in the costing of the reformer, the shift converter, the steam generators and the carbon dioxide removal subsystem, necessary for the production of the required hydrogen, are described. In addition, a brief description is given of the cost assumptions for the blowers used for the circulation of air through the cathode components of the fuel cell modules.

Steam-Methane Reformer: A schematic of a typical steam-reforming unit is shown in Figure 4.20. The reforming furnace is gas-fired, and the convection tube banks are fabricated from carbon steel. The radiant tubes, made from stainless steel, operate at a pressure of 0.689 MPa (100 psi) absolute. High-Btu gas is preheated, desulfurized by passage through activated carbon beds, mixed with preheated steam and fed to the catalyst-filled furnace, which operates at 1033° to 1144°K (1400° to 1600°F). This reaction

$$CH_4 + H_2O \longrightarrow CO + 3H_2 \quad \Delta H_{1600°F} = 97{,}400 \text{ Btu/lb-mol } CH_4$$

is highly endothermic. As one lb-mol of methane produces 4 mols of hydrogen (after shift conversion) or 42.87 std m^3 (1,514 scf), the reformer heat duty, Q/P, may be calculated as shown:

$$Q/P = \frac{97{,}400 \text{ Btu}}{1{,}514 \text{ scf}} = 64.3 \text{ Btu/scf of } H_2 \text{ required}$$

FIGURE 4.20:   SCHEMATIC OF A STEAM-METHANE REFORMER SYSTEM

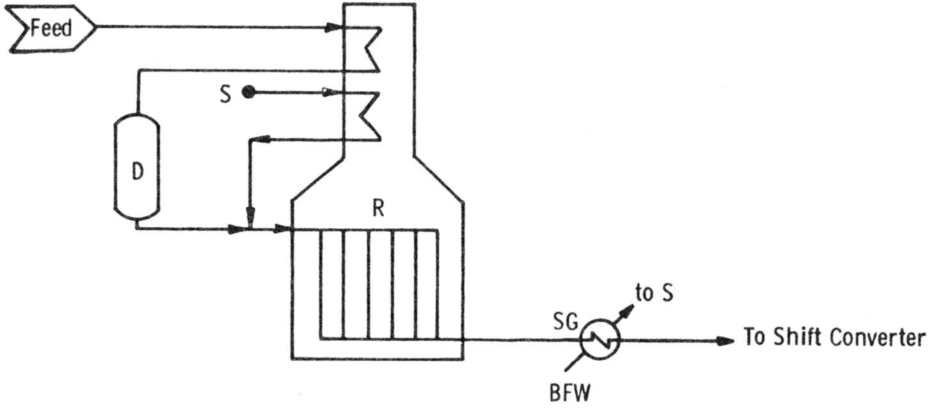

D: De-sulfurizer   S: Steam feed to reformer

SG: Steam-Generator   BFW: Boiler feed water

R: Gas-Fired Steam Reformer

The reformer heat duty for all six hydrogen production levels is presented in Table 4.17. Shown also are the total quantities of high-Btu gas required, based on arguments presented earlier in this chapter.

Using these heat-duty values, the base cost of the steam-methane reformer may be calculated. The method of cost estimation for all fuel conditioning costs follows an approach outlined by Guthrie (44) which uses mid-1970 prices. The costs have been factored upward by the ratio of the average Marshall and Stevens indexes (48) for the second and third quarters of 1974 and 1970. The base costs in mid-1974 terms are shown in the last column of Table 4.17.

TABLE 4.17:   HEAT DUTY AND HIGH-BTU GAS REQUIREMENTS AND
BASE COSTS (MID-1974) OF STEAM-METHANE REFORMERS FOR
HYDROGEN PRODUCTION IN FUEL CELL POWER PLANTS

Hydrogen Production Rate x 10^{-3},		Reformer Heat Duty x 10^{-6},	High-Btu Gas Usage x 10^{-3},	Base Cost x 10^{-3},
ton/d	scf/hr	Btu/hr	scf/hr	dollars
34.5	530	34	200	220
39.5	610	39	230	250
138	2,100	140	800	550
158	2,400	160	920	590
345	5,300	340	2,000	1,100
395	6,100	390	2,300	1,200

These base costs must be adjusted upward by a multiplier which is the sum of factors to allow for furnace type, radiant tube material and pressure. This yields purchased equipment base costs, which are shown in the last column of Table 4.18.

TABLE 4.18: BREAKDOWN OF ADJUSTMENTS TO BASE COSTS (MID-1974) OF STEAM-METHANE REFORMERS FOR HYDROGEN PRODUCTION IN FUEL CELL POWER PLANTS

Base Cost x 10^{-3}, dollars	Furnace Type	Tube Material	Tube Pressure	Total	Purchased Equipment Base Cost x 10^{-3}, dollars
220	1.345	0.345	0.05	1.740	380
250	1.345	0.345	0.05	1.740	440
550	1.342	0.341	0.05	1.733	950
590	1.342	0.341	0.05	1.733	1,020
1,100	1.341	0.340	0.05	1.731	1,900
1,200	1.341	0.340	0.05	1.731	2,100

The total direct cost of the reformer includes the purchased equipment base cost computed earlier, plus the materials and labor required for installation. Following Guthrie (44), installation materials (less concrete) typically average 25.4%. These are summarized in Table 4.19. The total cost shown does not include indirect costs associated with construction overhead, engineering, interest during construction, etc.

TABLE 4.19: REFORMER COSTS (MID-1974) FOR FUEL CELL POWER PLANTS

System Type	Fuel Cell Rating, MW	Reformer Costs x 10^{-3}, dollars			
		Purchased Equipment	Installation Material	Total Material	Site Labor
Alkaline	25	380	97	480	110
Acid	25	440	110	550	130
Alkaline	100	950	240	1,200	280
Acid	100	1,020	260	1,300	300
Alkaline	250	1,900	480	2,400	550
Acid	250	2,100	530	2,600	610

Shift Converter: After leaving the steam reforming unit, the gases are cooled to between 644° and 700°K (700° and 800°F) and passed over a water-gas shift catalyst to convert the carbon monoxide component to carbon dioxide and hydrogen by the reaction

$$CO + H_2O \longrightarrow CO_2 + H_2$$

A schematic of the shift conversion unit is shown in Figure 4.21. The shift converter is a pressure vessel loaded with an appropriate catalyst. In the four smallest sizes a length-to-diameter ratio of four is assumed. If a space velocity of 0.556 s^{-1} (2,000 hr^{-1}) is assumed, the volume of catalyst required may be calculated. The overall volume of the vessel is considered to be twice that of the catalyst.

The costing technique again follows that of Guthrie (44), with an escalation factor of 1.34 to convert mid-1970 costs to those of mid-1974. In the penultimate column of Table 4.21, the base costs (mid-1974) shown are for a carbon steel vessel operating at 0.345 MPa (50 psi) absolute. The costs include shell and two heads, nozzles and runways, skirt, base ring and lugs. The shift converter, however, should operate from 2.76 to 3.45 MPa (400 to 500 psi) absolute for compatibility with the steam-methane reformer. All costs at 0.345 MPa (50 psi) absolute were multiplied by 2.8 to allow for high-pressure design costs. These costs are shown in the last column of Table 4.22.

FIGURE 4.21:   SCHEMATIC OF SHIFT CONVERSION UNIT FOR FUEL
               CELL POWER PLANTS

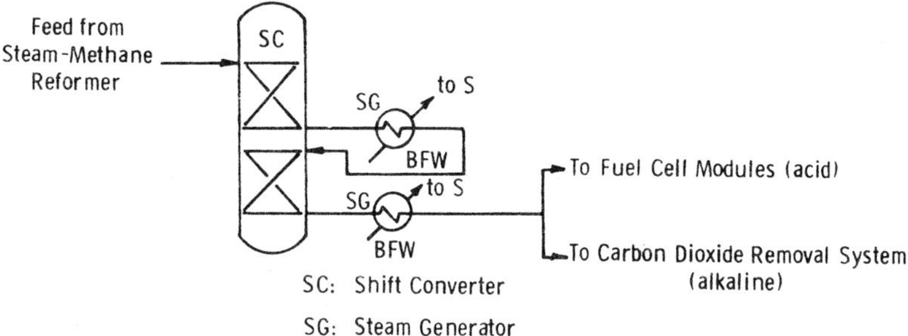

Installation costs of vertical pressure vessels are typically 3.05 times the base
costs (44). The total material and site labor costs of the shift converter units
for the different ratings of acid and alkaline fuel-cell power plants are shown in
Table 4.20.

TABLE 4.20: SHIFT-CONVERTER VOLUMES, DIMENSIONS AND BASE
             COSTS (MID-1974)

Hydrogen Required, ton/d	Volume, ft^3 Catalyst	Volume, ft^3 Vessel	Vessel Dimensions, ft Length	Vessel Dimensions, ft Diameter	Base Costs x 10^{-3}, dollars
34.5	270	540	23	5.5	36
39.5	320	640	24	6	45
138	1,100	2,200	35	9	77
158	1,250	2,500	39	9	81
345	2,700	5,400	70	10	140
395	3,200	6,400	80	10	180
1,400	11,000	22,000	90	18	610
1,600	13,000	26,000	100	18	640

TABLE 4.21: SHIFT-CONVERTER COSTS (MID-1974) FOR FUEL CELL
             POWER PLANTS

System Type	Fuel Cell Rating, MW	Shift Converter Costs x 10^{-3}, dollars Base	Installation Material	Total Material	Site Labor
Alkaline	25	36	20	56	21
Acid	25	45	25	70	26
Alkaline	100	77	43	120	45
Acid	100	81	46	130	47
Alkaline	250	140	79	220	81
Acid	250	180	100	280	100
Alkaline	900	610	340	950	350
Acid	900	640	360	1,000	370

TABLE 4.22: STEAM REQUIREMENTS AND GENERATOR COSTS (MID-1974) FOR FUEL CELL POWER PLANTS

Hydrogen Production Rate x 10^{-3} ton/d	scf/hr	Steam Rate x 10^{-3}, lb/hr	Steam Generator Base Cost x 10^{-3}, dollars
34.5	530	19	66
39.5	610	22	77
138	2,100	75	199
158	2,400	85	230
345	5,300	190	350
395	6,100	220	420
1,400	21,000	750	1,500
1,600	24,000	850	1,600

Steam Generators: An excess of steam, generally three to five times the stoichiometric requirement, is employed in the steam-methane reformer. Steam is raised in three main locations: [1] between the reformer and shift converter, in the cooling of the fuel gases from 1144° to 700°K (1600° to 800°F), as shown in Figure 4.20; [2] by the use of the heat rejected by the shift converter; and [3] in cooling the hydrogen-rich fuel gas from 700° to 464°K (800° to 375°F), the temperature of operation of the acid fuel-cell modules. The steam generators of [2] and [3] are shown schematically in Figure 4.21.

For the purposes of calculation, it will be assumed that 3 mols of steam are required in the reformer for every 4 mols of hydrogen reaching the fuel cell modules. The steam requirements are, therefore, 0.570 kg of steam/std m^3 of hydrogen (0.0356 lb of steam/scf of hydrogen). Table 4.22 presents the steam rates and steam generator base costs (mid-1974) as a function of hydrogen production. The costing method again follows that of Guthrie (44). The total material and site labor costs are shown in Table 4.23 for all four ratings of acid and alkaline fuel cell power plants.

TABLE 4.23: STEAM GENERATOR COSTS (MID-1974) FOR FUEL CELL POWER PLANTS

System Type	Fuel Cell Rating, MW	Base	Installation Material	Total Material	Site Labor
Alkaline	25	66	8.4	74	20
Acid	25	77	9.9	87	24
Alkaline	100	190	24	210	58
Acid	100	230	29	260	70
Alkaline	250	350	45	400	110
Acid	250	420	54	470	130
Alkaline	900	1,500	140	1,700	450
Acid	900	1,600	200	1,800	470

(- - Steam Generator Costs x 10^{-3}, dollars - -)

Carbon Dioxide Removal System: Leaving the shift converter, the fuel gas steam is at a temperature of 700°K (800°F) and a pressure of 0.483 MPa (70 psi) absolute. The carbon dioxide-carbon monoxide ratio is approximately 50. The gas composition is ~70% hydrogen, ~10% steam and the balance carbon oxides and inerts. The steam may be cooled and the steam condensed to yield a gas mixture of 80% H_2, 17% carbon oxides and 3% inerts. This fuel gas may be fed directly to the acid fuel-cell modules.

In the alkaline system, however, carbon dioxide removal must still be accomplished in order to protect the potassium hydroxide electrolyte. The process considered for this application is shown schematically in Figure 4.22. It consists of the Lurgi Rectisol Process, which uses refrigerated methanol. The total capital investment for a Rectisol System capable of stripping carbon dioxide and hydrogen sulfide from 4.16 kg-mols/s (33,000 lb-mols/hr) of fuel gas in the Bituminous Coal Research Bi-Gas Process (49) was $23.5 million (mid-1970) (50). When factored upward by 1.34 to convert to mid-1974 costs, the total capital investment required is approximately $0.582 per kg-mol/s ($950 per lb-mol/hr) of fuel gas.

FIGURE 4.22: SCHEMATIC OF CARBON DIOXIDE REMOVAL SYSTEM
FOR ALKALINE FUEL-CELL POWER PLANTS

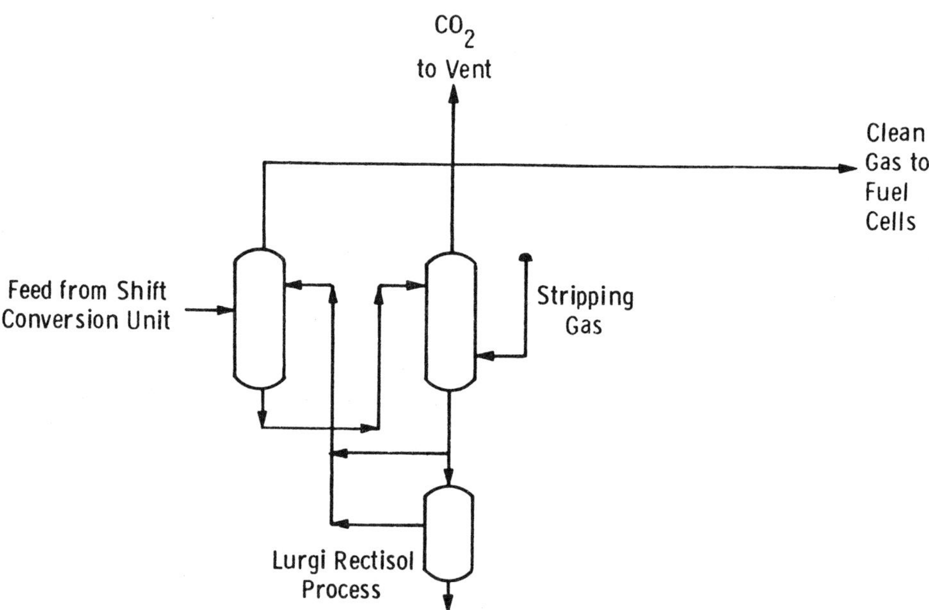

If it is assumed that the direct installed cost of the scrubbing system represents 28.5% of the total capital investment (52) and that the material-to-site labor ratio is equal to three, approximate estimates of the equipment and site labor costs for the Rectisol Process may be derived. These are shown for the four ratings of alkaline fuel cell power plants in Table 4.24. A linear relationship is assumed between scrubber cost and volume of gas scrubbed. Because of the additional load on the scrubbing system of the 900-MW plant due to the use of medium-Btu gas as the fuel, the costs are multiplied by 1.35.

Two of these scrubbing units are required per power plant, as carbon dioxide must also be removed from air. Although air contains approximately four orders of magnitude less carbon dioxide than does the fuel gas, the costs associated with scrubbing air were assumed to be the same as those given in Table 4.24.

TABLE 4.24: ESTIMATES OF MATERIAL AND SITE LABOR COSTS FOR FUEL GAS SCRUBBING IN AN ALKALINE FUEL CELL POWER PLANT

Fuel Cell Rating, MW	Hydrogen Required, ton/d	Fuel Gas Flow Rate, lb-mols/hr	Material Costs x 10^{-3}, dollars	Site Labor Costs x 10^{-3}, dollars
25	34.5	1,760	420	140
100	138	7,200	1,700	560
250	345	17,600	4,200	1,410
900	1,400	72,000	23,000	7,700

In general, the equipment cost was assumed to be a linear function of the quantity of gas passed through the process. For a 50% utilization of oxygen in air, the quantity of throughput air would be approximately a factor of eight greater than that of fuel gas. The advantages associated with the lower concentration of carbon dioxide in air, however, were assumed to approximately negate the 8 to 1 flow penalty. The treatment above, though very crude and allowing for no economy of scale, yields reasonable costs for scrubbing equipment to remove carbon dioxide from fuels and air. The value of $22/kW for air scrubbing in this closed-cycle process is acceptable when compared with $5/kW for the much simpler process involving the use of potassium hydroxide on a once-through basis (8).

Air Blowers: The costing of air blowers for the fuel cell subsystem was performed following the method of Guthrie (44). The air blower cost, C, in mid-1974 dollars is given by

$$C = \$(9)(1.34)(\text{Required airflow rate in scfm})^{0.68}$$

With the assumption of 50% utilization of the oxygen in the throughput air, a value of $19,000 is calculated for the air blowers in the 25-MW phosphoric acid fuel cell power plant. A site labor cost equivalent to 33% of the equipment cost was arbitrarily assumed. The material and site labor costs for all power plants were derived similarly.

Possible errors in these estimates arising from the underestimation of the pressure drop through the fuel cell modules are likely to have a negligible impact on the total capitalization of the fuel cell power plant. A total material and site labor cost of $25,000 for the 25-MW acid system air blowers is trivial when compared with, for example, a cost of $3.8 million for the fuel cell subsystem.

Appendix 2—Power-Conditioning Subsystem

Power-conditioning subsystems are necessary in fuel cell power plants for the conversion of the dc output of the fuel cell modules to 60 Hz ac power. NASA has specified that the 25-MW plant (25-, 100-, 250- and 900-MW ratings apply to fuel cell subsystems. The overall plant ratings are less, due to inverter subsystem losses and other parasitic losses in the fuel cell power system.) should deliver power at 69 kV to the distribution net, but the output from the other three sizes—100-, 250- and 900-MW—should be at 500 kV. The following is a discussion of inverter subsystems and transformers which are necessary to accomplish the required power conversion, and the resultant implications for the selection of fuel cell module sizes.

In general, there are at least seven power conversion schemes that can be considered for this application. These are: [1] chopper-inverter; [2] inverter; [3] buck-boost inverters; [4] complementary inverters; [5] HF link; [6] hybrid (HF link plus simple line-commutated inverter), and [7] force-commutated inverter. In a recent study at Westinghouse Research Laboratories (51), scheme 7 was found to be optimum from an economic and technical standpoint. Force-commutated systems not only operate with lower losses (4.5 to 5% versus 5 to 6% for the more conventional line-commutated systems) but also offer considerable operational advantages:

(a) They will ride through a system fault.

(b) It is possible to control their behavior with respect to reactive power demand and deliver independently of real power, thus conferring great stability on the power conversion system.

(c) They will start and run into a passive load.

Force-commutated inverter systems, however, suffer from a major limitation in that the state of the art indicates an upper limit of approximately 2 kV output voltage. Because of transformer primary-current considerations, this constraint limits use of this type of inverter system to the 25-MW case.

For the larger power plants (100-, 250- and 900-MW) a line-commutated inverter system must be used. Systems 1 through 3 are typical line-commutated schemes and represent relatively simple extensions of high voltage dc (HVDC) technology. Of these three, the buck-boost inverter is optimum from a cost standpoint. Further, it is close to being the most efficient and has no operational disadvantages compared to the other line-commutated schemes.

Most significantly, power factor improvement is achievable with this scheme. This results in a considerable savings in cost because auxiliary power-factor correction equipment is not needed. Although this inverter system can run, in theory, into a passive load, the stability of this operational mode is questionable in practice unless a synchronous capacitor is used to supply reactive VA.

Another disadvantage of line-commutated inverter systems, which applied also to this scheme, is that the dc side must be quickly interrupted after a system fault in order to protect the overall system. From this standpoint, line-commutated schemes compare unfavorably with force-commutated systems, which, as noted above, are able to ride through system faults.

The basic inverter unit, in the force- and line-commutated systems above, is the 3-phase bridge or Graetz connection. This is illustrated in Figure 4.23. It should be noted that the ac outputs of the two inverters are out of phase by $\pi/6$. The transformer required for step-up of the inverters' output to distribution voltages levels should have a twin core structure. For the line-commutated inverter scheme, wye-delta and delta-delta windings are specified, while in the force-commutated inverter case, one core should involve a wye-open wye winding plus a delta tertiary, and the other a delta-wye winding. These specifications, plus the poor power factor at which the transformers operate, result in transformer costs which are approximately double those of conventional transformers of the same kVA rating (52).

FIGURE 4.23: BLOCK DIAGRAM OF POWER CONVERSION EQUIPMENT REQUIRED FOR FUEL-CELL POWER PLANTS

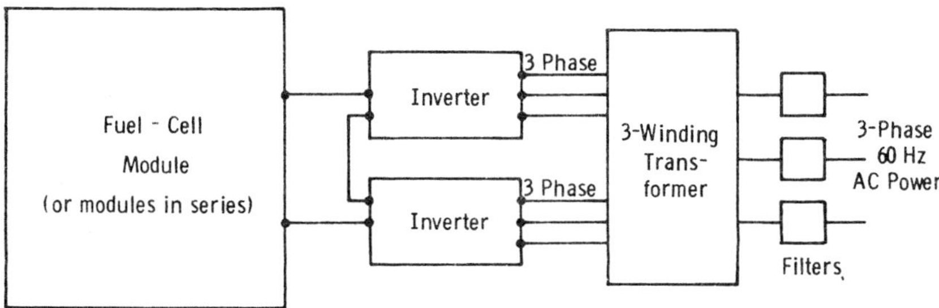

Table 4.25 outlines the relationship between the dc and three-phase ac currents and voltages for the force- and line-commutated systems, where α is the firing angle delay of the line-commutated inverter, and ϕ is the phase-shift angle of the force-commutated inverter.

TABLE 4.25: THE RELATIONSHIPS BETWEEN ELECTRICAL INPUT AND OUTPUT OF INVERTER SYSTEMS

System Parameters	Inverter System Type	
	Force-Commutated	Line-Commutated
Voltage	$V_{ac} = \left[\dfrac{\sqrt{6}}{\pi}\right] V_{dc}$	$V_{ac} = \left[\dfrac{\pi}{3\sqrt{2}}\right] \cos\alpha \, V_{dc}$
Current	$I_{ac} = \left[\dfrac{\pi}{3\sqrt{2}}\right] (\cos\phi)(I_{dc})$	$I_{ac} = \left[\dfrac{\sqrt{6}}{\pi}\right] I_{dc}$
Power	$\sqrt{3} \, (V_{ac})(I_{ac})(\cos\phi)$	$\sqrt{3} \, (V_{ac})(I_{ac})(\cos\alpha)$

The sizes, weights and selling prices of inverter systems for the four power plant ratings are shown in Table 4.26. All the data shown are based on the results of a previous study of inverter systems for 25-MW fuel cell power plants. The plan area/MW (51) for the buck-boost systems at the various rating levels were calculated by multiplication of the plan area requirement at the 25-MW level by the cube root of the ratio of the ratings of the larger system to the 25-MW unit. This factoring upwards ensures adequate clearance at the higher ratings, thus minimizing problems associated with creepage paths. The weights are taken from the above-mentioned study (51). The selling price data are derived from estimates of $60/kW and $55/kW for 25-MW line-commutated and force-commutated systems respectively, and from an estimate of ~$65/kW for the buck-boost system at the 900-MW level. The values for the 100- and 25-MW systems were derived by linear interpolation. The increases in capital costs ongoing to the larger plant sizes is due to higher costs of filters and the greater need for grading networks for the valves.

TABLE 4.26: SIZES, WEIGHTS AND COSTS OF INVERTER SYSTEMS FOR
FUEL CELL POWER PLANTS

Fuel Cell Rating, MW	Inverter Type	Plan Area, ft^2/MW*	Weight, lb/MW	Selling Price, $/MW**
25	Force-commutated	29	7,900	55,000
100	Buck-boost	23	4,800	60,000
250	Buck-boost	32	4,800	61,000
900	Buck-boost	48	4,800	65,000

*8-foot total height excluding bushings, for 25-MW; approximately 30 feet for 100-, 250- and 900-MW plants because of 500 kV tie.
**Prices include cost of single transformation. If double transformation is required, the costs shown must be increased by $10,000/MW.

Because of the lack of demonstrated reliability of fuel cell modules in extended service, it is not unreasonable to specify for all fuel cell power plants that a fault on the dc side should result in outages of not more than 10% of the rated capacity. Thus, in all cases, ten banks of fuel-cell modules are envisioned, each with its own inverter pair and transformer. In general, only one set of filters and switch-gear will be used on the high sides of the transformers.

In the 25-MW plants, each of the banks of fuel cell modules may be run at ~2200 A dc (1.15 kV dc). Single transformation from ~900 V ac to the required 69 kV ac level is achievable. In the 900-MW plants, ten banks of fuel cell modules are envisioned, each operating at 2900 A dc and ~31 kV dc. The ~23 kV ac output of the inverter pair permits single transformation to the required level of 500 kV ac.

Single transformation, however, is not possible without a significant inverter cost penalty in the 100-MW plants. Under the 10% outage constraint, each bank would contain 10 MW of fuel cell modules. As the ac output from the inverters must be at least 16 kV to facilitate single transformation to 500 kV ac, the fuel cell dc voltage requirements could be as high as ~27 kV, assuming a worst-case inverter power factor of 0.8. This would result in a dc current input to the inverters of ~370 A, which is too low for the inverters to be economical (51).

Rather than pay inverter costs more than twice those for systems operating at ~2200 A dc, it is more reasonable to pay the additional price for double transformation. This is estimated (52) to be ~$10,000/MW. An intermediate ac voltage of 34.5 kV will optimize the system with a dc voltage in each 10-MW bank of 4.6 kV (2200 A dc).

Similar considerations apply to the 250-MW plant. The 25-MW banks would supply less than 1 kA dc under the worst-case condition mentioned above. Once again, double transformation is economically more attractive than paying for increased inverter capability. Here, an intermediate ac voltage of 69 kV is envisioned with a dc voltage of ~11.5 kV (2200 A dc) in each 25-MW bank.

For the purposes of this study, an efficiency of 95.5% is assumed for the power-conditioning subsystems of all 25-MW power plants, based on the use of a force-commutated inverter system. Line-commutated inverter systems must be employed in the 100-, 250- and 900-MW power plants. Here, an efficiency of 95% is assumed. Double transformation, however, is required in all 100- and 250-MW power plants for the reasons outlined above. The second transformer is assumed to have an efficiency of 99.5%, so the net efficiency of the power-conditioning subsystem of the 100- and 250-MW plants is (95%)(0.995) or 94.5%.

The site labor costs are based on recent experience (mid-1974) of the Westinghouse Electric Corporation with the installation of a VAR generator (a device to inhibit voltage flickering in an arc furnace) in a steel-making facility of Akron Steel (53). The installation of this equipment, which is very similar in character to the power-conditioning subsystems described above, involved a site labor of approximately $3/kW. This value has been employed in this study.

The equipment, site labor and total costs of the power-conditioning subsystems for the four fuel cell power plant ratings are presented in Table 4.27. The costs for the 100- and 250-MW power plants reflect the cost of double transformation. All these designs and costs are based on a maximum input variation of the dc current and voltage to the power-conditioning subsystem of 5%. Point 9 in all four fuel cell power systems explores the effect of the variation of the voltage degradation at constant power for 5 to 15% in 25-MW plants. The power-conditioning subsystem must, therefore, have the capability of handling this greater variation of current and voltage characteristic of fuel cell subsystems.

The cost of the power-conditioning equipment may be assumed to be proportional to the dc current (51). At constant power the system with 15% voltage degradation would have to carry 20% more current than the system with a 5% voltage degradation. The costs of the power-conditioning subsystem for AC9, AL9, MC9 and SE9 were estimated by adding 20% to the value for the base case.

TABLE 4.27: VENDOR, SITE LABOR AND TOTAL COSTS (MID-1974) OF POWER-CONDITIONING SUBSYSTEMS FOR FUEL CELL POWER PLANTS

Fuel Cell Rating, MW	- - Power-Conditioning Costs x 10^{-3}, dollars - -		
	Equipment	Site Labor	Total
25	1,380	75	1,455
100*	7,000	300	7,300
250*	17,500	750	18,250
900	59,000	2,700	61,700

*Costs include additional transformation step at $10,000/MW.

Appendix 3—Oxygen Plants for Fuel Cell Power Systems

A fuel cell plant may be integrated with an oxygen plant as well as with a coal gasification plant. If oxygen is supplied to both the coal gasification unit (for production of medium- or high-Btu gas) and to the fuel cell plant, the total oxygen cost will be of similar magnitude to the coal cost.

Oxygen requirements depend on the type and size of the fuel cell and source of the coal, and are roughly 0.8 and 1.4 Mg of oxygen per Mg of coal, respectively, for the gasification and fuel cell plants. Thus, substitution of oxygen for air at the fuel cell cathode would have to give a considerable improvement in the power density at a given efficiency to justify the high oxygen costs involved.

The cost of oxygen from an on-site plant as operated by the utility is a strong function of plant size and the cost of energy for the compressor. Requirement for a 25-MW fuel cell would be about 3.15 kg/s (300 ton/d) of oxygen. The maximum size plant design commercially available is 21 kg/s ($\sim$2,000 tons/d) of oxygen, so that large fuel cell plants (e.g., 900 MW) would require multiple plants of this size. Only minor cost reductions would be realized for multiple plants on the same site, beyond the single 21 kg/s (2,000 ton/d) plant size.

The cost of oxygen produced by different size commercial plants (utility operated) was estimated (54) on the basis of the approximate commercial plant cost and operational data supplied by Air Products Corporation (55). The costs are adjusted to July 1974 dollars. Table 4.28 shows oxygen costs for two plant sizes, 3.15 and 21 kg/s (300 and 2,000 tons/d) and for an electric motor driven compressor or a turbine driven compressor. The calculations assume that the turbine driven compressor is operated on steam generated from the free waste heat from the fuel cell and that the electric motor driven compressor uses electricity from the fuel cell costing 8.3 mills/MJ (30 mills/kWh), which would appear to be about the minimum cost that could be expected from a 315 Ms (10 year) life optimized fuel cell from the present study. A recirculating water cost of $0.026/m^3 ($0.10/1,000 gal) is assumed. Also included are 18% per year capital based charges for depreciation, interest, local taxes, etc., plus 2% per year for maintenance.

TABLE 4.28: OXYGEN PRODUCTION COSTS (330 USE DAYS PER YEAR)

	300 tons/d	2,000 tons/d
Motor-driven plants		
Capital charges at 18%	1,008,000	3,412,800
Maintenance at 2% of investment	112,000	379,200
Power at 3.0¢/kWh	946,600	5,132,000
Water at 10¢/1,000 gal	63,100	342,100
Labor at $30,000/man year	180,000	240,000
Total	2,309,700	9,506,100
$/ton	23.33	14.41
Turbine-driven plants		
Capital charges at 18%	1,064,200	3,603,200
Maintenance at 2% of investment	118,200	400,400
Power at 3.0¢/kWh	17,900	121,200
Water at 10¢/1,000 gal	248,800	1,344,400
Labor at $30,000/man year	180,000	240,000
Steam, free	0	0
Total	1,629,000	5,709,200
$/ton	16.45	8.65

It should be noted that the costs do not include the cost of the steam generating equipment which is used to remove waste heat from the fuel cell for the turbine-driven compressor plants, and that oxygen costs for electric motor-driven compressor plants will vary with the cost of electricity produced by the fuel cell.

Electric-driven compressors are assumed for Point 5 for the 25-MW acid, alkaline and high-temperature solid electrolyte fuel cells, and the fuel cell electrical output is derated by the energy required. A turbine-driven compressor using steam generated by the waste heat is assumed for the molten carbonate Point 5 (25-MW dc fuel cell) and Point 17 (250-MW dc fuel cell), so the electrical output is not derated for those system points.

REFERENCES

(1) J.P. Ackerman, L.E. Link and J.J. Barghusen, *Assessment Study of Devices for the Generation of Electricity for Stored Hydrogen,* First Report (Draft Copy), Argonne National Laboratory, Argonne, Illinois 60439; August 1975.

(2) W. Leuckel, Power Systems Division, United Technologies Corporation; comment in discussion at July 7 and 8, 1975 Workshop of Assessment Study for the Generation of Electricity for Stored Hydrogen, Argonne National Laboratory.

(3) L.J. Nuttall, *Solid Polymer Electrolysis Fuel Cell Status Report,* Record of the Tenth Intersociety Energy Conversion Engineering Conference, August 18-22, 1975; pp. 210-7.

(4) J.W. Harrison in Part II—Appendix of Reference (1).

(5) H. Bohm and K. Maas, *Methanol/Air Acidic Fuel Cell System,* 9th Intersociety Energy Conversion Engineering Conference Proceedings, August 26-30, 1974; pp. 836-40.

(6) J.O'M. Bockris and S. Srinivasan, *Fuel Cells: Their Electrochemistry,* McGraw-Hill, New York, 1969; Chapter 10.

(7) R. Roberts, *Preliminary Draft—Needs and Recommendations for a National Fuel Cell Program,* proposed for Division of Conservation, Energy Research and Development Administration, by Mitre Corporation, July 21, 1975; pp. 7-8.

(8) G. Ciporos, in Part II—Appendix of Reference (1).

(9) A.P. Fickett, Electric Power Research Institute, in Alkaline Fuel Cell Discussions at the EPRI Catalyst Workshop, January 1975.

(10) J.K. Truitt, *15 kW Hydrocarbon-Air Fuel Cell Electric Power Plant Design,* Contract DA-44-009-AMC-1806(T), January 1968.

(11) L.G. Marianowski, *JP-4 Fueled Molten Carbonate Fuel Cells,* Contract DA-44-009-AMC-1456(T), July 1966.

(12) L.G. Marianowski et al, *JP-4 Fueled Molten Carbonate Fuel Cells,* U.S. Army Contract DA-44-009-AMC-1456(T), January 1968.

(13) *Assessment Study of Devices for the Generation of Electricity from Stored Hydrogen,* Part II—Appendix, August 1975, by J.P. Ackerman, J.J. Barghusen and L.E. Link, Argonne National Laboratory. Contribution from United Technologies entitled *Use of Fuel Cells to Generate Electricity from Hydrogen.*

(14) G.H.J. Broers, *High Temperature Galvanic Fuel Cells,* Thesis, University of Amsterdam, 1958.

(15) G.H.J. Broers and H.J.J. von Ballegoy, "Journees Int. Etude Piles Combustibles," *Comptes Rendus III,* Bruxelles, 77-86 (1969).

(16) A.B. Hart and G.J. Womack, *Fuel Cells,* Chapman and Hall, Ltd., London (1967).

(17) C.G. von Fredersdorff, "An Outline of the Economics of a Domestic Fuel Cell System," in *Fuel Cells,* Vol. 2, J. Young, Editor, Reinhold Publishing Company, New York, 1963.

(18) EPRI Meeting, High Temperature Fuel Cell Discussion Notes, comments by B. Baker and Houghtby.

(19) Y.L. Sandler, *J. Electrochemical Society,* 109, 1115-18, 1962.

(20) R.W. Hardy, W.E. Chase and J. McCallum in *Hydrocarbon Fuel Cell Technology,* B. Baker, Editor, Academic Press, 1965.

(21) J.O'M. Bockris and S. Srinivasan, *Fuel Cells: Their Electrochemistry,* McGraw-Hill Co., 1969, pp. 622-3.

(22) J.M. King, *Advanced Fuel Cell Technology for Utility Applications,* Record of the Tenth Intersociety Energy Engineering Conference, pp. 237-40, August 18-22, 1975.

(23) M. Anbar, D.F. McMillen and R.D. Weaver, *Electrochemical Power Generation Using a Liquid Lead Electrode as a Catalyst for the Oxidation of Carbonaceous Fuels,* Record of the Tenth Intersociety Energy Engineering Conference, pp. 48-55, August 18-22, 1975.

(24) T.L. Markin, *Limiting Problems in the Development of a High Temperature Solid Oxide Electrolytic Fuel Cell,* Proceedings of the 18th International Symposium on Power Sources, Brighton, England; September 1972, Paper 3, pp. 31-42.

(25) Final Report—Project Fuel Cell, Research and Development Report No. 57, Office of Coal Research, Department of the Interior, Washington, D.C., 1971.

(26) D.H. Archer, L. Elikan and R.L. Zchradnik, "The Performance of Solid-Electrolyte Cells in Batteries on $CO-H_2$ Mixtures; A 100-Watt Solid-Electrolyte Power Supply," in *Hydrocarbon Fuel Cell Technology,* Edited by B.S. Baker, Academic Press, New York, 1965; pp. 51-75.

(27) E.F. Sverdrup, C.J. Warde and A.D. Glasser, "A Fuel-Cell Power System for Central-Station Power Generation Using Coal as a Fuel," from *Electrocatalysis to Fuel Cells,* G. Sandstede (Editor), University of Washington Press, 1972; pp. 255-77.

(28) C.J. Warde and A.O. Isenberg, in Part II—Appendix of Reference (1).

(29) S.G. Abers, B.S. Baker, R.D. Pasquale and I. Michelko, *Phosphoric Acid Fuel Cell Stack Development,* Record of the Tenth Intersociety Energy Conversion Engineering Conference, August 18-22, 1975, pp. 218-21.

(30) K.V. Kordesch and R.F. Scarr, *Thin Carbon Electrodes for Acidic Fuel Cells,* Conference Proceeding, 7th Intersociety Energy Conversion Engineering Conference, September 25-29, 1972; pp. 12-19.

(31) W.R. Wolfe, Jr., K.B. Keating, V. Mehra and L.H. Cutler, *A New Class of Fuel Cell Anode Catalysts,* 8th Intersociety Energy Conversion Engineering Conference Proceedings, August 13-16, 1973; pp. 92-5.

(32) T.G. Schiller and A.P. Meyer, *1.5-kW Fuel Cell Power Plant,* Contract DAAK 02-C-0518, U.S. Army Mobility Equipment, Research and Development Center, Fort Belvoir, Virginia, May 20, 1971.

(33) Reference (2), p. 524.

(34) "Assumptions, Models and Correlations for Task I," for *Study of Advanced Energy Conversion Techniques for Utility Applications Using Coal and Coal-Derived Fuels,* compiled by D.T. Beecher, Project Manager, Westinghouse Electric Corporation; January 21, 1975; pp. 2-9.

(35) O.J. Adlhart, and A.J. Hartner, *Plastic-Bonded Electrodes,* Proceedings of the 21st Power Sources Conference, 1967, pp. 4-6.

(36) G.T. Lee, J.D. Leslie and H.M. Rodikohr, "The Cost of Hydrogen Made from Natural Gas," *Hydrocarbon Processing and Petroleum Refiner,* 42 (9) 125-8 (1963).

(37) March Monthly Progress Narrative, *Study of Advanced Energy Conversion Techniques for Utility Applications Using Coal and Coal-Derived Fuels,* compiled by D.T. Beecher, Project Manager, Westinghouse Electric Corporation, April 14, 1975; p. 42.

(38) "JANAF Thermochemical Tables," 2nd Edition, *NSRDS,* NBS 37, June 1971.

(39) K.V. Kordesch, Private Communication, April, 1975.

(40) Englehard Industries, Murray Hill, N.J.

(41) R.H. Perry and C.H. Chiltan, *Chemical Engineers Handbook,* pp. 19-67, McGraw-Hill Book Company, New York, 1973.

(42) Westinghouse Electric Corporation, Price List No. 1252, p. 19.

(43) Reference (25), pp. 17-36.

(44) K.M. Guthrie, *Process Plant, Estimating, Evaluation and Control,* Craftsman Book Company of America, 1974.

(45) *Economic Evaluation of a 200 MW Coal-Burning Fuel-Cell Power Plant,* Final report

prepared by the Institute of Gas Technology for Westinghouse Electric Corporation, January 1969.

(46) J. Werssbart and R. Ruka, "A Solid Electrolyte Fuel Cell," *J. Electrochem. Soc.,* 109 (8) 723-6 (1962).

(47) K.V. Kordesch, *Carbon Electrodes,* Conference Proceedings, Fuel Cell Catalysis Workshop, EPRI Special Report, SR-13, August 1975, pp. 101-6.

(48) *Chemical Engineering,* (McGraw-Hill); March 22, 1971, p. 170, and March 17, 1975, p. 120.

(49) *Engineering Study and Technical Evaluation of the BCR Two-Stage Super-Pressure Gasification Process,* R&D Report No. 60, Office of Coal Research, Department of the Interior, Washington, D.C., 1971.

(50) W.P. Hegarty and B.E. Moody, "Evaluating the Bi-Gas SNG Process," *Chemical Engineering Progress,* 69 (3), pp. 37-42, March 1973.

(51) P. Wood, Power Electronics Dept., Westinghouse Research Laboratories, Pittsburgh, Pa., private communication.

(52) Power Transformer Division, Westinghouse Electric Corporation, Sharon, Pa.

(53) Switchgear Division, Westinghouse Electric Corporation, Trafford, Pa.

(54) E.J. Vidt, Chemical Engineering Systems Dept., Westinghouse Research Laboratories, Pittsburgh, Pa., private communication.

(55) J.F. Strecansky, Air Products and Chemicals, Allentown, Pa., private communication.

Acknowledgements

The fuel cell task was accomplished in its entirety at the Westinghouse Research Laboratories. C.J. Warde led the effort, while assuming responsibility for the parametric assessment of the acid and alkaline power systems. The parametric assessments of the molten carbonate and solid electrolye power systems were performed by R.J. Ruka and A.O. Isenberg, respectively.

Others making contributions were:

L.E. Brecher, who provided information on fuel conditioning equipment perforance and cost.

J.T. Brown and E.S. Busselli, who developed and applied costing procedures.

F.J. Sisk, who provided information on steam-bottoming plants.

E.J. Vidt, who developed information on oxygen plant costs.

P. Wood, who provided information on the necessary power condition apparatus, and its performance and cost.

C.T. McCreedy and S.M. Scherer of Chas. T. Main, Inc. of Boston, who prepared the balance of plant description and costing, site drawings, and provided consultation on plant island arrangements and plant constructability.

FUEL CELLS
FOR PUBLIC UTILITY APPLICATIONS
GENERAL ELECTRIC STUDY

In 1976 the General Electric Company prepared a report (NASA CR 134948, Volume II, Part 3) that contained a section relating to an evaluation of specific low and high temperature fuel cells. This section by J.C. Corman and R.B. Fleming is editorialized below.

FUEL CELLS—LOW TEMPERATURE

The parametric variations for Case 1 through 15 fuel cells studied are given in Table 5.4 (page 163).

Description of Cycle

Figure 5.1 is a schematic of the base case for low-temperature fuel cells. The base case was a solid polymer electrolyte (SPE) fuel cell. This case used high-Btu gas as the fuel entering the plant, but this fuel was converted to H_2 before it entered the fuel cell. In the fuel conversion process, the shift reactor converts CO to CO_2 (with the addition of water). The methanator converts the small remaining amount of CO to CH_4 because even small quantities of CO are harmful to the performance of the SPE fuel cell.

The H_2 fuel (containing 3% by volume CH_4) enters the anode side of the fuel cell, where most of the H_2 is consumed. The fuel purge, containing mostly CH_4 (on a mass basis), is returned to the reformer to satisfy part of the thermal energy requirement of the reformer. Except for the use of fuel purge, no integration was assumed between the fuel cell and the fuel conversion system.

Air is used as the oxidizer on the cathode side of the cell. The air passes through a blower and into a humidifier-cooler, where it is preheated and saturated to the correct water vapor pressure for use in the fuel cell. The air passes through the fuel cell, where O_2 is consumed and product water is added; the air is then discharged to the atmosphere. The coolant stream, which is water in the case of the SPE cells, is cooled by evaporation of a fraction of the water and by warming the cooler air stream. Thus the humidifier-cooler serves to remove most of the fuel cell waste heat.

FIGURE 5.1: LOW-TEMPERATURE FUEL CELL (BASE CASE)

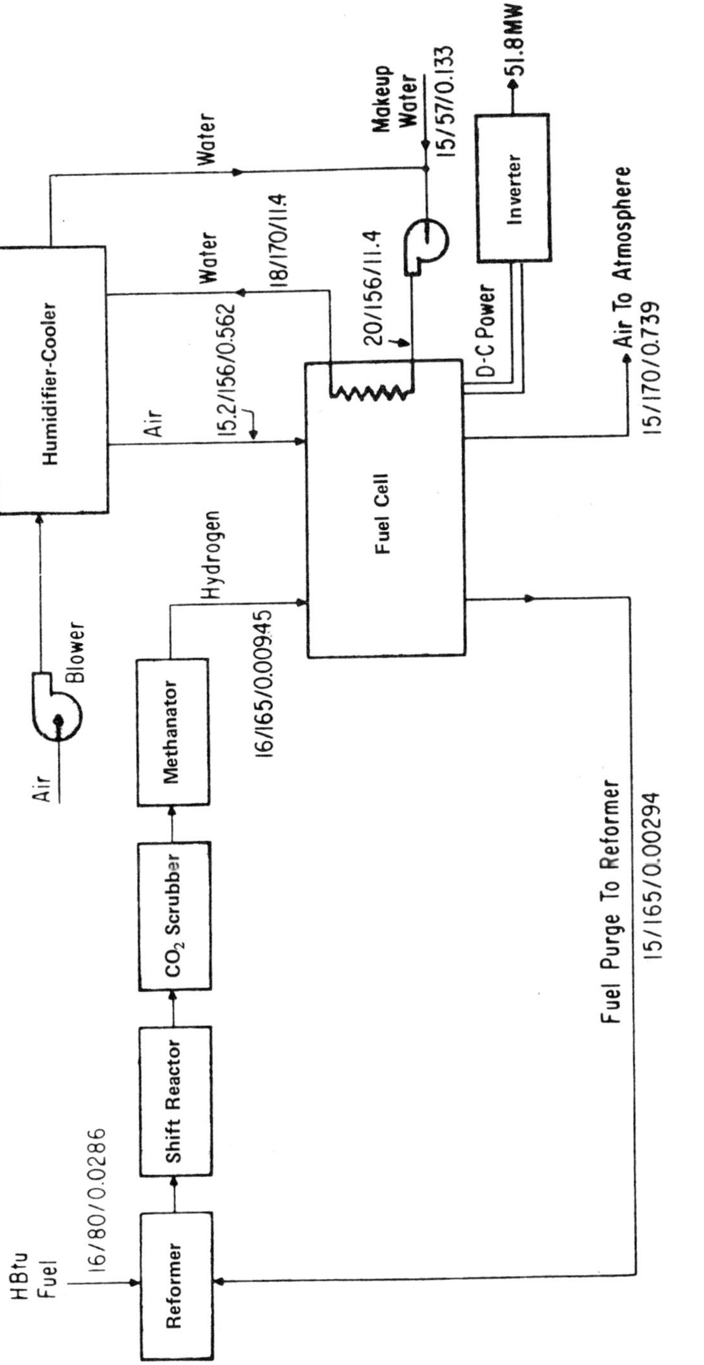

NOTE: Pressure(Psia)/Temperature(°F)/Flow Rate(×10⁶lb/hr)

High-Btu gas was selected as the fuel for the base case, as high-Btu gas is a possible future fuel to be pipelined to fuel cell plants.

SPE Fuel Cells: The base case and most of the parametric variations from the base case were with SPE type cells. Parametric variations included the substitution of hydrogen for the high-Btu fuel. This eliminated the need for the fuel conversion system, as hydrogen was assumed to be piped into the power plant from a remote hydrogen plant. Since the piped-in hydrogen was assumed to be dry, a humidifier had to be added to the incoming fuel stream.

Another variation for the SPE cell was the substitution of oxygen for air as the oxidizer (Case 8). Case 8 is unique in that the operating pressure and temperature were increased and the mass flow of oxygen was the stoichiometric rate (there was no purge of the oxidizer stream from the cathode compartment to the atmosphere).

In that case, the coolant water was cooled by flashing a portion of the cooling water stream, and all of the product water, into low pressure steam [about 59 psia (407 kN/m^2) with a saturation temperature of 292°F (418°K)]. This steam, which represents most of the fuel cell waste heat, is available for integration with the hydrogen plant, if it could be located at the power plant site. A potential exists for utilizing the waste heat from a low-temperature fuel cell, thereby reducing the cost of electricity. This possibility was not explored in the parametric variations. The base case current density, temperature, and electrolyte thickness were chosen as typical of the SPE cells. Variations above and below these values were examined in the parametric variations.

Phosphoric Acid Fuel Cells: Four parametric variations were calculated for phosphoric acid cells (Cases 12 through 15). For the phosphoric acid cases, the methanator shown in Figure 5.1 was eliminated, as the phosphoric acid cell is less sensitive to carbon monoxide.

The phosphoric acid cell was assumed to operate at 375°F (464°K) and near atmospheric pressure. To achieve good performance levels, the acid concentration in the matrix was held at 98%. This requires a vapor pressure of about 6 psia (41 kN/m^2) in the reactant streams; consequently, as shown in Figure 5.1, a humidifier was used, followed by a heater to provide the correct temperature. The coolant was changed to an organic fluid to permit operation near atmospheric pressure at 375°F (464°K). This permits thin-walled cooling passages within the fuel cell and reduces costs (compared with using water as a coolant).

Analytical Procedure and Assumptions

Hydrogen Purity: The purity of hydrogen produced from high-Btu gas in an on-site reformer (as in the base case, Figure 5.1) was assumed to be 96% hydrogen by volume, on a dry basis, and saturated with water at 165°F (374°K). The hydrogen produced in a remote plant and piped into the power plant was assumed to be completely dry, and 98% hydrogen, at a temperature of 59°F (288°K).

SPE Fuel Cells: The performance data used for the SPE cell were generated using the best demonstrated cell resistance. This has the effect of optimizing cell performance within the demonstrated capability. No allowances for diffusion losses were included; these losses could amount to as much as 0.5% of the required fuel flow rates. Air flow rates were established at 2.5 times the stoichiometric oxygen requirement. Above this rate no further performance improvement is noted on test. This value may be found to be in excess of optimum when the blower energy is examined more closely.

The hydrogen purge rates were set to hold the minimum hydrogen mol fraction within the cell at 48% on a dry basis. An examination of test data with a wide variation in mol fraction at the inlet shows that the mol fraction of hydrogen may be reduced in the cell by a factor of two without affecting measured cell performance. The concentration of carbon monoxide in the fuel in the order of 10 ppm was assumed to have no effect on performance of the SPE cells; this has been confirmed by test data.

A primary assumption made for the SPE cell (also for the phosphoric acid cell) is that there will be no improvement in the performance over the best that has been experimentally demonstrated. The only assumption of improvement over present day practice is a decrease in platinum catalyst loading. Present-day minimum catalyst loadings (total of both anode and cathode loadings) are in the range of 1.2 to 1.5 g/ft^2 (13 to 16 g/m^2) while some of the performance data for this study were from SPE cells with 8 g/ft^2 (86 g/m^2).

The assumed catalyst loading for this study was taken to be 0.2 g/ft^2 (2.2 g/m^2). This assumption is based on platinum surface areas now available (about 20 m^2/g), compared with the maximum surface areas that have been produced experimentally on substrates (about 150 m^2/g). The reduction to 0.2 g/ft^2 is considered feasible, without a degradation in performance or life, assuming that further research and development in this area is carried out. The effect of changes in catalyst loading on costs is covered in the following section, "Design and Cost Basis." The efficiency of the inverter (including transformer losses) was assumed to be 98.2% for output power levels less than 60 MW, to account for increased transformer losses in smaller sizes.

Phosphoric Acid Fuel Cells: Data that were available on phosphoric acid fuel cells were from General Electric Company tests that were performed in 1968 and earlier. It is understood that more recent proprietary development by other organizations has improved both the performance and life from these earlier tests. Because precise recent data were not available, performance characteristics of the SPE cell were used, even though that performance is somewhat better than what is understood to be the performance now forecast for phosphoric acid cells.

The analytical procedures used were similar to those used for the SPE except that the reactants were humidified to a partial pressure permitting operation of the cells at 98% acid concentration. No provision was made for methanation as it is understood that 0.5% CO can be tolerated at 375°F (464°K). No performance penalty was taken for the effect of the CO. Considering the 1968 General Electric data and the very low purge rates used in this study, this is an optimistic assumption. Further development tests may well show that the CO may have to be reduced by methanation or partial oxidation. Catalyst loading for the phosphoric acid cells was assumed to be the same as for the SPE cells.

Design and Cost Basis

Hydrogen and Oxygen Costs: For this study, the cost of hydrogen piped in from a remote plant was $2.53/million Btu ($2.40/billion J). For this piped-in hydrogen the composition was taken to be:

	------- Composition, % -------	
	By Volume	By Mass
H_2	98	84.27
CH_4	1.6	10.95
N_2	0.4	4.78
	100.0	100.00

The higher heating value of hydrogen is 61,031 Btu/lb (142 MJ/kg), and of methane is 23,890 Btu/lb (55.6 MJ/kg), giving a higher heating value of the mixture of 54,047 Btu/lb (125.7 MJ/kg). This value was used in determining the cost of piped-in hydrogen.

Rather than making a detailed analysis of how the anode purge gas could be used to satisfy part of the heat requirements of the reformer, a cost credit for the purge gas was allowed as follows. The dry-basis composition of this purge gas is:

	Composition, % by mass
H_2	8.84
CH_4	57.48
N_2	33.67

and the higher heating value of this mixture is 19,130 Btu/lb (44.50 MJ/kg). This heating value was multiplied by a cost credit of $2.00/million Btu ($1.90 per billion J) and by the dry-basis purge flow rate to calculate the purge flow cost credit. For Case 8, where oxygen was used as the oxidizer, the oxygen cost used was $9.00/ton ($9.92/Mg). This cost was provided by NASA.

Reformer System: The reformer system consisted of the four elements (reformer, shift reactor, CO_2 scrubber, and methanator) shown in Figure 5.1. Capital costs for the reformer system were determined from the data in Table 5.1, provided by the Foster Wheeler Energy Corporation.

TABLE 5.1: CAPITAL COSTS FOR REFORMER SYSTEM

Capacity		On-Site Investment ($ millions)	Off-Site Investment* ($ millions)	Total Investment ($ millions)
Standard ft^3/day	lb/hr			
25×10^6	5,427	6.25	1.25	7.50
50×10^6	10,850	9.98	2.00	11.98
100×10^6	21,710	17.29	3.46	20.75

*Includes water treatment and waste disposal equipment, initial charge of catalyst and chemicals, etc.

A reformer system with a capacity of 100 x 10^6 standard ft³/day (2.8 x 10^6 standard m³/day) is about the largest plant that can be built; plants larger than that would be built in modules. For purposes of this study, interpolations were made using data from Table 5.1. For the phosphoric acid cell cases, the reformer system costs were reduced by 8% to account for the fact that the methanator is not needed to remove carbon monoxide for the phosphoric acid cells. The efficiency of the reformer system (ratio of Btu/hr of product out to Btu/hr of high-Btu gas in) was calculated to be 0.70.

Fuel Cell: The cost of the fuel cell stack was expressed on the basis of dollars per square foot of active cell electrolyte area. For the two types of low temperature cells studies, the overall fuel cell stack costs were: (a) For the SPE cell, $15.12/ft² ($163/m²) and (b) For the phosphoric acid cell, $14.40/ft² ($155/m²). These costs do not include catalyst cost but do include materials, manufacturing, and assembly labor for the following elements of the fuel cell stack:

(a)   Electrolyte
(b)   Anode
(c)   Cathode
(d)   Cooling passages
(e)   Frame
(f)   End plates
(g)   All other parts that make up the fuel cell stack

Table 5.2 gives data on electrolyte areas, and sizes and weights of the fuel cell stack.

TABLE 5.2: DATA FOR FUEL CELL STACK

Case Number	Total Active Area of Cell Electrolyte (ft²)	Cost per Unit Area ($/ft²)	Total Cost of Fuel Cell Stack* ($ millions)	Volume of Fuel Cell Stack (ft³)	Weight of Fuel Cell Stack (lb)
1	315,600	15.12	4.77	2362	100,000
2	157,800	15.12	2.39	1181	50,000
4	315,600	15.12	4.77	2362	100,000
5	315,600	15.12	4.77	2362	100,000
6	315,600	15.12	4.77	2362	100,000
7	252,200	15.12	3.81	2000	80,000
8	827,200	15.12	12.5	6200	300,000
9	735,200	15.12	11.1	5501	240,000
10	309,100	15.12	4.67	2350	100,000
11	324,000	15.12	4.90	2500	104,000
12	315,600	14.40	4.54	2950	145,000
13	315,600	14.40	4.54	2950	145,000
14	733,500	14.40	10.56	6900	335,000
15	252,200	14.40	3.63	2400	116,000

* Not including catalyst

Catalyst Cost: The cost of platinum catalyst was assumed to be $5.50/g, a cost near the 1974 market value of platinum. The platinum loading was taken to be 0.2 g/ft^2 (2.1 g/m^2) of active electrolyte area (this loading is the sum of the anode and cathode loadings). The total catalyst cost can be calculated by taking the product of the surface area from Table 5.2, the cost of $5.50/g and the loading of 0.2 g/ft^2.

Inasmuch as the platinum loading of 0.2 g/ft^2 has not been demonstrated experimentally but rather is a projection from present practice, it is useful to predict the effect of changes in catalyst loadings. Figure 5.2 shows the effect of loading on the catalyst capital cost and on cost of electricity. Two cases are shown. Case 1 is for a relatively low power density [net output: 152 W/ft^2 (1640 W/m^2)] and Case 8 is for a high density [243 W/ft^2 (2620 W/m^2)]. From the figure it can be determined for Case 1 that an increase in loading from 0.2 to 1.2 g/ft^2 will increase the cost of electricity by about 1.2 mills/kWh.

Catalyst and Electrolyte Replacement: The useful life of a low-temperature fuel cell is generally limited by degradation of the catalyst or electrolyte. After a period of some years, the electrolyte and catalyst must be replaced. The estimated costs for replacement are given below.

The replacement cost is divided into two elements. First, there is the reprocessing of the platinum. In this process there is a processing charge and a loss of platinum, which is estimated to be 2% of the platinum submitted for reprocessing. The estimated costs are: processing, $0.46/g; loss (0.02 x $5.50/gram), $0.11/g; for a total reprocessing cost of $0.57/g. This reprocessing cost is estimated to be the same for the SPE and the phosphoric acid cell.

The second cost element is the materials and labor charge to disassemble the cells, replace the electrolyte and other parts, and reassemble the cell. The estimated costs for the SPE cells are: electrolyte material, $2.15/ft^2; labor, $5.00 per ft^2; for a total of $7.15/ft^2 ($77.00/m^2. It was not possible to estimate accurately the labor and materials costs but a crude analysis led to about the same cost for the phosphoric acid cells as for the SPE cells. The catalyst and electrolyte replacement costs are therefore estimated to be $0.57/g of platinum, plus $7.15/ft^2 ($77.00/m^2) of active electrolyte area, for both types of cells.

The period of time between replacements is difficult to estimate because of the lack of experimental data over long periods of time under operating conditions. For the phosphoric acid cell the period of 40,000 hours of operation was chosen, as this is believed to be the goal of present development. For the SPE cell a period of 100,000 operating hours was selected because long-term tests have been conducted at 180°F (355°K) for up to 34,000 hours, with no sign of performance deterioration.

For those cases operating near 180°F, the extension of time by a factor of three should be realistic. For the higher temperature Case 8 (300°F or 472°K), there was less justification for selecting 100,000 hours. However, there have been life tests at this elevated temperature for up to 800 hours, using a newly developed electrolyte material. In these tests, degradation of the electrolyte polymer (the normal failure mode for earlier electrolyte materials) was carefully monitored and no products of degradation were found.

FIGURE 5.2: EFFECT OF CHANGE IN CATALYST LOADING

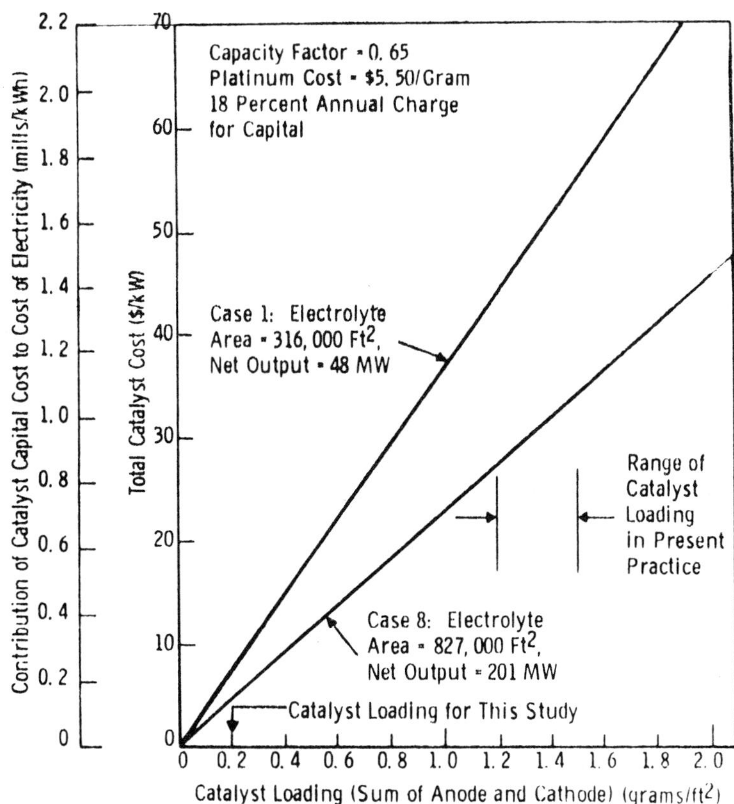

In order to determine the effect of changes in replacement period, Figure 5.3 was prepared. This figure can be used to determine, for example, the cost effect of reducing the 100,000 hour period for Case 8 to some lower number, say to 30,000 hours. At a 30,000-hour replacement period, the contribution of catalyst and electrolyte replacement to the cost of electricity will rise to 1.0 mils/kWh, compared with 0.3 mils/kWh at the assumed period of 100,000 hours for Case 8.

Oxygen Case: In Case 8, oxygen was used as the oxidizer. One reason for this is that the fuel cell plant is assumed to be located near the plant that converts coal to hydrogen, and this hydrogen plant also needs an oxygen supply. Thus, one oxygen plant could supply both the hydrogen plant and the fuel cell power plant.

A unique characteristic of Case 8 is that it is the only one in which steam can be produced conveniently and in large quantities. In the SPE cell, there is no vapor pressure suppression (as in the phosphoric acid cells), and product water collects in the cell at the cell operating conditions of 115 psia (793 kN/m²) and 300°F (422°K). This water can be collected and flashed into steam at about

292°F (418°K) and a saturation pressure of about 59 psia (407 kN/m²). In addition, the separate stream of cooling water circulating through the cell can be partially flashed to steam to join the product water steam. The water not flashed to steam is returned to the cooling water loop. Thus, almost all of the waste heat from the fuel cell appears as the latent heat of steam, which could be used in fuel conversion or other processes.

FIGURE 5.3: EFFECT OF FUEL CELL LIFE

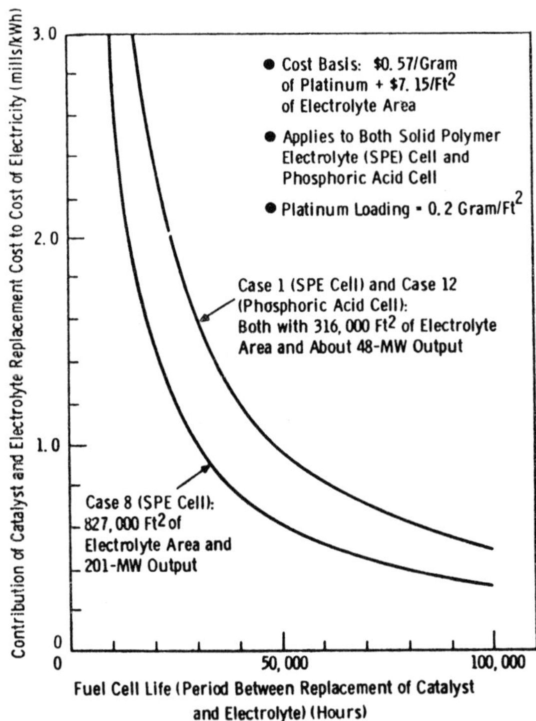

A complete integration of the fuel cell and the hydrogen plant was beyond the scope of this contract. However, some opportunity for integration does exist, and an approximate evaluation indicated that about one-third of the steam produced could be used in the hydrogen plant. To approximate the cost savings of integration, it was assumed that a cost credit could be allowed for one-third of the steam produced by the fuel cell.

The credit allowed for each 1,000 pounds of steam used was 1.35 multiplied by the fuel cost in dollars per million Btu. This is a typical figure for industrial steam, saturated and at pressures under 100 psi (690 kN/m²). For a fuel cost of $2.00/million Btu, the credit amounted to $2.70/1,000 lb of steam ($5.95 per 1,000 kg), and this is the figure that was used.

Table 5.3 shows the effect of allowing this steam credit. The table also shows a cost breakdown, and illustrates the effect of an increase in catalyst loading from the 0.2 g/ft² (2.2 g/m²) assumed in the study to 1.2 g/ft² (13 g/m²) that is typical of the minimum loading for present commercial units. It can be seen from the table that the steam credit is relatively large, and that integration between the fuel cell and the hydrogen plant is essential from a cost standpoint. If a use could be found for any part of the remaining two-thirds of the steam, an even further cost savings could be realized. The steam credit was allowed only in Table 5.3, and not in the other cost data presented later.

TABLE 5.3: FUEL CELL COST BREAKDOWN

(Case 8, Solid Polymer Electrolyte [SPE] Cell with Oxygen)

	Results for Case 8	Effect of Change in Platinum Loading	Effect of Cost Credit by Using 1/3 Steam Generated by Fuel Cell
Platinum Loading (g/ft²)	0.2	1.2	0.2
Capital Charge, Catalyst (Mill/kWh)	0.14	0.86	0.14
Capital Charge, Other (Mills/kWh)	7.56	7.56	7.56
Fuel Cost (Hydrogen) (Mills/kWh)*	15.93	15.93	15.93
Oxygen Cost (Mills/kWh)	3.67	3.67	3.67
Maintenance and Operating Charge (Mills/kWh)	4.10	4.10	4.10
Credit for Steam (Mills/kWh)**	0.0	0.0	(3.58)
Totals (Mills/kWh)	31.40	32.12	27.82

*Includes credit of 0.83 mill/kWh for fuel purge flow returned
**Steam credit of $2.70/1000 pounds of steam supplied

Current Inverters: Costs for the dc and ac inverters were based upon present solid state technology that has been developed for high voltage dc power transmission projects. Total inversion equipment costs in 1974 dollars, including installation, are as shown on the following page for various plant ratings. This equipment cost includes arrestors, valves and control equipment, converter, transformer, auxiliary power and motor control center, capacitors, smoothing reactors, etc. Voltages are assumed to be 600 V on the dc side, and 230 kV on the ac side (except for the nominal 25 MW system, for which the ac side is at 69 kV).

Plant Rating in MW	Dollars per kW at High Voltage Terminal
25	69
50	58
200	44

Results

Results for the study of low-temperature fuel cells are tabulated in Table 5.4, which includes the major cycle input parameters. Capital cost distributions are given in Table 5.5. A summary giving major cycle characteristics for the low temperature fuel cell base case is given in Table 5.6. Auxiliary losses and power outputs are shown in Table 5.7. A number of observations can be made from the results shown in Table 5.4.

When high-Btu gas is used as the fuel, the overall energy efficiency is extremely low (12.7% for the base case), because of the double penalty of converting coal to high-Btu gas, followed by converting high-Btu gas to hydrogen.

Further, the cost of fuel is very high (32.9 mills/kWh for the base case) because of the high cost of high-Btu gas. When hydrogen is used as the fuel, the overall energy efficiency rises, and the fuel cost drops (see Cases 6 and 13, for example).

The highest overall energy efficiency (31.1%) and the lowest cost of fuel (19.6 mills per kWh) were obtained with hydrogen and oxygen (Case 8). (Note that the energy efficiency does not include the energy required to produce the oxygen, but the fuel cost does include the cost of oxygen; this is consistent with the approach taken in the open-cycle MHD system.) If the steam produced in the hydrogen-oxygen cell can be utilized in the hydrogen plant, the cost of electricity could be reduced to 27.8 mills/kWh, as shown in Table 5.3.

The fuel cell costs of electricity, for those cases where hydrogen was the fuel, were characterized by very low capital costs and very high fuel costs. This would normally place the fuel cell in a peaking plant category from an economic standpoint. However, it should be pointed out that these economics are a consequence of assuming the hydrogen to be purchased over the fence at a certain cost per Btu.

Therefore, the capital charges associated with the equipment that converts coal to hydrogen are included in the fuel costs, and consequently appear to be a variable cost. If the capital cost of the hydrogen plant were included with the capital cost of the power plant, and if the fuel cost were only the cost of coal to the hydrogen plant, then the cost of electricity would not change, but the cost distribution between capital and fuel would be more characteristic of a base load plant.

There was very little cost difference between the SPE and the phosphoric acid cells; for example, the costs of electricity of comparable Cases 6 and 13 were 36.2 and 36.9 mills/kWh, respectively. Because of a sizeable advantage in cost of electricity, the hydrogen-oxygen case with an SPE cell (Case 8) is recommended for further study.

TABLE 5.4: PARAMETRIC VARIATIONS FOR TASK I STUDY FUEL CELLS–LOW TEMPERATURE

Parameters	1*	2	4	5	6	7	8	9	10	11	12	13	14	15
Power output (MWe)	48	24	48	48	48	48	201	48	48	48	47	47	47	47
Coal and conversion process	Ill. #6	Ill. #6	Mont.	N.D.	Ill. #6	Ill. #6	Ill. #6	Ill. #6	Ill. #6	Ill. #6	Ill. #6	Ill. #6	Ill. #6	Ill. #6
	HBtu	HBtu	HBtu	HBtu	H_2	H_2	H_2 (on site)	H_2	H_2	H_2	HBtu	H_2	H_2	H_2
Oxidizer	Air	Air	Air	Air	Air	Air	O_2	Air	Air	Air	Air	Air	Air	Air
Fuel cell type	SPE	SPE	SPE	SPE	SPE	SPE	SPE	SPE	SPE	SPE	Phos	Phos	Phos	Phos
Current density (A/ft^2)	250	250	250	250	250	350	300	100	250	250	250	250	100	350
Operating temperature (maximum) (°F)	170	170	170	170	170	170	300	170	170	170	375	375	375	375
Electrolyte thickness (inches)	0.005	0.005	0.005	0.005	0.005	0.005	0.005	0.005	0.002	0.010	0.020	0.020	0.020	0.020
Actual powerplant output (MWe)	48	24	48	48	48	48	201	48	48	48	47	47	47	47
Thermodynamic efficiency (%)	0	0	0	0	0	0	0	0	0	0	0	0	0	0
Powerplant efficiency (%)	25.2	25.1	25.2	25.2	38.3	34.6	51.1	41.3	39.1	36.7	29.8	37.9	40.8	33.9
Overall energy efficiency (%)	12.7	12.7	12.7	12.7	23.3	21.1	31.1	25.2	23.9	22.4	15.0	23.1	24.9	20.7
Coal consumption (lb/kWh)	2.50	2.50	3.01	3.91	1.36	1.50	1.02	1.26	1.33	1.41	2.10	1.37	1.27	1.53
Plant capital cost ($ million)	30	16	30	30	16	14	49	26	16	16	27	15	25	14
Plant capital cost ($/kWe)	634	687	634	634	325	295	242	542	329	335	570	317	527	287
Cost of electricity, capacity factor is 0.65														
Capital (mills/kWh)	20.1	20.7	20.1	20.1	10.3	9.3	7.7	17.1	10.4	10.6	18.0	10.0	16.7	9.1
Fuel (mills/kWh)	32.9	33.0	32.9	32.9	21.2	23.4	19.6	19.6	20.7	22.1	28.6	21.4	19.9	23.9
Maintenance and operating (mills/kWh)	4.7	5.1	4.7	4.7	4.7	4.7	4.1	5.4	4.7	4.8	5.5	5.5	7.1	5.3
Total (mills/kWh)	57.7	59.9	57.7	57.7	36.2	37.4	31.3	42.1	35.9	37.5	52.1	36.9	43.7	38.3
Sensitivity														
Capacity factor is 0.50 (total mills/kWh)	65.0	67.8	65.0	65.0	40.6	41.5	34.8	48.6	40.3	41.9	58.8	41.2	50.0	42.3
Capacity factor is 0.80 (total mills/kWh)	53.2	54.9	53.2	53.2	33.5	34.9	29.2	38.1	33.1	34.7	48.0	34.2	39.7	35.8
Capital Δ is 20% (Δmills/kWh)	4.0	4.3	4.0	4.0	2.1	1.9	1.5	3.4	2.1	2.1	3.6	2.0	3.3	1.8
Fuel Δ is 20% (Δmills/kWh)	6.6	6.6	6.6	6.6	4.2	4.7	3.2	3.9	4.1	4.4	5.7	4.3	4.0	4.8
Estimated time for construction (yr)	2	2	2	2	2	2	2	3	2	2	2	2	2	2
Estimated date of 1st commercial service (yr)	1986	1986	1986	1986	1986	1986	1992	1985	1985	1985	1982	1982	1982	1982

*Base case

Note: HBtu = high Btu; Ill. = Illinois; Mont. = Montana; N.D. = North Dakota; Phos = phosphoric acid; and SPE = solid polymer electrolyte.

TABLE 5.5: CAPITAL COST DISTRIBUTIONS FOR LOW-TEMPERATURE FUEL CELL

	Case Number													
	1	2	4	5	6	7	8	9	10	11	12	13	14	15
Major components														
Prime cycle														
Fuel cell stack, MM$	4.8	2.4	4.8	4.8	4.8	3.8	12.5	11.1	4.8	4.9	4.5	4.5	10.6	3.6
Humidifier-cooler, MM$	0.1	0	0.1	0.1	0.1	0.1	0	0	0	0.1	0.1	0.1	0	0.1
Catalyst, MM$	0.3	0.2	0.3	0.3	0.3	0.3	0.9	0.8	0.3	0.4	0.3	0.3	0.8	0.3
(loading g/ft^2)	(0.2)	(0.2)	(0.2)	(0.2)	(0.2)	(0.2)	(0.2)	(0.2)	(0.2)	(0.2)	(0.2)	(0.2)	(0.2)	(0.2)
Furnace and fuel processing														
Fuel processing, MM$	10.5	5.8	10.5	10.5	0	0	0	0	0	0	8.3	0	0	0
Fuel preheater-humidifier	0	0	0	0	0.3	0.3	0	0.3	0.3	0.3	0	0	0	0
Subtotal of major components, MM$	15.7	8.4	15.7	15.7	5.5	4.4	13.4	12.2	5.5	5.6	13.2	4.9	11.5	3.9
Balance of plant														
Cooling tower, MM$	0	0	0	0	0	0	0	0	0	0	0	0	0	0
dc to ac inverters, MM$	3.0	1.8	3.0	3.0	3.0	3.0	9.0	3.0	3.0	3.0	3.0	3.0	3.0	3.0
All others, MM$	1.9	1.0	1.9	1.9	1.9	1.9	7.3	2.3	2.0	2.0	2.0	2.0	2.3	2.0
Site labor, MM$	0.4	0.2	0.4	0.4	0.4	0.4	1.4	0.5	0.4	0.4	0.5	0.5	0.5	0.5
Subtotal of balance of plant, MM$	5.3	2.9	5.3	5.3	5.3	5.3	17.7	5.8	5.5	5.5	5.5	5.5	5.9	5.5
Contingency, MM$	4.2	2.3	4.2	4.2	2.2	1.9	6.2	3.6	2.2	2.2	3.7	2.1	3.5	1.9
Escalation costs, MM$	2.8	1.5	2.8	2.8	1.4	1.3	5.7	2.4	1.5	1.5	2.5	1.4	2.3	1.3
Interest during construction, MM$	2.3	1.3	2.3	2.3	1.2	1.1	5.6	2.0	1.2	1.2	2.1	1.1	1.9	1.0
Total capital cost, MM$	30.3	16.4	30.3	30.3	15.6	14.1	48.6	26.0	15.8	16.0	27.0	15.0	25.0	13.6
Major components cost, $/kWe	328.6	352.4	328.6	328.6	114.8	93.4	66.9	254.6	114.0	117.8	278.7	103.5	241.7	83.1
Balance of plant, $/kWe	110.9	123.7	110.9	110.9	110.9	111.4	88.5	121.0	114.0	114.7	116.2	116.2	123.4	116.2
Contingency, $/kWe	87.9	95.2	87.9	87.9	45.1	40.9	31.1	75.1	45.6	46.5	79.0	44.0	73.0	39.9
Escalation costs, $/kWe	58.9	63.9	58.9	58.9	30.3	27.5	28.2	50.4	30.6	31.2	53.0	29.5	49.0	26.7
Interest during construction, $/kWe	48.5	52.5	48.5	48.5	24.9	22.6	27.9	41.4	25.1	25.6	43.5	24.2	40.3	22.0
Total capital cost, $/kWe	634.8	687.7	634.8	634.8	325.9	295.7	242.5	542.5	329.3	335.7	570.4	317.4	527.4	287.9

TABLE 5.6: SUMMARY SHEET FUEL CELLS–LOW-TEMPERATURE BASE CASE

Cycle Parameter

Net power output (MWe)	48
Coal type	Illinois No. 6
Prime cycle	
Oxidizer	Air
Fuel cell type	Solid polymer electrolyte
Current density (A/ft^2)	250
Operating temperature ($^\circ$F)	170
Electrolyte thickness (inch)	0.005

Performance and Cost

Powerplant efficiency (%)	25.2
Overall energy efficiency (%)	12.7
Plant capital cost ($ x 10^6)	30
Plant capital cost ($/kWe)	634
Cost of electricity (mills/kWh)	57.7

Natural Resources

Coal (lb/kWh)	2.50
Water (gal/kWh)	
Cooling (H_2 plant cooling tower)	0.05
Processing (H_2 plant)	0.31
Fuel cell cooling and air humidification	0.33
Land (acres/100 MWe)	8.3

Environmental Intrusion

	lb/10^6-Btu Input	lb/kWh Output	Btu/kWh
SO_2	0	0	
NO_x (from H_2 process)	0.1	5.7×10^{-4}	
HC	0	0	
CO	0	0	
Particulates	0	0	
Heat to water			0
Heat, total rejected			10,128
Wastes			None

Major Component Characteristics

Major Component	Size (ft) W x L (or D) x H	Weight (lb) (x 10^6)	Cost ($ x 10^6)	Units Required	Total Cost ($ x 10^6)	$/kW Output
Fuel cell stack	10 x 40 x 6	0.10	4.8	1	4.8	100
Humidifier-cooler	10 x 40 x 10	0.02	0.1	1	0.1	2.1

TABLE 5.7: POWER OUTPUT AND AUXILIARY POWER DEMAND FOR BASE CASE AND PARAMETER VARIATIONS: FUEL CELLS—LOW TEMPERATURE

	\.\.\. Case Number \.\.\.													
	1	2	4	5	6	7	8	9	10	11	12	13	14	15
Prime cycle power output, MW	52.1	26.0	52.1	52.1	52.1	52.1	208.3	52.1	52.1	52.1	52.1	52.1	52.1	52.1
Bottoming cycle power output, MW	0	0	0	0	0	0	0	0	0	0	0	0	0	0
Furnace power output, MW	0	0	0	0	0	0	0	0	0	0	0	0	0	0
Balance of plant auxiliary power required, MW	3.4	1.7	3.4	3.4	3.4	3.6	4.6	3.2	3.3	3.6	3.8	3.8	3.8	3.8
Furnace auxiliary power required, MW	0	0	0	0	0	0	0	0	0	0	0	0	0	0
Transformer losses, MW	0.4	0.2	0.4	0.4	0.4	0.4	1.0	0.4	0.4	0.4	0.4	0.4	0.4	0.4
Inverter losses, MW	0.5	0.3	0.5	0.5	0.5	0.5	2.1	0.5	0.5	0.5	0.5	0.5	0.5	0.5
Net station output, MW	47.8	23.8	47.8	47.8	47.8	47.6	200.5	48.0	47.9	47.6	47.4	47.4	47.4	47.4

FUEL CELLS—HIGH TEMPERATURE

Description of Cycle

While most of the fuel cell effort was devoted to low-temperature cells, a brief investigation was made of high-temperature, solid electrolyte fuel cells. Figure 5.4 shows a schematic of the high-temperature fuel cell base case. In this system, the low-Btu gas fuel is preheated before it enters the cell. After passing through the anode side of the cell, the gas passes on to a combustor that provides hot gas to the boiler and reheater for the bottoming cycle.

The cathode side of the cell is supplied with air from a blower and air preheater. Hot air leaving the fuel cell is cooled in the fuel preheater and then joins the anode stream in the combustor. The bottoming cycle uses conventional temperatures of 1000°F (811°K) for superheating and reheating, and a pressure at the turbine inlet of 3515 psia (24.2 MN/m²).

Three variations of this base case were studied (the parametric variations are listed later in the Results section). The first variation was a change in the type of coal supplied to the gasifier. The next two variations were made to determine the effect of changes in the current density and the electrolyte thickness.

Analytical Procedure and Assumptions

The solid electrolyte for the high-temperature fuel cell was zirconia (ZrO_2). Cell operating temperature was 1832°F (1273°K). In order to maintain a temperature near this level throughout the cell, a large amount of air was circulated through the air side of the cell (see Figure 5.4). At this temperature, no catalyst is needed.

The low-Btu gasifier providing fuel to the cell is basically the same as the other gasifiers in this study. One difference is that the gasifier was free standing; that is, there was no integration between the fuel cell system, or its bottoming cycle, and the gasifier.

The high-temperature fuel cell requires low-pressure gas. Because the gasifier operates at elevated pressures, the low-Btu gas was expanded through a turbine at the end of the gasification process. This expansion cooled the gas, and moisture had to be removed before the gas left the gasifier.

As shown in Figure 5.4, the fuel was preheated by high-temperature air leaving the fuel cell. Because of the high temperature, a heat storage regenerator with a refractory matrix was used, similar to the units described under sections of this report on open-cycle MHD and closed-cycle inert gas MHD. Within the preheater, the fuel changed to a new equilibrium composition at the high-temperature discharge.

The stream leaving the anode side of the fuel cell is not completely depleted of fuel; enough hydrogen and carbon monoxide remain to burn in a combustor for the bottoming cycle. For the cases studied, the constituents leaving the fuel side of the fuel cell and entering the combustor were as shown below.

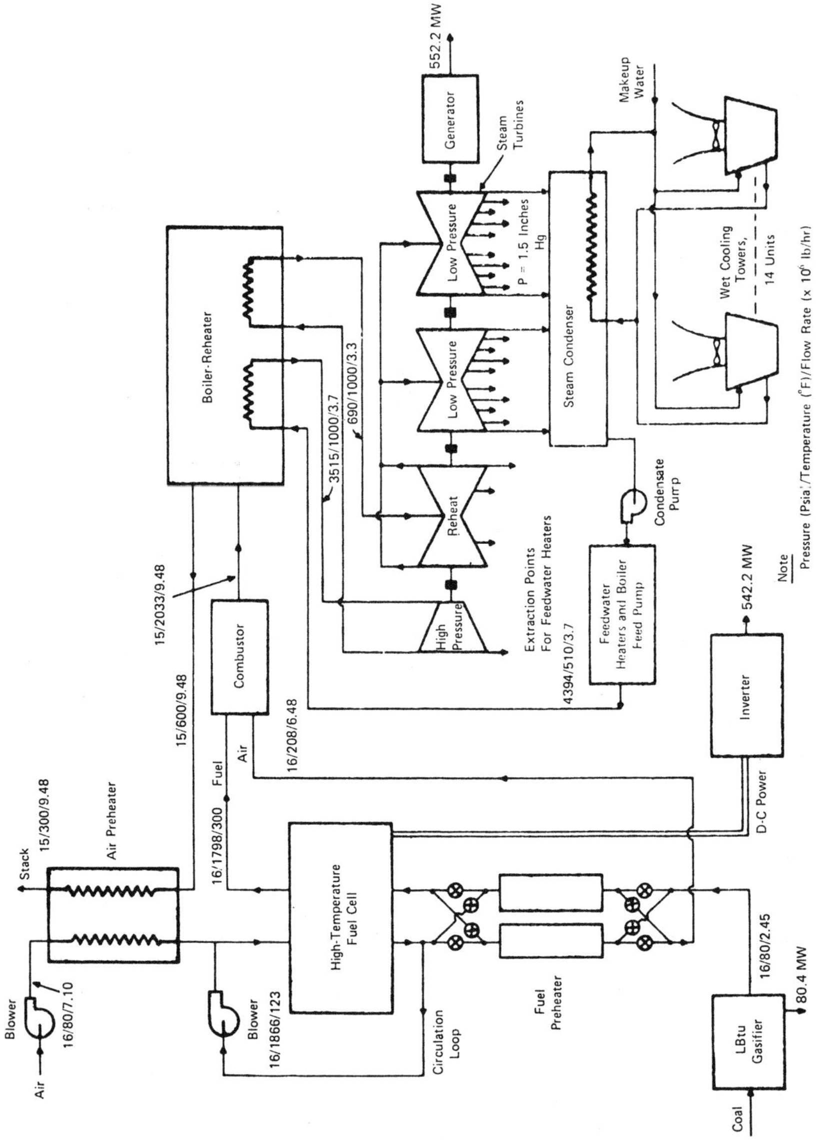

FIGURE 5.4: HIGH-TEMPERATURE FUEL CELL BASE CASE

Constituent	Percent by Mass
H_2	0.9
N_2	39.1
H_2O	16.2
CO	15.2
CO_2	28.6
	100.0

Performance data for the high-temperature fuel cell were obtained primarily from General Electric tests. Reference (1) describes the work on which the test data were based and gives a number of references to literature used in this study.

Design and Cost Basis

The following materials were assumed for the cost estimates: porous tube support, ZrO_2; fuel electrode, Ni and zirconia; interconnections, cobalt chromite; electrolyte, calcia-stabilized zirconia; and air electrode, indium oxide doped with tin. Cost data were taken from Reference (2). This document gave a most probable cost of \$13.42/ft^2, which was estimated to escalate to \$15.83/ft^2 (\$170/m^2) by mid-1974.

The performance estimates of Reference (2) showed much higher performance than was estimated from the General Electric tests [Reference (1)]. For example, the power density used in Cases 1 and 2 was 117 W/ft^2 (1260 W/m^2) (based on the General Electric data), while the Reference (1) data projected a power density of 688 W/ft^2 (7410 W/m^2). If this higher density could be achieved, a sizable reduction in fuel cell capital cost could be realized.

Results

Results for the study of high-temperature fuel cells are tabulated in Table 5.8, which includes the major cycle input parameters. A breakdown of capital costs is given in Table 5.9. A summary giving major cycle characteristics for the high-temperature fuel cell base case is given in Table 5.10. Auxiliary losses and power outputs are shown in Table 5.11.

Several observations can be made from the results shown in Table 5.8. The overall energy efficiency was moderate (a maximum of 34.3%). Increasing the current density to 700 A/ft^2 (7,500 A/m^2) caused a large drop in efficiency. The principal drawback appeared to be the very high capital cost (\$974/kW for the base case). Largely because of this high cost, the cost of electricity was also relatively high (45 mills/kWh for the base case).

The largest contributions to the capital cost were made by the gasifier (\$202 million for the base case) and the balance of plant (\$226 million). The balance of plant was costly principally because of the large amount of high-temperature piping.

The high-temperature fuel cell case that is recommended for further study is Case 1. A variation on that case that should be considered is the integration of the high-temperature fuel cell system with the low-Btu gasifier.

TABLE 5.8: PARAMETRIC VARIATIONS FOR TASK I STUDY (FUEL CELLS–HIGH TEMPERATURE)*

Parameters	Case Number			
	1**	2	3	4
Power output (MWe)	1,112	1,111	632	824
Coal and conversion process	Ill. #6	Mont.	Ill. #6	Ill. #6
	LBtu	LBtu	LBtu	LBtu
Oxidizer	Air	Air	Air	Air
Current density (A/ft^2)	200	200	700	700
Electrolyte thickness (inches)	0.020	0.020	0.020	0.005
Steam bottoming cycle				
Turbine inlet temperature (°F)	1000	1000	1000	1000
Turbine inlet pressure (psig)	3,500	3,500	3,500	3,500
Reheat temperature (°F)	1000	1000	1000	1000
Maximum feedwater temperature (°F)	510	510	510	510
Heat rejection (inch Hg)	WCT***	WCT	WCT	WCT
	15	15	15	15
Actual powerplant output (MWe)	1,112	1,111	632	824
Thermodynamic efficiency (%)	0	0	0	0
Powerplant efficiency (%)	31.5	34.3	24.5	27.9
Overall energy efficiency (%)	31.5	34.3	24.5	27.9
Coal consumption (lb/kWh)	1.00	1.11	1.29	1.14
Plant capital cost ($ million)	1,083	1,087	575	719
Plant capital cost ($/kWe)	974	978	910	861
Cost of electricity, capacity factor is 0.65				
Capital (mills/kWh)	30.8	30.9	28.8	27.2
Fuel (mills/kWh)	9.2	8.5	11.8	10.4
Maintenance and operation (mills/kWh)	5.0	5.0	4.8	4.7
Total (mills/kWh)	45.0	44.4	45.4	42.3
Sensitivity				
Capacity factor is 0.50 (total mills/kWh)	55.7	55.2	55.5	51.9
Capacity factor is 0.80 (total mills/kWh)	38.3	37.7	39.1	36.4
Capital Δ is 20% (Δmills/kWh)	6.2	6.2	5.8	5.4
Fuel Δ is 20% (Δmills/kWh)	1.8	1.7	2.4	2.1
Estimated time for construction (yr)	6	6	5	5
Estimated date of 1st commercial service (yr)	1998	1998	2005	2005

*Zirconia, 1832°F operating temperature.   **Base Case.   ***WCT = wet cooling tower.

TABLE 5.9: CAPITAL COST DISTRIBUTIONS FOR HIGH-TEMPERATURE FUEL CELLS

	Case Number			
	1	2	3	4
Major components				
Prime cycle				
Fuel cell stack, MM$	74.1	74.1	10.2	14.6
Air preheater, MM$	2.2	2.2	1.6	1.9
Bottoming cycle				
Steam boiler, MM$	16.5	16.5	10.9	13.0
Steam turb-gen, MM$	20.3	20.3	20.3	20.3
Fuel processing				
Gasifier including boost				
steam turb-comp, MM$	202.0	204.0	157.0	175.0
Fuel preheater, MM$	9.6	9.6	6.9	8.0
Subtotal of major components, MM$	324.7	326.7	206.9	232.8
Balance of plant				
Cooling tower, MM$	3.5	3.5	3.5	3.5
dc to ac inverters, MM$	19.6	19.6	4.2	11.5
All other, MM$	154.3	154.3	75.6	107.1
Site labor, MM$	48.9	48.9	25.9	35.2
Subtotal of balance of plant, MM$	226.3	226.3	109.2	157.3
Contingency, MM$	110.2	110.6	63.2	78.0
Escalation costs, MM$	189.1	189.8	90.6	111.9
Interest during construction, MM$	232.9	233.8	105.5	130.2
Total capital cost, MM$	1,083.2	1,087.1	575.4	710.2
Major components cost, $/kWe	292.1	294.0	327.4	282.4
Balance of Plant, $/kWe	203.5	203.6	172.8	190.9
Contingency, $/kWe	99.1	99.5	100.0	94.7
Escalation costs, $/kWe	170.1	170.7	143.4	135.7
Interest during construction, $/kWe	209.5	210.3	166.9	157.9
Total capital cost, $/kWe	974.5	978.1	910.5	861.6

TABLE 5.10: SUMMARY SHEET FUEL CELLS–HIGH-TEMPERATURE BASE CASE

Cycle Parameter

Net power output (MWe)	1,112
Coal type	Illinois No. 6
Prime cycle	
Oxidizer	Air
Current density (A/ft^2)	200
Electrolyte thickness	0.020
Bottoming cycle (steam)	
Turbine inlet temperature (°F)	1000
Turbine inlet pressure (psig)	3,500
Reheat temperature (°F)	1000
Heat rejection	Wet cooling tower

Performance and Cost

Thermodynamic efficiency (%)	–
Powerplant efficiency (%)	–
Overall energy efficiency (%)	31.5
Plant capital cost ($ x 10^6)	1,083
Plant capital cost ($/kWe)	974
Cost of electricity (mills/kWh)	45.0

Natural Resources

Coal (lb/kWh)	1.00
Water (gal/kWh)	
Total	0.277
Cooling	0.187
Processing	0.090
Makeup	0
NO$_x$ suppression	0
Stack gas cleanup	0
Land (acres/100 MWe)	0.36

Environmental Intrusion

	lb/10^6-Btu Input	lb/kWh Output	Btu/kWh	lb/kWh	lb/day
SO$_2$	0.2	0.0022			
NO$_x$	0	0			
HC	0	0			
CO	0	0			
Particulates	–	–			
Heat to water			2,080		
Heat, total rejected			7,420		
Wastes					
Ash				0.113	3.02 x 10^6
Sulfur				0.038	1.01 x 10^6

Major Component Characteristics

Major Component	Unit or Module Size (ft) (W x L (or D) x H)	Weight (lb) (x 10^6)	Cost ($ x 10^6)	Units Required	Total Cost ($ x 10^6)	$/kW Output
Prime cycle						
Fuel cell stack	8 x 293 x 30	1.58	7.41	10	74.1	–
Air preheater	–	2.42	2.2	1	2.2	–
Fuel preheater	–	–	9.6	1	9.6	–
Bottoming cycle						
Furnace-boiler	80 x 77 x 82	8.43	16.5	1	16.5	–
Steam turbine-generator	30 x 174 x 25	4.60	20.3	1	20.3	–

TABLE 5.11: POWER OUTPUT AND AUXILIARY POWER DEMAND FOR BASE CASE AND PARAMETRIC VARIATIONS: FUEL CELLS–HIGH TEMPERATURE

	. . . . Case Number.			
	1	2	3	4
Prime cycle power output, MW	550.5	550.5	80.6	269.0
Bottoming cycle power output, MW	555.0	555.0	550.0	552.0
Furnace power output, MW	80.8	80.8	59.0	67.7
Balance of plant auxiliary power required, MW	63.3	63.4	53.3	57.3
Furnace auxiliary power required, MW	0	0	0	0
Transformer losses, MW	5.9	5.9	3.6	4.4
Inverter losses, MW	5.5	5.5	0.8	2.7
Net station output, MW	1,111.6	1,111.5	631.9	824.3

REFERENCES

(1) White, D.W., *Progress in High-Temperature, Zirconia-Electrolyte Cell Technology at General Electric,* Report 68-C254, Corporate Research and Development, General Electric Company, Schenectady, New York, July 1968.

(2) *Project Fuel Cell Final Report,* R&D Report 47, Office of Coal Research, Department of the Interior, Contract 14-01-0001-303, Westinghouse Electric Corporation, Pittsburgh, Pa., 1970.

FUEL CELL POWER PLANT EVALUATION

The National Aeronautics and Space Administration contracted with both the Westinghouse Electric Corporation and the General Electric Company to study ten advanced energy conversion systems for central-station, base-load electric power generation using coal and coal-derived fuels. The fuel cell section authored by Westinghouse (NASA CR-134941, Vol. XII) is reported in the fourth chapter of this book. The General Electric study (NASA CR-13948, Vol. II, Part 3) can be found in the fifth chapter. A comparative evaluation of the fuel cell power plant was made by Marvin Warshay in February 1976, as part of NASA Report TM X-71855, and is editorialized below.

INTRODUCTION

Fuel cells offer the potential for high-efficiency generation of electricity. These electrochemical devices are not Carnot-cycle limited as are conventional thermodynamic cycles. In particular, fuel cells that operate at high temperatures ($1200°$ to $1800°F$) fit well into the ECAS study framework. They are quite suitable as large base-load utility power plants. In fact, their high-quality waste heat is best utilized in large systems. Waste heat can be used either by a bottoming cycle or by a coal gasifier, or in some cases by both. Further, close integration can effectively improve the energy efficiency of the coal conversion process.

All fuel cells require clean fuels. However, high-temperature fuel cells do not require fuels to be clean within extremely narrow tolerances. Consequently, their fuel processing requirements are not very stringent. Finally, the increase in the reaction rates brought about by high temperatures helps the fuel cell to approach its potential for high efficiency. All these aspects of the high-temperature fuel cell systems lead to energy conversion systems with very high overall efficiencies.

Low-temperature fuel cells are less suited for the primary ECAS utility application, that is, base-load power generation from coal-derived fuels. First of all, the requirements for clean fuel are much more stringent for low-temperature fuel cells than for high-temperature fuel cells. Secondly, there is less potential for utilization of waste heat in the power plant at the low temperatures. Third, the rate processes are slower at low temperatures; polarization losses are significant when operating at desirable current density levels. All three of these conditions reduce the efficiency of a low-temperature fuel cell power plant system.

However, greater potential for low temperature fuel cell utility applications is foreseen for non-base-load service. Outside the context of ECAS, low temperature fuel cells are associated with dispersed power generation, peaking or intermediate service, load following, efficient use of natural gas or petroleum products, transmission savings, total energy savings through onsite waste heat utilization, modularity advantages, high efficiency at wide load variation, the hydrogen economy, etc. These potential fuel cell applications are discussed later. Previous reports on fuel cell power plants have indicated their potential for high efficiency. The present ECAS (Energy Conversion Alternatives Study) study not only further documents this, but also adds cost dimensions to the description.

LOW TEMPERATURE FUEL CELLS

General Electric Treatment

Two low temperature fuel cell systems were treated by G.E., the solid polymer electrolyte (SPE) system and the phosphoric acid system. The SPE fuel cell received the major emphasis, being chosen as the base case (Table 6.1) and treated in 10 out of 14 parametric variations. In the base case, high Btu (H Btu) gas was used as the fuel, but hydrogen was used in over half of the parametric variations. The oxidizer was air in all but one case, in which oxygen was used.

TABLE 6.1: GENERAL ELECTRIC BASE-CASE PARAMETERS FOR LOW TEMPERATURE FUEL CELL POWER PLANT (CASE 1)

Power output, MWe	48
Coal	Illinois #6
Conversion process	HBTU
Oxidizer	Air
Fuel-cell type	SPE
Current density, A/ft^2	250
Operating temperature (maximum), $^\circ F$	170
Electrolyte thickness, in. (cm)	0.005 (0.0127)
Actual powerplant output, MWe	48

The choice of H Btu fuel was dictated by the selection of the 48-MWe (substation size) power plant. This small power plant would, in most cases, be used at a dispersed site rather than as a central station power plant. Undoubtedly the first gasified fuel to be transported through gas pipelines will be substitute natural gas, that is H Btu, rather than a hydrogen-rich fuel, which fuel cells prefer.

However, the H Btu gas presents fuel processing problems, especially for the SPE system. (See Figure 6.1 for a typical low temperature fuel cell schematic.) The base-case SPE fuel cell was operated at 250 A/ft^2 and 170°F; the phosphoric acid fuel cell operating temperature was 375°F. The range of current density variation in the study was 100 to 350 A/ft^2. A single central station sized power plant case (201 MWe; SPE; onsite hydrogen; oxygen) was included in the G.E. study.

FIGURE 6.1: G.E. LOW TEMPERATURE SOLID POLYMER ELECTROLYTE FUEL CELL POWER PLANT BASE CASE

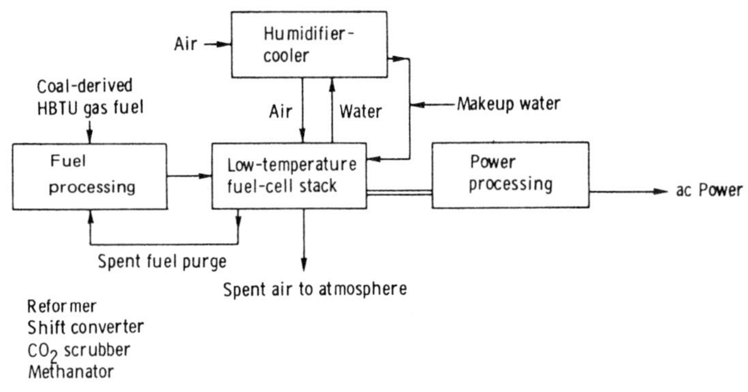

Westinghouse Treatment

Two low temperature fuel cell systems were treated by Westinghouse, the aqueous alkaline (KOH) and the aqueous phosphoric acid systems. (See Figure 6.2 for schematics of the systems.) An appropriate base case for each of the fuel cell systems was selected (Table 6.2). One characteristic that all base cases had, including the ones for high temperature fuel cells, was the nominal 25 MWe power plant size. This size selection is in keeping with the first fuel cell electric utility power plant, the FCG-1.

The assumed fuel cell useful life for both low temperature base cases was 10,000 hours. Parametric variations covered fuel cell life up to 100,000 hours. H Btu fuel and air oxidizer were used in the majority of the 32 low temperature cases. Methanol and intermediate Btu gas (designated medium Btu (M Btu) by Westinghouse but termed I Btu in the ECAS study) were other fuels investigated; oxygen was used as the oxidizer in three cases.

The operating temperatures were 158°F (70°C) for the KOH fuel cell system and 375°F (190°C) for the acid system. The KOH base case current density was 100 mA/cm^2, with parametric variations up to 250 mA/cm^2 investigated.

(Current densities are commonly reported in either A/ft² or mA/cm². The latter is 7% lower than the former.) The phosphoric acid system base case current density was 200 mA/cm² with parametric variations up to 400 mA/cm² investigated. These current densities were a direct reflection of the operating temperatures of the two fuel cell systems. Finally, to determine the effects of scale-up, several large power plants at a nominal 900 MWe size were investigated.

FIGURE 6.2: WESTINGHOUSE LOW TEMPERATURE ACID (H₃PO₄) AND ALKALINE (KOH) FUEL CELL POWER PLANT BASE CASES

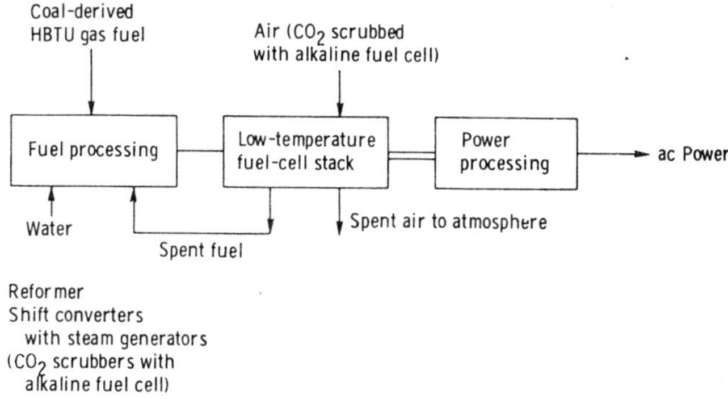

TABLE 6.2: WESTINGHOUSE BASE CASE PARAMETERS FOR LOW TEMPERATURE FUEL CELL POWER PLANTS (CASE 1)

	Aqueous acid (H_3PO_4) systems	Aqueous alkaline (KOH) systems
Power output, MWe	23.4	22
Fuel-cell rating, MW	25	25
Fuel	HBTU	HBTU
Oxidizer	Air	Air
Fuel-cell life, hr	10 000	10 000
Voltage degradation, percent	5	5
Temperature, °C	190	70
Electrolyte type	85 wt % H_3PO_4	30 wt % KOH
Electrolyte thickness, cm (in.)	0.05 (0.02)	0.05 (0.02)
Anode type	Pt/C	Pt/C
Anode catalyst loading, mg Pt/cm²	1.0	1.0
Cathode type	Pt/C	Ag/C
Cathode catalyst loading, mg Pt (or Ag)/cm²	1.0	5.0
Current density, mA/cm²	200	100
Average cell voltage, V	0.7	0.8

HIGH TEMPERATURE FUEL CELLS

General Electric Treatment

The G.E. ECAS contract had the major emphasis on low temperature fuel cells with only a small study of high temperature fuel cell systems. General Electric treated the high temperature (1832°F) zirconia solid electrolyte (SE) fuel cell in four cases, all of which included stream bottoming cycles. The fuel was low Btu (L Btu) gas. (See Table 6.3 for details of the base case.)

TABLE 6.3: GENERAL ELECTRIC BASE CASE PARAMETERS FOR HIGH TEMPERATURE FUEL CELL POWER PLANT (CASE 1)

Power output, MWe	1112
Coal	Illinois # 6
Conversion process	LBTU
Oxidizer	Air
Current density, A/ft^2	200
Electrolyte thickness, in. (cm)	0.02 (0.05)
Steam bottoming cycle:	
Turbine-inlet temperature, °F	1000
Turbine-inlet pressure, psig	3500
Reheat temperature, °F	1000
Maximum feedwater temperature, °F	510
Heat rejection method	[a]WCT
Condenser pressure, in. of Hg	1.5
Actual powerplant output, MWe	1112

[a]Wet cooling tower.

However, there was no integration of the L Btu gasifier with the power plant although the capital cost of the gasifier is included in the plant cost. (See Figure 6.3 for a schematic of the system.) Air was the oxidizer in all cases.

Two current densities were investigated, 200 and 700 A/ft^2. The electrolyte thickness was 0.020 inch in three out of four cases; for one case the effect of a fourfold reduction in electrolyte thickness (0.005 inch) was explored. The power plant outputs varied from 632 to 1,112 MWe.

FIGURE 6.3: SIMPLIFIED SCHEMATIC OF G.E. HIGH TEMPERATURE ZIRCONIA FUEL CELL POWER PLANT BASE CASE

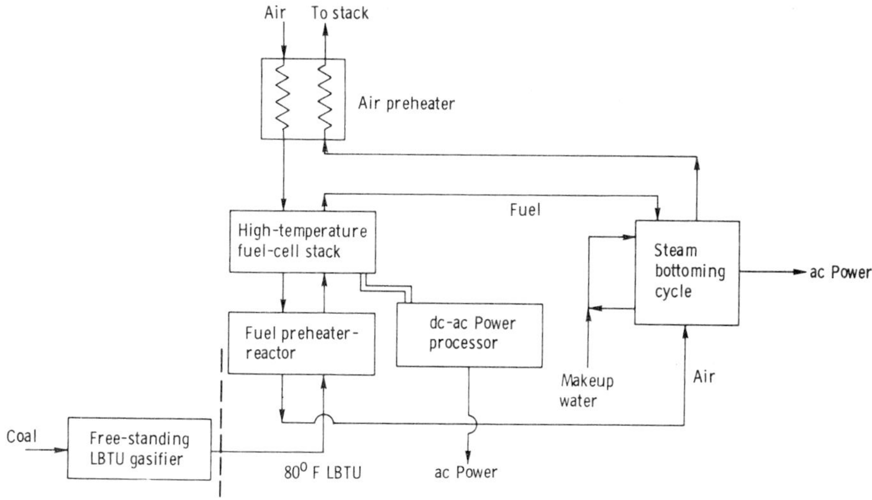

Westinghouse Treatment

Westinghouse examined two high temperature fuel cell systems, the molten carbonate (MC) system (1200° to 1381°F; or 650° to 750°C) and the zirconia solid electrolyte (SE) system (1652° to 2012°F; or 900° to 1100°C). See Figure 6.4 for schematics of the two systems.

FIGURE 6.4: WESTINGHOUSE HIGH TEMPERATURE MOLTEN CARBONATE AND ZIRCONIA FUEL CELL POWER PLANT BASE CASES

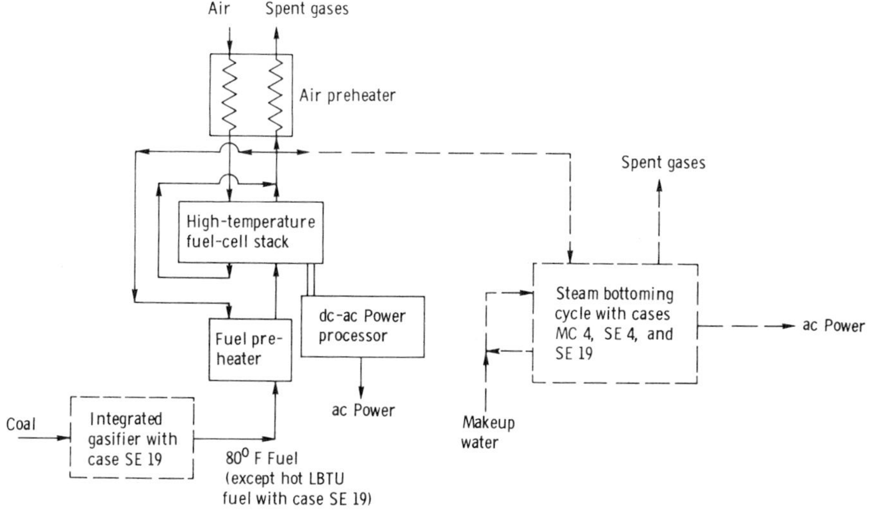

TABLE 6.4: WESTINGHOUSE BASE CASE PARAMETERS FOR HIGH TEMPERATURE FUEL CELL POWER PLANT (CASES 1)

	Molten carbonate systems	Solid electrolyte (SPE) systems
Power output, MWe	22.6	22.9
Fuel-cell rating, MW dc	25	25
Fuel	HBTU	HBTU
Oxidizer	Air	Air
Fuel-cell life, hr	10 000	10 000
Voltage degradation, percent	5	5
Temperature, $^{\circ}C$	650	1000
Electrolyte type	Paste of Li, Na, K, carbonates, and alkali aluminates	$\left(ZrO_2\right)_{1-x}\left(Y_2O_3\right)_x$
Electrolyte thickness, cm (in.)	0.1 (0.04)	0.004 (0.0015)
Anode type	Ni	$NiZrO_2$-cermet
Cathode type	Lithiated NiO	$In_2O_3/PrCoO_{3-x}$
Interconnection type	------------	Cr_2O_3
Interconnection thickness, cm (in.)	------------	0.002 (0.0008)
Current density, mA/cm^2	200	400
Average cell voltage, V	7	0.84

An appropriate base case was chosen for each system. (See Table 6.4 for details.) H Btu was the predominant fuel used for the 16 MC cases, while I Btu fuel predominated in the 20 SE cases. Additionally, I Btu gas and methanol were investigated in the MC systems. The zirconia SE investigation also covered H Btu and L Btu cases. Air was the oxidizer in all but two cases. In 13 out of 16 MC cases the current density was 200 mA/cm². It ranged from 150 to 300 mA/cm² in the remaining parametric variations. In the SE cases, 400 mA/cm² was the most common current density; the range covered 800 mA/cm². Thinner cell electrolytes were investigated in the Westinghouse zirconia SE study (0.002 and 0.004 cm) than were investigated by G.E.

The MC power plant sizes ranged to 1,255 MWe, a case that included a steam bottoming cycle. The SE cases, which covered sizes up to 1,164 MWe, included three integrated power plant cases. In one case the fuel cell and steam bottoming cycle were integrated, in another case the fuel cell and gasifier were integrated, and in the last case all three were integrated. The gasifier/fuel cell integration represented a version of Westinghouse's previously studied "Project Fuel Cell" concept. In this concept, the fuel cell is inside the gasifier to maximize heat and mass transfer from the fuel cell.

OVERALL COMPARISON

With the aid of Figure 6.5 an overall comparison of the results of the fuel cell portion of the parametric study can be made. The G.E. low temperature fuel cell system points are divided into two groups to indicate the significant effect of hydrogen fuel. Likewise, the Westinghouse high temperature points are divided into two groups to indicate the significant effect of using a steam bottoming cycle and/or integration with the gasifier. The general conclusions that can be drawn from inspection of Figure 6.5 are as follows:

(1)   With low temperature fuel cell power systems the use of hydrogen fuel in place of H Btu gas improves efficiency and lowers the cost of electricity (COE).

(2)   The Westinghouse estimates of low temperature fuel cell efficiencies (with no hydrogen fuel cases) were all higher than the G.E. overall efficiency estimates for H Btu/air. The Westinghouse estimates of COE for these same cases were either higher or lower than those of G.E.

(3)   The results indicate that, in general, high temperature fuel cell systems are more efficient than low temperature fuel cell systems.

(4)   Using a steam bottoming cycle and/or integration with the gasifier results in the highest efficiencies obtained with a high temperature fuel cell.

(5)   The estimates by Westinghouse of the high temperature fuel cell efficiencies for systems with a steam bottoming cycle are significantly higher than G.E.'s estimates.

Additional conclusions based on a closer examination of the results are discussed in the following section. In addition, the present general conclusions are more precisely stated and discussed. Finally, it is important to state that optimization was not a goal of Phase 1 of ECAS. Optimization could result in points with reduced COE.

FIGURE 6.5:   ENERGY EFFICIENCY—COST OF ELECTRICITY MAP FOR VARIOUS TYPES OF FUEL CELL POWER PLANTS

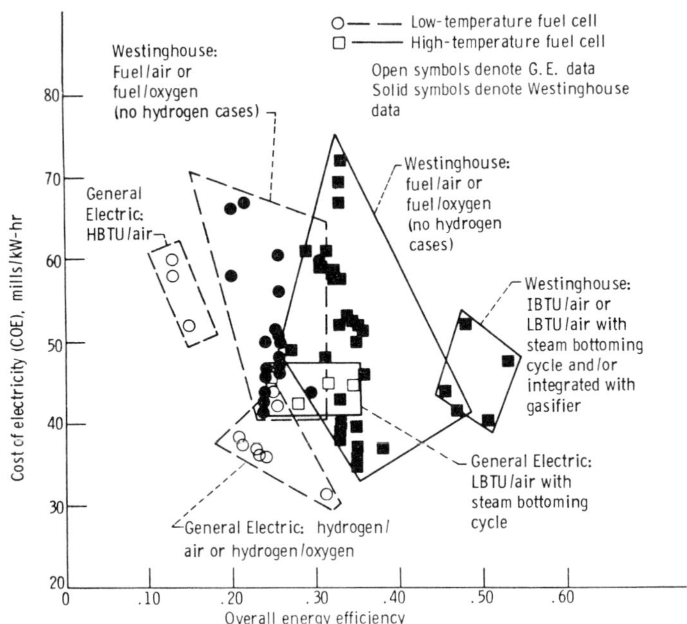

ASSESSMENT OF LOW TEMPERATURE FUEL CELLS

The general results obtained by both General Electric and Westinghouse in their studies of fuel cell systems are as follows.

General Electric Solid Polymer Electrolyte Systems

In G.E.'s treatment of low temperature fuel cells, major emphasis was placed on the SPE fuel cell system. The overall efficiency of the SPE system operating with H Btu/air at 170°F in only 12.7%. The fact that the fuel cell power plant efficiency is double that illustrates the heavy penalty the fuel cell pays for coal conversion to H Btu fuel.

Significant performance jumps to 23 to 25% occur when the SPE operates with hydrogen/air. This is no surprise because fuel cells operate best on hydrogen. Also the conversion efficiency to hydrogen is somewhat better than to H Btu gas. Nor is it surprising that the best efficiency, 31.1%, was registered by the 201-MWe system operating with hydrogen/oxygen at 300°F. The cost data for the SPE systems follow the same pattern as the performance efficiencies.

The higest cost electricity (55 to 60 mills/kW-hr) would be produced by
H Btu/air systems, the lowest (31 mills/kW-hr) by the most efficient hydro-
gen/oxygen system. In all cases, the fuel, cost which reflects power plant efficiency, is the major component of the COE.

For the hydrogen cases, both fuel and capital components of COE costs are
lower than for the H Btu cases. In fact, capital costs for the hydrogen cases
are sharply lower. The hydrogen-fueled converters do not require costly fuel
processing to convert the fuel to hydrogen-rich gas, as in the case of H Btu.
Operating and maintenance costs were not high for any of the SPE cases. The
cell life was assumed to be 100,000 hours for all fuel cells. This assumption is
considered very optimistic for the 300°F case.

General Electric Phosphoric Acid Systems

The results for the G.E. phosphoric acid fuel cell followed exactly the pattern
of the SPE results in COE, efficiency, and response to hydrogen or H Btu fuel.
However, a surprising G.E. result is the mere 4.7 percentage point power plant efficiency advantage for the higher temperature (375°F) phosphoric acid H Btu/air
system compared with the SPE system (170°F). Furthermore, the results indicate
no advantage in efficiency for the phosphoric acid system over an SPE system
using hydrogen fuel.

Westinghouse Phosphoric Acid Systems

For the phosphoric acid cases, comparisons can be made between the G.E. and
Westinghouse results. Westinghouse included a very broad parametric study
almost exclusively devoted to H Btu/air, with no hydrogen cases considered.
Several similarities as well as differences in the results of the two contractor
studies stand out.

The Westinghouse results indicate, as did G.E.'s results, that the fuel COE component is highest. However, the power plant efficiencies upon which fuel costs
hinge are significantly higher for Westinghouse than for G.E. The Westinghouse
power plant efficiencies for H Btu/air cases are approximately 36%, compared
with 29.8% for G.E. Also the Westinghouse overall efficiencies, at approximately
24%, are about 50% greater than G.E.'s (15.6%). This is reflective of both the
higher power plant efficiency and the higher gasifier conversion efficiency estimated by Westinghouse.

However, since both contractors used the NASA specified cost for over-the-
fence H Btu gas, a lower or higher conversion efficiency is not reflected in the
fuel COE in this parametric study. In actual detailed designs and in the ECAS
system cases integrated with L Btu and Btu gasifiers, the efficiency of the coal
conversion process does affect the fuel costs. Another significant difference
between the G.E. and Westinghouse H Btu/air results is in capital cost. The
Westinghouse capital cost estimates ($339/kWe to $463/kWe) are all lower than
for the G.E. H Btu/air base case ($570/kWe). The difference is largely a re-
flection of the much higher fuel processing costs for H Btu estimated by G.E.

Finally, a contrast exists between the G.E. oxygen case (hydrogen/oxygen SPE,
case 8) results and the Westinghouse low temperature fuel cell case results

(H Btu/oxygen phosphoric acid, cases 5 and 16). The Westinghouse H Btu/oxygen systems were more costly and less efficient than the Westinghouse H Btu/air systems. However, the G.E. hydrogen/oxygen system was more efficient and has a lower COE than any of the G.E. hydrogen/air systems. Part of the explanation lies in the two contractors' treatments of oxygen production.

Westinghouse elected to estimate the costs of producing oxygen on site; G.E. used over-the-fence oxygen at a NASA-specified cost of $9/ton. Doing the calculation the first way not only resulted in a higher oxygen price, but in a significant (from 36 to 29.6%) lowering of power plant efficiency. This accounts for the parasitic power required to run the oxygen plant. For further discussion of this point see the preceding chapter, a detailed discussion of G.E. hydrogen/oxygen SPE case 8.

Westinghouse Alkaline (KOH) Systems

In the Westinghouse treatment of the alkaline fuel cell, the more conventional reactant scrubbing technique of handling the electrolyte carbonation problem was selected. Neither the carbon dioxide rejecting buffered base technique nor the cyclic decarbonation technique was examined in the Westinghouse study.

The Westinghouse parametric treatment of the alkaline (KOH) fuel cell systems was also almost exclusively devoted to H Btu/air cases. (The only other case treated was an I Btu/air case.) Comparing these results with Westinghouse's phosphoric acid results indicates the following: Despite higher power plant efficiency the KOH system COE's are higher. This means capital costs are higher for the KOH cases. High capital costs result from (1) higher reactant processing costs to handle the carbonation problem and (2) the higher fuel cell stack costs resulting from the lower power density operation (due to the lower current density) of this lowest temperature fuel cell (158°F).

ASSESSMENT OF HIGH TEMPERATURE FUEL CELLS

General Electric Zirconia Solid Electrolyte System

The overall efficiencies for the G.E. zirconia SE fuel cells (27.9 to 34.3%) are surprisingly low for this highest temperature (1832°F) fuel cell system. As discussed in the next section, this is primarily attributable to zirconia fuel cell performance, not to lower than expected steam bottoming cycle performance.

In G.E.'s treatment of fuel cells, their projection of long life (40,000 to 100,000 hours) kept the operation and maintenance (O and M) COE to approximately 5 mills/kW-hr. By contrast, the Westinghouse O and M COE for fuel cells varies from 2 to over 30 mills/kW-hr. The reason is that in the Westinghouse treatment, fuel cell useful life (which is 10,000 hr for all four base cases) is allowed to vary up to 100,000 hr in the parametric variations.

Westinghouse Zirconia Solid Electrolyte System

The Westinghouse results indicate a much greater zirconia fuel cell efficiency in all cases than is estimated by G.E. Apparently, Westinghouse's thinner-cell concept results in much higher energy densities than the G.E. zirconia concept. The Westinghouse system with a steam bottoming cycle (case SE 19) has an overall efficiency of 47.8% compared with 31.5 to 34.3% for G.E. systems. The Westinghouse system operates at twice the current density (400 mA/cm^2) of the G.E. systems while generating approximately equal voltage (0.56 V compared with 0.58 V for G.E.).

Both the capital and fuel costs are lower for the Westinghouse cases. However, because of the long life assumption by G.E. (10 yr compared with 10,000 hr by Westinghouse), the O and M estimate by G.E. (5 mils/kW-hr) is much less than that estimated by Westinghouse (23 mills/kW-hr). This causes the total COE estimate made by Westinghouse to exceed by a small amount the COE for the G.E. case. The majority of the Westinghouse parametric cases use I Btu fuel because Westinghouse had greater confidence in it than in H Btu gas. Also, there are more technical data available on the effects of using I Btu gas in the zirconia fuel cell.

Westinghouse Molten Carbonate Systems

The most interesting Westinghouse molten carbonate system for base load application is the I Btu/air system (1200°F), which produces 1,255 MWe with the aid of a steam bottoming cycle (case MC 4). The power plant and overall efficiencies for this case are very attractive at 54.4 and 45.7%, respectively. For a similar zirconia SE system, power plant and overall efficiencies were 60.2 and 50.6%, respectively (case SE 4). For these two cases, the capital, fuel and O and M costs were very close to one another, with the O and M cost being the largest. This produced a COE of 40 to 45 mills/kW-hr. To what extent the fuel and O and M COE's can be reduced by a reasonable life projection beyond 10,000 hr is discussed later.

SIGNIFICANT TRENDS

The Westinghouse results show most clearly (Table 6.5) that efficiency increases with fuel cell temperature. In going from the low temperature phosphoric acid fuel cell to the high temperature molten carbonate system and finally to the very high temperature zirconia solid electrolyte system, the overall efficiency almost doubles. These results are for H Btu fuel.

Table 6.6 illustrates the enhanced overall efficiencies that result from using I Btu fuel instead of H Btu fuel and from using a steam bottoming cycle, in the case of the two high temperature fuel cells. Only the high temperature fuel cells have the quality of waste heat that can be used by a steam bottoming cycle, and then only with large (900 MWe) fuel cell power plants. (The increased size affects the phosphoric acid low temperature fuel cell power plant very little).

That most of the increased efficiency for the high temperature fuel cell systems comes from integration can be seen in Table 6.7. This table indicates that each of the high temperature systems gains approximately 15 percentage points in overall efficiency through using a steam bottoming cycle. (Not shown in this table but included in the Westinghouse results are data for both high temperature systems that indicate no efficiency improvements due to simple scale-up from 25 to 250 MWe.)

TABLE 6.5:   EFFECT OF FUEL CELL TEMPERATURE (TYPE) ON EFFICIENCY AND COST OF ELECTRICITY FOR 25-MWe POWER PLANTS WESTINGHOUSE RESULTS

Fuel-cell temperature, ^{o}F	Fuel-cell type[a]	Efficiency, percent		Cost of electricity,[b] COE, mills/kW-hr
		Powerplant	Overall	
375	H_3PO_4	35.5	23.9	50
1200	Molten carbonate	48.8	32.9	58
1832	ZrO_2 SE	69.7	46.9	42

[a]HBTU/air.

[b]Assumed 10 000-hr life for all fuel cells.

TABLE 6.6:   EFFECT OF FUEL CELL TEMPERATURE (TYPE) ON EFFICIENCY AND COST OF ELECTRICITY FOR 900-MWe POWER PLANTS WESTINGHOUSE RESULTS

Fuel-cell temperature, ^{o}F	Fuel-cell type[a]	Efficiency, percent		Cost of electricity,[b] COE, mills/kW-hr
		Powerplant	Overall	
375	H_3PO_4	34.8	29.3	44
1200	Molten carbonate[c]	54.4	45.7	44
1832	ZrO_2 SE[d]	60.2	50.6	40

[a]IBTU/air.

[b]Assumed 10 000-hr life for all fuel cells.

[c]Fuel cell with steam bottoming cycle; 1255 MWe total.

[d]Fuel cell with steam bottoming cycle; 1164 MWe total.

TABLE 6.7:   EFFECT OF STEAM BOTTOMING CYCLE ON HIGH TEMPERATURE FUEL CELL PERFORMANCE WESTINGHOUSE RESULTS

Fuel-cell temperature, ^{o}F	Fuel-cell type[a]	Overall efficiency, percent		Cost of electricity,[b] COE, mills/kW-hr	
		25-MWe systems	Nominal 900-MWe systems	25-MWe systems	Nominal 900-MWe systems
1200	Molten carbonate[c]	30.6	45.7	60	44
1832	ZrO_2 SE[d]	35.0	50.6	52	40

[a] IBTU/air.

[b] Assumed 10 000-hr life for all fuel cells.

[c] 1255 MWe total.

[d] 1164 MWe total.

Looking at these tables from the standpoint of COE, the results indicate great benefit for the high temperature fuel cells from using a steam bottoming cycle. There is no clear COE trend related to fuel cell temperature (Tables 6.5 and 6.6). The benefits of integration are again most clearly shown in Table 6.7. Cost of electricity is lowered more than 13 mills/kW-hr in both high temperature fuel cell systems. (The lower COE for phosphoric acid indicated in Table 6.6 stems principally from the lower price of I Btu fuel compared with H Btu fuel).

Table 6.8, the last in this set of Westinghouse results, provides some results for various modes of high temperature solid electrolyte system integration. The use of a steam bottoming cycle compared with integration with the gasifier seems to be an efficiency/COE trade-off; higher efficiency but also higher COE result from integration with the gasifier. In the third case, which has both integration with the gasifier and use of a steam bottoming cycle, it is not surprising that the L Btu fuel results in the lowest efficiency. What is surprising is that it does not also result in the lowest COE since L Btu should be the cheaper fuel. [However, there is no assurance that the contractor's estimate (which is not explicitly stated) of L Btu cost is consistent with the NASA specified costs for H Btu and I Btu fuels].

The G.E. results (Table 6.9) also show the efficiency increase in going from the low temperature fuel cells to the high temperature fuel cells; the overall efficiency increases almost 18 percentage points. This efficiency increase with temperature results from a comparison that is not entirely valid; the high temperature case uses a steam bottoming cycle and uses L Btu fuel from a freestanding gasifier instead of H Btu fuel. However, closer examination of the complete set of G.E. data indicates that the trend is very substantially correct.

TABLE 6.8: EFFICIENCY AND COST OF ELECTRICITY FOR THREE TYPES OF INTEGRATED, HIGH TEMPERATURE, SOLID ELECTROLYTE FUEL CELL POWER PLANTS WESTINGHOUSE RESULTS

Size (total MWe)	Fuel	Extent of integration	Efficiency, percent		Cost of electricity,[a] COE, mills/kW-hr
			Powerplant	Overall	
1164	IBTU	With steam bottoming cycle	60.2	50.6	40
219	IBTU	With gasifier[b]	53.2	53.2	48
1064	LBTU	With gasifier and steam bottoming cycle	47.8	47.8	52

[a]Assumed 10 000-hr life fuel cell.

[b]This is the most intimate integration, and is the "Project Fuel Cell" concept of Westinghouse.

While the trend of the G.E. efficiency data in Table 6.9 is in agreement with the Westinghouse results, the absolute values indicate several differences between the two contractors' results. These differences are as follows:

(1) Westinghouse used a higher conversion efficiency for H Btu production from coal than G.E. (67% against 50%).

(2) Westinghouse calculated a higher phosphoric acid power plant efficiency than G.E. (36% against 29.8%).

(3) Westinghouse calculated a higher overall efficiency for the high temperature solid electrolyte fuel cell system (47.7% against 31.5%).

From an analysis of G.E.'s and Westinghouse's treatments of the phosphoric acid system it appears that the power plant efficiency difference stems principally from G.E.'s lower fuel cell efficiency plus a lower degree of fuel cell-fuel processor integration. Close integration of the fuel cell with the fuel processor can increase the conversion efficiency substantially.

The explanation for the lower efficiencies obtained with the G.E. high temperature solid electrolyte concept lies in the use of much thicker cells (0.020 inch electrolyte) than in the Westinghouse concept [0.00158 inch (0.004 cm) electrolyte]. For this concept G.E. voltages varied between 0.18 and 0.58 V compared with a 0.51 to 0.84 voltage range for Westinghouse.

To complete the discussion of the G.E. results in Table 6.9, the lower COE value for the zirconia SE system merely reflects a savings due to the use of a steam bottoming cycle and is not possible with the low temperature fuel cells. This would be consistent with the conclusion based upon the Westinghouse results.

Comparing the COE values of G.E. and Westinghouse in Tables 6.5, 6.8 and 6.9 might lead to the specious conclusion that the values are close for both the phosphoric acid and zirconia cases. This is because the G.E. life assumptions were much longer than the 10,000 hr Westinghouse life assumptions. A closer look at the G.E. and Westinghouse data reveals higher capital costs in both G.E. cases but lower O and M costs. These data are examined in greater detail in the next section.

Table 6.10 illustrates the dual benefit of using hydrogen instead of H Btu fuel. Using hydrogen fuel results in higher efficiencies and in lower COE's for both the SPE and phosphoric acid low temperature fuel cell systems.

Higher overall efficiency is attributable to the absence of a fuel processor and to the higher conversion efficiency for producing hydrogen from coal. The lower COE's are the result of the elimination of the fuel processor cost.

TABLE 6.9: EFFECT OF FUEL CELL TEMPERATURE (TYPE) ON POWER PLANT EFFICIENCY AND COST OF ELECTRICITY GENERAL ELECTRIC RESULTS

Fuel-cell temperature, $^\circ$F	Fuel-cell type	Efficiency, percent		Cost of electricity,[a] COE, mills/kW-hr
		Powerplant	Overall	
170	SPE[b]	25.2	12.7	58
375	H_3PO_4[b]	29.8	15.0	52
1832	ZrO_2 SE[c]	31.5[d]	31.5	45

[a]Assumed 100 000-hr life for SPE system, 40 000-hr life for H_3PO_4 system, and 10-yr life with 10-percent/yr replacement for ZrO_2 SE system.

[b]HBTU/air; 48 MWe.

[c]LBTU/air; with steam bottoming cycle; 1112 MWe total.

[d]Fuel cost based upon coal rate to gasifier. Gasifier cost included in capital cost. However, powerplant not integrated with gasifier in terms of heat and mass conservation.

**TABLE 6.10: EFFECT OF FUEL TYPE ON EFFICIENCY AND COST OF
ELECTRICITY FOR LOW TEMPERATURE
FUEL CELL POWER PLANTS
GENERAL ELECTRIC RESULTS**

Fuel[a]	Fuel-cell type	Efficiency, percent		Cost of electricity,[b] COE, mills/kW-hr
		Powerplant	Overall	
HBTU	SPE	25.2	12.7	58
Hydrogen[c]	SPE	38.3	23.3	36
HBTU	H_3PO_4	29.8	15.0	52
Hydrogen[c]	H_3PO_4	37.9	23.1	37

[a]Air oxidizer.

[b]Assumed 100 000-hr life for SPE system, and 40 000-hr life
    for H_3PO_4 system.

[c]Hydrogen obtained from gasified coal.

DETAILED EVALUATION

The specific results obtained by General Electric and Westinghouse in their
studies of fuel cell systems are evaluated in detail here.

Solid Polymer Electrolyte

Hydrogen/Oxygen—300°F—201 MWe (G.E. Case 8): The detailed results for this
interesting case are listed in Table 6.11. The use of hydrogen fuel has obvious
benefits. Using hydrogen/oxygen at the elevated temperature of 300°F (up
from 170°F) results in the highest low temperature fuel cell power plant efficiency of 51.1% (31.1% overall) in the G.E. results. At 31 mills/kW-hr the
electricity produced by this 201 MWe power plant is also the lowest in cost
of all the fuel cells reported in the G.E. study. The capital cost is $242/kWe.
The potential also exists, as G.E. points out, for further cost reduction through
sale of steam, which can be produced from waste heat. The fuel cell operates
at a moderate pressure of 115 psi.

However, the preceding results are probably optimistic in three areas. First,
the performance penalty to account for the equivalent fuel required to manufacture oxygen will lower overall efficiency. (Efficiency is lowered by approximately 4.6 percentage points according to a NASA estimate based upon information obtained from oxygen manufacturers. NASA also estimated that the
effect upon COE would be to increase it from 31 mills/kW-hr to approximately
37 mills/kW-hr. In its writeup, G.E. mentions the fact that the use of the
NASA-specified $9/ton oxygen cost overlooks the impact upon efficiency).
Second, it is considered highly optimistic that an SPE polymer can be developed
to last 100,000 hr at 300°F. General Electric has extrapolated to 100,000 hr
life on the basis of an 800 hr test.

TABLE 6.11: GENERAL ELECTRIC VALUES OF ALL RELEVANT PARAMETERS FOR LOW TEMPERATURE FUEL CELL POWER PLANTS

Parameter	Case	
	8	12
Power output, MWe	201	47
Coal	Illinois #6	Illinois #6
Conversion process	Hydrogen (on site)	HBTU
Oxidizer	Oxygen	Air
Fuel-cell type	SPE	H_3PO_4
Current density, A/ft^2	300	250
Operating temperature (maximum), $^\circ F$	300	375
Electrolyte thickness, in. (cm)	0.005(0.0127)	9.020(0.05)
Actual powerplant output, MWe	201	47
Thermodynamic efficiency, percent	0	0
Powerplant efficiency, percent	51.1	29.8
Overall energy efficiency, percent	31.1	15.0
Coal consumption, lb/kW-hr	1.02	2.10
Plant capital cost, million dollars	49	27
Total direct costs, $/kWe	155	395
Plant capital cost, $/kWe	242	570
Cost of electricity (capacity factor =0.65):		
Capital, mills/kW-hr	7.7	18.0
Fuel, mills/kW-hr	19.6	28.6
Maintenance and operating, mills/kW-hr	4.1	5.5
Total, mills/kW-hr	31.3	52.1
Sensitivity:		
Capacity factor = 0.50 (total mills/kW-hr)	34.8	58.8
Capacity factor = 0.80 (total mills/kW-hr)	29.2	48.0
Capital Δ = 20 percent (Δmills/kW-hr)	1.5	3.6
Fuel Δ = 20 percent (Δmills/kW-hr)	3.2	5.7
Estimated time for construction, yr	3	2
Estimated date of commercial availability	1992	1982

If the SPE polymer life at 170°F is 100,000 hr, there is G.E. evidence which indicates that at 300°F the life must be much shorter. Below a 30,000 hr life, the O and M costs rise steeply.

Finally, the projected reduction of current SPE fuel cell costs from approximately $70 or $80/ft² to $16/ft² is believed optimistic. Achievement of this large a cost reduction while maintaining cell performance would require complete success in three major materials areas: the SPE polymer, the metallic screens, and the electrode catalyst.

Phosphoric Acid Electrolyte

High Btu Gas/Air—375°F (G.E. Case 12, 48 MWe and Westinghouse Case AC 12, 23 MWe): In terms of commercial interest, the phosphoric acid fuel cell systems operating on H Btu/air are of greatest current interest. Most similar to the United Technologies Corp. FCG-1 fuel cell power plant that is proposed for intermediate peaking service are G.E. case 12 and Westinghouse case AC 12. Data for these cases are shown in Tables 6.11 and 6.12.

As was discussed, earlier, the higher overall efficiency of the Westinghouse treatment (23.9% for Westinghouse case 12, compared with 15.0% for G.E. case 12) is attributed primarily to the higher fuel cell power plant efficiency estimate by Westinghouse coupled with their more efficient gasifier concept. The second important item is the capital cost.

The G.E. estimate of $570/kWe is much higher than Westinghouse's capitalization estimate of $371/kWe. Before the contingency, escalation, indirect, and interest during construction costs are added, the total direct cost (including site labor) are estimated at $395/kWe by G.E. and $276/kWe by Westinghouse.

The capital cost difference between the two contractors' estimates is attributable to the fuel processing cost difference. The G.E. estimate for fuel processing is $173/kWe; the Westinghouse estimate for their phosphoric acid system is only $38/kWe (both before contingency, escalation, and other costs).

The added cost of a carbon dioxide scrubber (about $14/kWe) in the G.E. concept and differences in cost accounting between G.E. and Westinghouse cannot account for this large a difference. Based upon information from manufacturers and upon engineering design data, it is believed that actual costs may be somewhere between the G.E. and Westinghouse estimates.

The major item in the COE in both treatments is the fuel cost portion, 29 mills/kW-hr for G.E. and 25 mills/kW-hr for Westinghouse. It is likely that the fuel cost portion of COE will remain high. The O and M portion of COE for Westinghouse could be reduced by increased fuel cell life.

Westinghouse case AC 12 uses a 10,000 hr life, while G.E. case 12 assumes 40,000 hr—a reasonable projection of the current state of the art. Making the 40,000 hr life assumption for Westinghouse case 12 would reduce their COE to approximately 40 mills/kW-hr.

TABLE 6.12: WESTINGHOUSE VALUES OF ALL RELEVANT PARAMETERS FOR LOW AND HIGH TEMPERATURE FUEL CELL POWER PLANTS

Parameter	Aqueous acid system (case 12)	Molten carbonate system (case 4)	Solid electrolyte system		
			Case 4	Case 18	Case 19
Power output, MWe	23.4	1255	1164	219	1064
Fuel-cell rating, MW dc	25	900	900	250	900
Fuel	HBTU	IBTU	IBTU	IBTU	LBTU
Oxidizer	Air	Air	Air	Air	Air
Fuel-cell life, hr	10 000	10 000	10 000	10 000	10 000
Voltage degradation, percent	5	5	5	5	5
Temperature, $^{\circ}$C	190	650	1000	1000	1000
Electrolyte type	85 wt % H_3PO_4	Paste of Li, Na, K, carbonates, and alkali aluminates	$(ZrO_2)_{1-x}(Y_2O_3)_x$	$(ZrO_2)_{1-x}(CaO)_x$	$(ZrO_2)_{1-x}(Y_2O_3)_x$
Electrolyte thickness, cm	0.05	0.1	0.004	0.002	0.004
Anode type	Pt/C	Ni	Ni-ZrO_2 - cermet	Ni-ZrO_2 - cermet	Ni-ZrO_2 - cermet
Anode catalyst loading, mg Pt/cm^2	0.3	------------	----------------	----------------	----------------
Cathode type	Pt/C	Lithiated NiO	$In_2O_3/PrCoO_{3-x}$	$In_2O_3/PrCoO_{3-x}$	$In_2O_3/PrCoO_{3-x}$
Cathode catalyst loading, mg Pt/cm^2	0.3	------------	----------------	----------------	----------------
Interconnection type	---	------------	Cr_2O_3	Cr_2O_3	Cr_2O_3
Interconnection thickness, cm	---	------------	0.002	0.002	0.002
Current density, mA/cm^2	200	200	400	800	400
Average cell voltage, V	0.7	0.7	0.66	0.68	0.56
Powerplant efficiency, percent	36.0	54.4	60.2	53.2	45.6
Overall energy efficiency, percent	24.2	45.7	50.6	53.2	47.7

(continued)

TABLE 6.12: (continued)

Parameter	Aqueous acid system (case 12)	Molten carbonate system (case 4)	Solid electrolyte system		
			Case 4	Case 18	Case 19
Total plant capital cost, million dollars:	8.64	569.90	539.05	205.46	893.28
Fuel processing equipment	0.645	2.30	2.41	52.36	75.37
Fuel-cell system	2.4	171.00	142.00	18.80	167.00
Steam turbine generator	0	11.72	11.52	0	0
Oxygen plant	0	0	0	0	0
Heat-recovery steam generator	0.086	20.20	15.62	0.211	11.80
Recuperator	0	4.84	18.4	4.20	28.00
Power conditioning	1.38	59.00	59.00	17.50	59.00
Capital cost, $/kWe					
Result breakdown:					
Total major-component cost, million dollars	4.51	269.06	248.95	93.87	341.17
Total major-component cost, $/kWe	193.74	227.74	215.40	429.58	327.96
Balance-of-plant cost, $/kWe	39.93	18.38	15.57	39.59	36.98
Site labor cost, $/kWe	42.70	29.47	33.35	137.54	89.08
Total direct cost, $/kWe	276.37	275.58	264.32	606.71	454.03
Indirect costs, $/kWe	21.78	15.03	17.01	70.14	45.43
Profit and owner costs, $/kWe	22.11	22.05	21.15	48.54	36.32
Contingency cost, $/kWe	12.44	22.05	21.15	36.40	38.59
Escalation cost, $/kWe	19.00	68.62	66.35	89.81	130.94
Interest during construction, $/kWe	19.46	79.04	76.43	96.71	153.38
Total capitalization, $/kWe	371.16	482.370	456.40	948.31	858.70
Cost of electricity, mills/kW-hr:					
Capital component	11.73	15.25	14.74	29.98	27.15
Fuel component	25.01	12.55	11.34	5.47	6.35
Operating-and-maintenance component	9.85	16.05	14.16	12.25	18.40
Total	46.68	43.85	40.24	47.70	51.89
Estimated time for construction, yr	1.5	5	5.0	3.0	5.5
Estimated date of commercial availability	1990	1990+	2000+	2000+	2000+

Molten Carbonate Electrolyte

1200°F—Integrated Steam Bottoming Cycle—1,255 MWe Total (Westinghouse Case 4): The overall efficiency of 45.7% (power plant efficiency, 54.4%) reduces the fuel cost (Table 6.12). At 13 mills/kW-hr the reduction is no more than half the fuel COE cost of low temperature fuel cells. It is further believed, that a complete integration of the molten carbonate (MC) system with the gasifier will yield further increases in efficiency.

The total capital cost estimate for the MC system is $482/kWe; the total direct cost is only $276/kWe. The capital cost influences the O and M portion, as well as the capital cost portion, of the 44 mills/kW-hr COE and is the primary influence in the Westinghouse calculation. NASA feels that a significant improvement in the current state of the art life of 10,000 hr (demonstrated in small scale fuel cells) is achievable. Correcting the O and M cost for a projected MC life of 30,000 to 50,000 hr reduces the COE to 34 to 32 mills/kW-hr. Because MC work is still in the small scale stage, conclusions (principally those involving cost) must be regarded as uncertain. For all fuel cell systems, the uncertainty in predicted efficiency is much less than the uncertainty in predicted costs.

Zirconia Solid Electrolyte

Intermediate Btu Gas/Air—1832°F (Westinghouse Case SE 4): Much of what was just said about the high temperature MC system applies to the higher temperature zirconia SE system (Table 6.12). For this case, the total direct costs and capitalization costs are $264/kWe and $466/kWe, respectively. The COE is 40 mills/kW-hr, and the overall efficiency is 50.6%.

For this system one could also calculate a reduction in O and M costs based on a 30,000 to 50,000 hr life, bringing the COE down to 34 to 32 mills/kW-hr. However, for this 1832°F system, no more than 1,000 hr life has been demonstrated on small scale cells. Therefore, a projection to 40,000 hr life is premature and the COE of 40 mills/kW-hr already represents a projection.

Intermediate Btu Gas/Air—1832°F—Integrated into Gasifier—219 MWe (Westinghouse Case SE 18): This treatment constitutes the Westinghouse "Project Fuel Cell" concept of zirconia SE fuel cell integration with the gasifier. By locating the fuel cell inside the gasifier, heat and mass transfer are maximized. This most intimate contact requires expensive alloy steel housings for the fuel cells to make them resistant to the corrosive gasifier environment. This results in high direct costs (including the highest site labor costs) of $607/kWe and a total capitalization cost of $948/kWe.

This system produces the highest overall efficiency of any fuel cell power plant system, 53.0%. This of course is reflected in the low fuel cost portion of COE. In comparing fuel components of COE it is important to realize that the 5 mills/kW-hr for this case (integrated with a gasifier) is based on coal. In the previous case the fuel component of COE was based upon an over the fence I Btu cost. However, it is instructive to compare the sum of the capital and fuel components of COE for each case. For case 4 this equals 26 mills/kW-hr, while for the present case, SE 18, it is 34 mills/kW-hr. The question is whether this difference is attributable to the expensive alloy steel costs in the present case or merely a difference between NASA specified I Btu costs and the contractor's cost estimate of producing I Btu gas from coal in an advanced gasifier. Further light is shed on this by looking ahead to the next case in Table 6.12.

These results are for an integrated gasifier, SE fuel cell, steam bottoming cycle (Westinghouse case SE 19). The fuel cell is not installed inside the gasifier. Therefore, expensive alloys are not required for the housing. The power output of 1,064 MWe is close to the 1,164 MWe of case SE 4. The fuel is L Btu gas, which probably accounts for a slightly lower overall efficiency, 47.8%, compared with case 4, 50.6%. The two cases are very close. Yet, the total capital and fuel COE components for case 19 equal 33 mills/kW hr compared with the 26 mills/kW-hr for case 4. This lends weight to the suspicion that the difference between cases 18 and 19 on one hand and case 4 on the other is associated with the NASA specified fuel costs versus the contractor's cost estimate of manufacturing clean gaseous fuels from coal. For case 18, the total COE is 48 mills/kW-hr. As was mentioned in the previous case SE 4 discussion, realistic projections of life (bringing COE down below 40 mills/kW-hr) beyond 10,000 hr must await further research and development work on this system.

Low Btu Gas/Air—1832°F—Bottoming Cycle—1,064 MWe (Westinghouse Case SE 19): To complete the discussion of this case begun in the previous section, its total direct and capitalization costs are $454/kWe and $859/kWe, respectively. These are lower than those of the previous case. However, due to higher O and M costs attributable to the 10,000 hr life fuel cell, the COE is 52 mills/kW-hr. The overall efficiency is 47.8%. The fuel manufacturing cost question and ultimate fuel cell life uncertainty, previously discussed, of course apply to this system as well.

Commerical Availability

As expected, the order of estimated dates of commercial availability follows the order of increasing fuel cell temperature, that is, the higher the fuel cell temperature the later the estimated availability date. This holds for both the General Electric and Westinghouse estimates. (See the individual data tables for these estimates).

SPECIAL FEATURES OF FUEL CELL POWER PLANTS

Fuel cells have a number of features, many of them unique, that have potentially important applications in the utilities industry. Many of these features can be translated into cost credits by utilities cost analysts. Only some of these features are discussed in this section since many of them are outside the scope of the ECAS study.

Fuel cells, especially low temperature systems, are very suitable for utility peaking and intermediate service. Advantage can be taken of the fuel cell's constant, or actually increased, efficiency at part load. (Characteristically, the fuel cell efficiency itself increases as the load decreases. However, where a fuel processor is part of the power plant, the net power plant efficiency increase at part load may be smaller). By using fuel cells to follow the changing load requirements, that is, bringing the fuel cell unit (or units) on at full rated load at times and at part load at other times, other utility energy convertors can be kept at their most efficient rated loads a higher percentage of the time. Substation sized fuel cell power plants dispersed throughout the utility service area can result in significant savings in transmission costs. The transmission costs are quite utility specific. However, one report estimates that the transmission savings would range between $60/kWe and $200/kWe. In addition, another report points out that the cost of underground transmission is 9 to 15 times that of overhead transmission.

Modularity is a feature that has several potential benefits. First of all, it provides flexibility. Substation fuel cells could be provided in various multiples of the basic module without sacrificing cost or efficiency. In fact a single module, probably considerably smaller than the 25 MW minimum size considered by the ECAS study, could be effectively utilized by a small municipal or rural power company.

Modularity also enables a utility to better match capacity growth requirements from year to year than with conventional power plants that have to be added in large units. To keep the $/kWe cost down, power plants have been growing in size in recent years. This ties up utility capital in unused capacity for a number of years until the growing power requirement matches the new capacity. In addition, utilities must be able to predict their growth rates perhaps 10 years in advance. Utilities reduce the impact of these difficulties by pooling power. Fuel cell power plant modularity provides them with another option.

A final important advantage of modularity is related to system reliability and availability. When one module out of a series of modules making up the fuel cell power plant is out of service, either because of scheduled maintenance or an unscheduled mishap, the other modules can continue to provide power. In fact the remaining fuel cell modules are usually capable of picking up some of the slack by providing significantly more than their rated capacities. There will be some loss in efficiency of course while the fuel cells are overloaded.

The final feature of fuel cells is their potential for total energy savings through onsite waste heat utilization. The present ECAS study illustrated, for high temperature fuel cell systems, the advantage of using waste heat in a steam bottoming cycle or in a gasifier. Also, for low temperature fuel cell systems some of the fuel cell stack waste heat can be used by the power plant fuel processor. But there is quite a lot of unused fuel cell stack waste heat. In their treatment of the SPE hydrogen/oxygen fuel cell (case 8) G.E. showed that a substantial credit could be realized by using steam generated from feul cell stack waste heat. This total energy savings concept is not limited to large power plants. Much smaller fuel cell power plants could be used to supply not only electric power, but also steam or hot water from fuel cell waste heat. This total energy concept could be suitable for industrial factories or even apartment complexes.

Finally, with respect to fuels, it will take a number of years before gasified or liquefied coal-derived fuels become available. In the mean time, those energy convertors that do not use coal directly will have to use natural gas or petroleum-derived fuels. Although both fuels are not in unlimited supply, they will be around for quite some time. In many cases, fuel cells make more efficient use of these fuels than many of the other energy convertors using or planning to use them.

CONCLUSIONS

The results of this study of fuel cell power plants indicate several general trends.

(1) Efficiency increases with fuel cell temperature, that is, proceeding from low temperature fuel cells, to the high temperature molten carbonate fuel cell, and finally to the very high temperature zirconia solid electrolyte system.

(2) Scale-up in power plant size only results in significant reduction in COE when it is accompanied by utilization of waste fuel cell heat through a steam bottoming cycle and/or integration with the gasifier.

Low Temperature Fuel Cell Systems

For the low temperature fuel cell systems, the following results were obtained:

(1) The inefficiency and cost of producing the required clean fuels are considerable.

(2) The use of hydrogen fuel results in the highest efficiency and lowest COE.

(3) Operating SPE fuel cells on H Btu/air at 170°F results in the lowest fuel cell power plant efficiency, 25.2%, and an overall efficiency of 12.7% as well as quite high COE (58 to 60 mills/kW-hour).

(4) On the other hand, operating SPE fuel cells at 300°F with hydrogen/oxygen results in considerable efficiency improvement, to 31.1% overall, and the lowest reported COE for any fuel cell system, 31 mills/kW-hr. However, accounting for the equivalent fuel requirement to manufacture oxygen would reduce overall efficiency to an estimated 26.5% and raise COE to approximately 37 mills/kW-hr. In addition, the basic COE costs probably reflect an optimistic projection of cell costs and SPE polymer life at 300°F.

(5) The phosphoric acid fuel cell system, which is closest to commercialization for intermediate peaking service, was treated by both G.E. and Westinghouse. Their results differed chiefly in gasifier and fuel cell power system efficiencies and in fuel processor costs. For a phosphoric acid, 40,000, hour fuel cell life system, the COE based upon Westinghouse estimates is approximately 40 mills/kW-hr compared with 52 mills/kW-hr estimated by G.E.

(6) Alkaline fuel cell costs are higher than those for phosphoric acid fuel cell systems, for two reasons. First, additional reactant processing is required to guard against carbonation of the alkaline electrolyte. Second, because lower power densities are obtained with the alkaline fuel cell at 158°F than with the phosphoric acid fuel cell at 375°F, alkaline fuel cell costs are higher. For the alkaline system in the 10,000 to 30,000 hour life range the COE was 50 to 61 mills/kW-hr.

High Temperature Fuel Cell Systems

For the high temperature fuel cell systems, the following results were obtained:

(1) The most promising molten carbonate case, utilizing a steam bottoming cycle, offered high overall efficiency, 45.7%. A reasonable projection of life from the present 10,000 hr to the 30,000 to 50,000 hr range results in a projected total COE for this molten carbonate system of 32 to 34 mills/kW-hr.

(2) The zirconia solid electrolyte system offers the potential for highest overall efficiency of any fuel cell system, 53.0%.

(3) For the zirconia systems, large fuel cell power plants utilizing a steam bottoming cycle and/or integrated with the gasifier produce electricity at high overall efficiency, 47.8 to 53% for the cases studied. The fuel cell steam bottoming cycle power plant offered the best combination of efficiency (50.6% overall) and COE (40 mills/kW-hr) of the zirconia systems. A projection of life beyond 10,000 hr (as was suggested in the case of the molten carbonate fuel cell) accompanied by a lower COE is not warranted for the zirconia fuel cell at this time since it already represents a substantial life projection.

For fuel cell power plant systems, confidence in predicted efficiencies is greater than confidence in predicted costs. In many cases, projections are based upon small laboratory sized units. While there is a high degree of confidence in the estimates of fuel cell system efficiency, the uncertainty in costs of coal-derived fuels makes the fuel portion of COE uncertain. However, with the upward trend in fuel costs, the emphasis that has been placed upon power plant efficiency in the ECAS study would seem to be quite reasonable.

Finally, it appears that the potential for low temperature fuel cell utilities application lies in dispersed sites to provide peaking or intermediate service, where advantage of fuel cells' special features can be taken. Fuel cells' special features are:

(1) to increase the overall efficiency of a utility's energy conversion equipment

(2) to reduce transmission costs

(3) to provide, through modularity, a means of better matching capacity with growth requirements

(4) to provide improved system reliability and availability

(5) to provide the fuel cell features in units small enough to meet the needs of very small utilities

(6) to provide total energy savings through onsite waste heat utilization

MARKETING CONSIDERATIONS

In 1976, Argonne National Laboratory conducted a short study that was published in June 1976 under the title of *Fuel Cell Benefit Analysis* (ANL/ES-51). The study was conducted by Samuel H. Nelson and John P. Ackerman under the direction of Lloyd R. Lawrence, Jr. This report indicates that fuel cell systems have the potential of saving 275,000 barrels of oil per day by 1985, and over \$1 billion per year in lower electric costs in the same time period. The report is editorialized below.

SUMMARY

Fuel cell technology has been developed to the point where efficient and practical generators can be introduced into the market place by 1980, but only if government support is available. The salient features of these generators are high efficiency, highly desirable environmental characteristics, flexibility of operation, availability in essentially any size, rapid installation, remote dispatch, and an ability to recover waste heat without any loss in efficiency.

Fuel cells are quite versatile and can be used for electric utility operation, integrated energy systems, and to provide industrial process heat and electricity. They also possess a substantial export potential. Benefits from government support of fuel cells, based upon economic competitiveness, are estimated below:

Range of Fuel Cell Installation, 1985 and 1990

1985	Low	Base	High
Mwe's installed	10,300	17,400	33,800
Associated Energy Savings (bbl oil/day)	95,000	174,000	279,000
Value of units exported (cumulative, in billions of dollars)	3.1	3.6	9.3

(continued)

200

1990	Low	Base	High
Mwe's installed	23,200	38,200	83,300
Associated Energy Savings (bbl oil/day)	228,000	380,000	562,000
Value of units exported (cumulative, in billions of dollars)	6.2	7.4	18.6

The above energy savings assume improvements in competing technologies. Compared to present technologies energy savings would in the base case be equivalent to 228,000 bbl of oil per day in 1985 and 494,000 in 1990. The discounted present value, at 10%, of dollars saved by fuel cells in the Northeast region, base case, and of exports to 1990 (assuming net exports are half of total exports) exceed $1,500,000,000 (in 1975 dollars).

HISTORICAL OVERVIEW OF FUEL CELLS AND PROGRAM

The history of fuel cells can be traced back to the experiments in 1839 by Sir William Grove, but it was not until the U.S. space program in the 1960s required a highly efficient, reliable electrical generator of very high energy density that fuel cells were put to practical use. The research community, fostered by the space program, naturally foresaw the terrestrial application of fuel cells, but the level of capital expenditure and effort required to develop fuel cell generators using carbonaceous fuels discouraged all but one or two companies from significant development programs.

Moreover, during the late sixties and early seventies, the current pressing need for energy independence and more efficient and pollution-free use of energy resources was not as clearly and generally recognized as it is now. Fuel cell development was continued, however, by a few organizations, most notably, United Technologies Corporation. Thus, the option of bringing the benefits of fuel cells to bear on energy problems remains open. Exercising this option will require federal support, and this study addresses potential benefits from such support.

FUEL CELL CHARACTERISTICS

There are a great variety of possible fuel cell types, but this study is concerned only with the first-generation, acid-electrolyte systems. It is possible to begin commercial use of generators of this type in the late 1970s with appropriate government support. Appendix 1 contains descriptions of other promising fuel cell types. Note that a portion of the benefit of first-generation fuel cell commercialization is embodied in its effect on these more advanced types.

Operating Characteristics

The acid-electrolyte fuel cells operate on gaseous or light liquid-hydrocarbon fuels with good efficiency and essentially no adverse environmental effects.

They are highly flexible in response to changing loads, and can be rapidly installed in sizes from 40 kW on up. The target specifications of the United Technologies FCG-1 system are given in Table 7.1.

TABLE 7.1: TARGET SPECIFICATIONS

Rating*[†]	26 Mw
Heat rate (end of life)*	9000 Btu/kw-hr @ 6-20 Mw 9300 Btu/kw-hr @ 26 Mw
Lifetime*	40,000 hrs
Cooling*	Dry air or water
Water required*	None
Noise*	Acceptable in residential area
Fuel*[✦]	Straight run naphtha or gas

Emissions (lb/10^6 Btu heat input)	Fuel cell	Federal standards for gas-fired central station
Particulates**[††]	2.9×10^{-6}	2.9×10^{-1}
NO_x**[††]	$1.3\text{-}1.8 \times 10^{-2}$	$2 \quad \times 10^{-1}$
SO_2**[✦✦]	2.3×10^{-5}	$8 \quad \times 10^{-1}$

Cost***	$200-300/kw
Operator requirements*	Remote dispatch
Startup time*	4-6 hr
Response to load change*	Very rapid (0.1 sec or less)
Usable reject heat*	24% of fuel energy at maximum temperature of 165°C and 33% at temperature below 100°C

---

*
 UTC specification.

[†]
Composed entirely of independent 4.8-Mw modules. It is possible to have a module size as low as 40 kw without sacrificing performance characteristics.

[✦]
Can be extended to clean coal gas, methanol, or light aliphatic liquids with minimum difficulty.

**
 York Research Corp., Y-7309, (April 1970).

[††]
Federal standards as of 8/17/71.

[✦✦]
For oil-fired central station--there is no requirement for gas-fired central station.

***
 Estimate, installed cost in 1975 dollars.

Environmental and Siting Characteristics

The environmental characteristics of fuel cells make them well suited to nearly any location. These attributes are of great value in such applications as replacement of urban plants and in environmentally sensitive areas like Southern California. They are also essential if the fuel cells are to operate so as to make use of reject heat, since competing technologies (e.g., gas turbines) are less attractive environmentally. (This application requires siting at the location where the heat is to be used.) Modules of 40-kW and 4.8-MW size are being developed for initial marketing. Inasmuch as cost and efficiency are relatively insensitive to size in this range and above, intervening sizes also can be manufactured if demand warrants.

Response to Varying Load

For certain applications, it is desirable to provide a rapidly varying output of power to match load requirements. The phosphoric acid systems being developed can match a demand that varies by more than half their rated output within 0.1 sec. The efficiency of fuel cells improves slightly as power output is reduced from 100 to about 25% of rated load. Other generating devices lose efficiency to a greater or lesser degree at part load.

For example, the simple-cycle gas turbine responds well to varying load but loses efficiency rapidly at partial output. The combined-cycle power plant retains efficiency rather well down to around 50% of rated output, but only if a period on the order of several hours is available to change from 100 to 50% output and back. One point to note is that fuel cell generators require about 4 to 6 hours for warm-up to operating temperature. The implication of this point is that fuel cells would be "idled" at about 15% of rated power output, which achieves the same efficiency as at rated power, thus providing some power at all times.

Siting and Installation

Because of the desirable environmental characteristics of fuel cells, a wide variety of sites can be used. No cooling water is required, so all that is needed is a concrete pad and output-power-line connection. The system is designed to be truck-transportable and to require minimum installation labor.

Operation and Maintenance

The 26-MW, fuel cell generator is designed for remote dispatch; no operators are required. Routine maintenance can be done by a traveling crew. Periodic replacement of zinc oxide, sulfur traps in the fuel processor, and lubrication of blowers and pumps are in this category. The fuel cell modules must be replaced periodically.

Waste Heat

If an application (such as heating and air conditioning or generation of low-pressure steam) can be found for the exhaust heat of the fuel cell system, an additional quarter of the fuel energy (above the electrical output) is available as heat from 230° to 330°F and still another third as lower temperature heat. Again, their favorable environmental characteristics facilitate the use of fuel cells for waste heat recovery or in integrated energy systems.

AREAS OF APPLICATION

Electric Utility Uses

Fuel cells are clearly very well suited to peaking and intermediate uses in electric utilities, particularly where high efficiency and freedom from adverse environmental effects are desirable. Table 7.2 summarizes costs and efficiencies projected for peaking turbines, coal, and oil-fired intermediate generators, combined cycles, and fuel cells.

TABLE 7.2: OPERATING CHARACTERISTICS AND CAPITAL COSTS FOR FUTURE GENERATING UNITS[a]

Unit Type	Size (Mw)	Lead Time (yr)	Installed Costs ($/kw)	Operating & Maintenance Cost Fixed ($/kw yr)	Variable (mills/kw-hr)	Heat Rate (Btu/kw-hr)[b]	Outage Rate (%)
Coal (Base)	600–1500	6	500	4.0	1.0	10,200	28.5
Oil (Base)	600–1500	5	330	3.6	.35	9,800	28.5
Oil (Int.)	400–600	5	300	3.2	.2	11,300	25
Combined Cycle	150–250	3	210	.9	3.7	9,000	30
Gas Turbine	50–200	2	140	.26	5.0	12,000 13,000[b]	23
Fuel Cell	26	2	(200–300)	.26	3.0	9,000	8

[a]Approximately 1987. Operating characteristics are those expected and not design conditions. These are estimates based upon the current literature.

[b]Depending on whether the utility faces a winter or summer peak.

Heat Recovery and Integrated Energy Systems Applications

Two factors make fuel cells ideally suited to these applications: their benign environmental characteristics, and their availability in appropriate sizes. As stated earlier, fuel cell exhaust is essentially free of SO_2, NO_x, and particulates, and the noise levels are acceptable. Fuel cells have been installed in apartment buildings in sizes as small as 12.5 kW, and improved 40-MW units are being tested. Small (40-kW) units are somewhat (50%) more expensive than large (MW-size) generators, but escalation of cost in small sizes is not nearly as great as with gas turbines, for example.

ASSUMPTIONS

As far as possible, this study will compare alternatives providing the same amount of energy. This analytical form is consistent with Steiner's theoretical conclusions, "If the list of services is the same, the benefits will be equal. In this case, benefit measuring is totally unnecessary and comparative cost provides necessary and sufficient conditions for choice" (1).

In evaluating fuel cell potential, a number of assumptions are made. These are: that the scope of this study does not extend beyond 1990, that the fuel cell must compete with the most likely alternative, and that initial market penetration will be in those areas most favorable to fuel cells.

BENEFIT ANALYSIS

The rate of introduction of fuel cells is highly dependent on the level of government funding. Therefore, two scenarios were evaluated: [1] with no government funding and [2] with large-scale funding to bring fuel cells to commercial realization as soon as possible. The first-case benefits are easy to evaluate as there are none; without some level of government support, it is extremely doubtful that fuel cells will become commercially viable. In case [2], the 56 26-MW units on tentative order would be delivered by 1981 and subsequent units would be available for purchase. Large-scale funding also moves forward the introduction of second-generation fuel cells to the mid to late 1980s. Benefit analysis is done only for case [2].

Utility Applications

Carrying out this analysis involves the following assumptions about the utility market:

[1] That oil is a permissible fuel, and that the natural gas is either (a) unavailable or (b) available at a premium price compared to oil.

[2] That the growth rate of capacity additions for 1976 to 1990 is about 5% per year, with capacities for specific years being as shown in Table 7.3. This is below the historic growth rate of 7%/yr for four reasons: (a) lower population growth rates, (b) rising or constant real prices as opposed to declining real prices, (c) the impact of changes in the rate structure and (d) the impact of conservation efforts by consumers (see Appendix 2 for an evaluation of the effect of a higher electric capacity growth rate on demand for fuel cells).

TABLE 7.3: NATIONAL ELECTRIC GENERATING CAPACITY

Year	Capacity Mw in Thousands
1975	509.4
1980	631.0
1985	751.1
1990	967.9

(This projection is very close to that by *Electrical World,* p 46, Sept. 15, 1975.)

Source: Reference (2)

[3] That the capacity level of peakers (internal combustion and gas turbine generating units) will be as shown in Table 7.4.

TABLE 7.4: PEAKER CAPACITY

Year	Capacity Mw in Thousands
1975	44.8
1980	52.0
1985	64.1
1990	85.2

Source: Reference (2)

The replacement and new capacity additions for the decade of the 1980s will be as shown in Table 7.5.

TABLE 7.5: PEAKER CAPACITY ADDITIONS—NEW AND REPLACEMENT

Period	Replacement Capacity (Mw in Thousands)	Added Capacity (Mw in Thousands)
1981–85	2.8	12.1
1986–90	3.7	21.1

Source: Reference (2)

The breakdown by type of replacement capacity is assumed equivalent to that existing in 1974, 11.5 % internal combustion and 88.5% gas turbines, while for added capacity, the internal combustion portion is about 3.5% (3).

[4] That environmental restrictions currently in force will not be significantly changed.

With this basic background, the utilization of fuel cells was estimated, under the assumption that the cost of fuel cells was no greater than that of the next best alternative. This involved making some assumptions, too, about how a fuel cell would actually operate in a utility system; namely, that:

[1] The fuel cell operates primarily as a peaking unit;

[2] The 15% of rated fuel cell capacity continuously on line would be operated as base load units;

[3] The capital charge rate (which includes interest, amortization, and taxes) is 17%;

[4] The required reserve ratio on fuel cells is 10%. This reflects the cell stack redundancy in the fuel cell section, which can maintain full power with several stacks off-line,

and the modularity of fuel cells;

[5] Fuel cells are sited at substations, and therefore have zero transmission line losses as opposed to about 3% losses for units at remote locations;

[6] There is no credit for the spinning reserve application of fuel cells;

[7] The useful life of all devices is 30 years;

[8] Kilowatt-hours of delivered electricity (the benefits) are equal;

[9] Fuel cells are run as peakers when gas turbines would otherwise be on line and as base load generators the rest of the time. Turbine peakers have a load factor of 11.9% (1,000 hours of operation).

With these assumptions plus local fuel prices (see Appendix 3) fuel cell use was estimated for 5 regions, the Northeast, Midwest, South, Southwest, and West. Whenever fuel cell costs were no greater than alternate systems, the assumption was made that they would achieve full penetration due to their many advantageous features. Several regions were divided into areas with expected winter and summer peaks because of the effect of ambient air temperature on gas turbine efficiency. In addition, Southern California was analyzed separately from the remainder of the West because of its unique characteristics. Each region was assumed to replace units in the proportion held in 1974 and to hold about the same share of the new capacity market as their share of total generation in 1975 (see Appendix 4).

Based upon this analysis (see Appendix 5) fuel cell capacity would be 5.4 GW in 1985 and 11.9 in 1990. The associated energy savings is equal to 11,100 bbl of oil per day in 1985 and 22,900 bbl in 1990. Considering the range of capital cost for fuel cells, installation could vary from 10.8 to 1.4 GW in 1985 and from 24.6 to 1.4 GW in 1990. The associated energy savings range from 23,000 to 2,900 bbl/day in 1985 and from 51,800 to 2,900 bbl/day in 1990. These estimates cover only the private utility intermediate peaking component of utility fuel cell demand. There are several other markets: replacement of older units, publicly owned power and cooperatives, explicit consideration of transmission and spinning reserve credits, and the load-growth-dependent market.

The replacement generating station market is a difficult one to estimate. By assuming replacement of all units greater than 40 years of age, an estimate of 28,000 MWe was made (United Technologies Corp.). Another estimate is closer to 50,000 MWe (2). The primary condition of replacement is that new units can be purchased and operated for less than the operational costs of units currently on line. With the exception of units that either have fallen apart or are in a nonviable location, the determination of replacement is economic. At what point does new capacity operating for the same amount of time, or some combination of new capacity that covers this time period, become less costly than existing units?

Assessing the costs involves knowing the operating characteristics of each old unit, the financial shape of the utility, its load curve, expected load growth, and expected plant; such detailed analysis is clearly beyond the scope of this study. However, it can be ascertained that older units fall into three categories, intermediate, peaking, and superpeaking. The superpeaking units are very old, have high heat rates, are seldom run, but can be put on line very quickly. Examples of such units are Blount St. 1 and 2 of Madison Gas and Electric Co. These units, because of their characteristics, are replaced only upon breakdown, since for such short operating time, new capacity is not economical.

The old peaking units are one determinant of new gas turbine capacity and, therefore, were included in the earlier analysis. In the intermediate load range, fuel cells will be used where they are (1) less expensive than the existing units and (2) the lowest cost intermediate unit available. These cost factors rule out the fuel cell in all regions except the Northeast and California, since elsewhere coal-fired units are economically dominant and the number of older units in California is insignificant. Information provided by United Technologies plus examination of plant data provided by the Federal Power Commission indicate that the market involved is about 5,000 MWe.

Units in this market have heat rates as low as 13,000 Btu with the highest identifiable heat rate at about 19,000 Btu. The fuel cell, if available, would economically replace units of 13,000, or better, at the lowest point of the capital cost spectrum and units of 14,900, or better, at the highest point. The estimated range of installation, therefore, is from 3,400 to 5,000 MW, with the most likely being 4,500 MW (based on an estimated load factor of 34% or 3,000 hr/yr and maintenance costs of 4.0 mills/kWh, which is typical for units identified as needing replacement although not operated as peaking units).

Associated fuel savings for 1990 ranges from 36,800 to 26,500 bbl/day, with 34,000 as the best estimate. (For units above 15,000 Btu/kWh, it would be economical to use combined cycles, and therefore fuel savings are considerably reduced. If the units replaced are the alternative, savings range from 45,900 to 35,600 bbl/day. The units are assumed to be evenly distributed across the heat rate spectrum.) For 1985 the range is 18,400 to 13,300 bbl/day, with 17,000 as the best estimate. Publicly owned power and rural cooperatives provided 10.7% of generation and 18.3% of sales in 1972. Most of this, 8.9% of generation and 13.0% of sales, is provided by publicly owned companies.

The great bulk of both sales and generation is provided by the largest publicly owned utilities. These systems are all large enough to be considered coequal with the private utilities, and one of them, Los Angeles Department of Water and Power, is included in the earlier analysis.

Some of these utilities do not generate power but purchase it from TVA for Tennessee utilities and BPA in Washington; i.e., they represent marketing of federal government power. Also these utilities have a much higher proportion of hydropower, with another five having hydro as their only power source. These systems are atypical and can be expected to expand slowly, if at all. Of the systems remaining the previous analysis indicates a range of fuel cell installation from 0 to 169 MW in 1985 and 0 to 494 MW in 1990 with the best estimate being 52 MW in 1985 and 156 MW in 1990. Associated energy savings range from 0 to 400 bbl/day in 1985 and 0 to 1,000 in 1990, with the best estimate 130 bbl/day in 1985 and 330 in 1990.

The smaller public power systems have been joining together with private companies and cooperatives to build large base load units. An example is St. George, Utah, which is purchasing 62 MW of both Warner Valley 1 and 2 (250-MW, coal-fired units for 1979 and 1980)(3). St. George currently has 7 MW of capacity run at an annual load factor of 11% (4). The cooperatives also have banded together, often at the state level, and have formed systems large enough to purchase 265 MW of a plant. In addition there are considerable amounts of purchased power, principally from the federal government. But there will still be the need for small units. Examination of data on municipals reveals two key characteristics: [1] equipment load factors are from 20 to 60% and [2] purchase power is about 50% of power distribution (5).

These systems are also quite small. The principal competition for this market is the diesel generator, which is economical in small sizes. The diesel also has a reasonably low heat rate, but as its size increases so does its cost. Considering the other alternatives, and the extremely small size, less than 20 MW for some systems, the estimated demand in this market for fuel cells is 0 to 1,000 MW by 1985 and 0 to 3,000 MW by 1990. Associated energy savings range up to 1,800 bbl/day in 1985 and up to 5,400 in 1990.

The question of spinning reserve and transmission line credits has received some consideration, but no clear-cut answers. Adequate assessment of the magnitude of such credits requires a system-by-system analysis. Stopping short of this approach involves relying on estimates. Spinning reserve credit has been estimated at about $10/kW (6). This credit is obviously highest where systems use high-cost fuel, because it is the utilization of intermediate capacity not otherwise on line while baseload units are backed off that is responsible for this cost. Spinning reserve use results therefore in fuel savings, since backing units off raises the heat rates, albeit only marginally. There are also substantial transmission savings for fuel cells instead of remote-sited baseload units, although these savings are small vs gas turbines except in highly congested areas.

This savings can range from $15 to $75 per kW, considering that transmission capital expenditures are much higher in congested corridors, while in vacant areas, such as the Southwest, land acquisition and undergrounding of lines are not yet problems.

Introducing these credits has some impact in the peaking baseload market upon units installed. The units installed by 1985 now range from 3.7 to 11.0 GW and by 1990 from 7.4 to 25.0 GW. Associated energy savings range from 7,600 to 23,400 bbl/day in 1985 and from 15,200 to 52,600 bbl/day in 1990.

Fuel cell deployment is dramatically affected by the average national load growth. If it is 6.3%, as Appendix 2 suggests is possible, then fuel cell energy savings could increase by up to 22,900 bbl/day in 1985 and 42,100 in 1990 and installations 12.0 GW in 1985 and 19.2 in 1990. There are also local effects; e.g., if a utility underestimates growth and load match it may experience an imminent need for capacity. The present means of meeting this need is the use of gas turbines. But since in such circumstances they would operate considerably longer than usual, often more than 2,000 hours, fuel cells should consequently dominate this market. Estimations are that from 300 to 1,000 MW will be installed for this purpose by 1985 and from 600 to 2,000 MW by 1990. Associated energy savings are from 800 to 2,700 bbl/day in 1985 and from 1,600 to 5,400 bbl/day in 1990.

Fuel cells fit into many markets due to their high reliability, good environmental features, modularity, and unique ratio of heat rate to load pattern. The total fuel cell installation for utility purposes ranges from 5,700 to 27,800 MW, with the base estimate at 9,100 MW for 1985. For 1990 the range is from 11,400 to 54,700 MW, with the base estimate at 18,400 MW. Associated energy savings range from 21,700 to 70,200 bbl/day in 1985 and from 43,300 to 142,600 bbl/day in 1990. The base estimates are 31,400 bbl/day in 1985 and 64,400 in 1990.

Examination of the economic potential of the fuel cell in a cost-benefit mode reveals sizeable benefits. Although fuel cell development costs are unavailable, the base cost savings ($9.50/kW-yr) for the Northeast region were calculated for the 1980 to 2015 period, assuming a 10% discount rate. Even with this high discount rate, the cumulated benefit to society in 1975 dollars is $253,000,000. (This is the highest rate used for cost-benefit studies; several other and lower rates are also in use.)

Integrated Energy Systems

The fuel cell is ideally suited to this market. It can be made economically at small sizes, can provide heat without loss of efficiency, and is highly reliable. (Based upon over 200,000 hours of actual operation, fuel cells should be available over 90% of the time.) Systems to provide hot water heating and space conditioning from waste heat recovery can use as much as 80% of the energy input to a 40-kW unit and 91% to a 26-MW unit, based upon higher heating value (7). The waste heat available is about 54% of input with about half as water at 160°F and half as steam at temperatures as high as 330°F (7).

Diesel engines and gas turbines also may be used in this way, with overall efficiency similar to that of the small fuel cell. However, there is no mention of such units of less than 100 kW being used (7)(8)(9)(10). The 40-kW units are estimated to cost about $500/kW installed, including heat recovery equipment, with costs per kW declining with increasing size. In the 40-kW to 100-kW range, the fuel cell will have as its competition advanced fossil-fueled furnaces and conventional electric service. Currently, furnaces for this size level achieve efficiencies of 65 to 80%.

However, Amana Corporation has introduced a gas furnace-air conditioner with an 84% efficiency (11). They plan to market the gas furnace, independent of the air conditioner, by 1978. The high efficiency is due to a very efficient heat exchanger and low heat losses through the stack resulting from external placement. This results in a 20% energy savings according to a study performed for Amana by the Institute of Gas Technology (IGT). Assuming this type of unit, and typical electric utility system efficiency, then in a 16-unit apartment in Hartford, Connecticut, there is about a 10% energy savings from a fuel cell with heat recovery (based upon information provided by United Technologies Corp.).

The residential-commercial market from 1978-1990 requires the equivalent of 100,000 MW of capacity. About 40% of this market is suitable for integrated energy systems. A complete assessment of this market is not possible because of the wide range of electric prices paid by consumers and because the load level is so locationally dependent.

However, in areas with high fuel and electricity costs, some market penetration of integrated systems is expected by 1985 and more substantially by 1990. By 1985 up to 1,000 MWe of capacity in the 40 to 100 kW range will be installed while 500 MWe in the 100-kW range will be in place. This capacity could increase to 3,000 MWe in each category by 1990. In the 100-kW range, the fuel cell offers no great cost or fuel savings over its alternatives (which will have a share of this market), but it does offer considerable environmental benefits. Assuming an annual load factor of 34%, the energy savings from integrated energy systems range up to 1,400 bbl/day in 1985 and 4,200 in 1990.

Process Steam

In 1972, industry required 17.6 quadrillion Btu of fossil fuel (12). About 30% of this, 5.1 quadrillion Btu, was either for process heat at no more than 330°F or for space heat (assuming constant utilization across temperature range). The combined demand is expected to reach 6.25 quadrillion Btu by 1985 (12) and 6.8 by 1990, at the same rate of growth. Because fuel cells can provide up to 54% of input power as useful heat at up to 330°F without any diminution of electric output, they face, therefore, a large potential market.

At present, most process heat is generated either in package boilers or by back-pressuring steam turbines. Package boilers operating on gas are about 75% efficient; on oil, about 70%. Back pressure units are much more capital-intensive. A large unit of this type would be equivalent to about 100 MWe (i.e., converting the extracted steam to electricity). The heat rate runs about 11,500 Btu/kWh, because the plant operates at lower temperature and pressure and with less regeneration than central electric stations (13). Smaller units do not do quite so well. There are two other, less widely used, means of providing process steam; extraction from central station steam electric plants, and from heat recovery of gas turbines.

Estimations are that extraction steam can save up to 30% of the process steam user's fuel demand (13) which, however, involves placing the process plant out at the power plant. Many users cannot afford such relocation, and even if they can, there are serious institutional problems. These involve regulation, long-term contracts, impacts on regional development, and provision of public services. [See (13) for a more complete discussion of institutional problems for matching large process steam users.] Gas turbines with waste heat recovery currently achieve 70 to 80% efficiencies (information provided by James Burroughs, Dow Chemical Co., based upon Dow's experience at an installation in Sudbury, Ont.). Given expected improvement, 80% efficiency in the 1980s seems reasonable, although the reliability of the gas turbine does not match that of the fuel cell.

One important unconventional competition to the fuel cell in this market is the templifier. The templifier employs the same principle as the heat pump, but by using a much larger compressor and working at higher temperature levels, it can produce hot water for industry. For industrial uses, a number of good heat sources are available, such as, condenser cooling water from electric power plants, cooling pond water, cooling tower water, warm water effluent from plant process, and overhead vapors from distillation processes. The templifier can heat water to 180°F from a source at 82°F and to 225°F from a source at 115°F, and still maintain a COP of 3. Its current limit is a top temperature of 230°F; however, if suitable refrigerant fluids were available this could be raised to 400°F. The space heating component of the process heat market is provided by units similar to those in residential-commercial applications. Thus maximum current efficiency over actual operating conditions is about 75%, and hence equivalent to a package boiler.

If the fuel cell were to be used in this market, it would often provide more electricity than the industrial user needs. This possibility implies a joint arrangement with electric utilities, with the fuel cell being baseloaded due to the high (about 90%) load factor for process steam use. These units will be competitive for baseload generation for two reasons: [1] capital costs are relatively low, hardly above the cost for all electric fuel cells, since process steam provision merely involves replacing an air cooling system with a water heat exchanger and [2] the utility will perceive a low effective heat rate since it can sell steam for up to what would be the cost of the lowest cost alternative. This is particularly true in high-fuel-cost regions.

The institutional problems of siting are not great to the process heat user, and seem, given past experience, to be a barrier of little consequence. The fuel cell is the lowest cost device in many cases and its nearest competitor may be either gas turbines, templifiers, or package boilers, depending on local and historical conditions. Figure 7.1 illustrates the system energy savings of fuel cells vs these alternatives when all power output of the fuel cell is equalized to the alternative system. These savings range from 44 to 130 Btu per 100 Btu of fossil fuel input to industry.

Because of these large potential savings, the fuel cell will penetrate those portions of the process heat market where the templifier is unsuitable due to lack of a good heat source or need for slightly higher temperatures; where the gas turbine is too expensive; and where package boilers are not in place or can be economically replaced. This penetration is estimated at from 7 to 13% of the available market in 1990 and from 3 to 7% by 1985, with base estimates at 10% and 5%, respectively. This leads to installation of from 4,560 to 10,550 MWe by 1985 and from 11,760 to 22,600 MWe by 1990, with respective base estimates of 7,580 and 16,800 MWe.

Assuming savings are determined by the least cost alternative, they range from 73,400 to 207,000 bbl/day in 1985, with the base estimate at 142,100. If the most energy efficient alternative is used these savings range from 37,500 to 87,500 bbl/day, with the base estimate at 62,500. Using present technology the savings would be as large as from 110,100 to 256,800 bbl/day, with the base estimate 183,300.

For 1990 expected energy savings range from 185,000 to 415,000 bbl/day, with the base estimate at 314,000. If the most energy efficient alternative is used, these savings range from 94,500 to 175,500 bbl/day with the base estimate 135,000. With present technology, these savings range from 277,000 to 507,500 bbl/day, with the base estimate 396,000.

FIGURE 7.1: FUEL CELL ENERGY SAVINGS COMPARED WITH ALTERNATIVE POWER SUPPLY SYSTEMS*

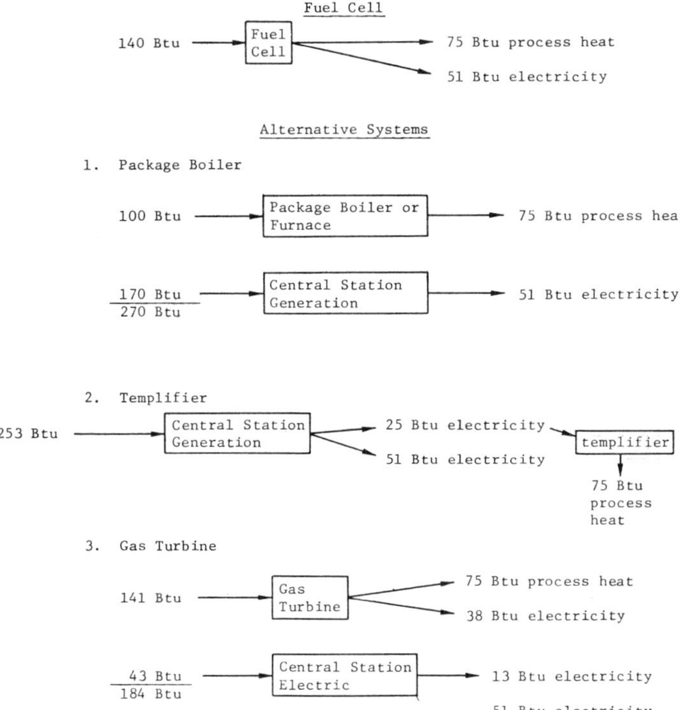

*Electricity generated at 30% efficiency. This is the 1974 average for generation and since efficiency has been falling for a decade, may be an overstatement.

Source: Reference (3)

Export Market

The foreign market for fuel cells is quite large. The fuel cell fits into four markets: [1] developed countries where the price differential between oil, coal, and uranium is small (e.g., Japan); [2] developed countries where environmental considerations are important (e.g., the Netherlands); [3] developing nations attracted by the small unit size and ability to run without operating personnel; and [4] oil- and gas-rich nations where the low capital cost, heat rate, and unit size, as well as ability to operate without trained personnel make fuel cells attractive (e.g., Iran). In total this market is estimated at about two times the USA utility market, with 25% of the export market being to Communist bloc nations. (Estimate based upon personal communications with Mr. Paul Farris, United Technologies.) The export market involves both the initial equipment sale, at $170 to $270/kW manufactured cost and a resale market half again as large, based on a 30-year operating life and $100/kW in charges for fuel cell power-section replacements.

Based upon the earlier estimates of fuel cells in the U.S. utility market, the foreign market for fuel cells ranges from 22,800 to 109,400 MW, with the best estimate 36,800 MW for the period 1890 to 1990. The total lifetime export potential of first-generation fuel cells to 1990 is, therefore, from $9.4 to $29.7 billion, and base cost estimate is some $11.1 billion. Because there is no competition, the foreign market and the USA export market may be considered equal. That is, an export potential exists of from $9.4 to $29.7 billion. Some portion of the export market reflects capture of market shares of USA manufactured prime movers. There is also a secondary effect, through currency revaluation, that reduces other exports and increases imports.

But the remainder reflects the net gain in exports. This net gain is reflected, in turn, in an increase in jobs for U.S. citizens and in taxes paid to governmental agencies. To illustrate the latter effect, take the base case and assume a 1990 cutoff date. In this case, total export value is about $7.4 billion. Assume that only half is net export, that each $30,000 yields a manufacturing job paying $12,000, that the tax rate in 1975 dollars stays constant, and that there are no associate jobs generated. (Actually, there will be some associated service jobs; how many, is locationally dependent. Clearly though, none is too few.) These assumptions, particularly the latter, are quite conservative. Yet the result is an average of 25,000 added jobs per year and increased federal income tax and social security revenues of $485 million (in 1975 dollars) for the decade of the 1980s.

From a benefit/cost standpoint the entire net export represents benefits. Limiting the analysis only to the decade of the 1980s, again assuming only half of exports are net, and using a 10% discount rate, current benefits in the base case are in excess of $1.25 billion.

CONCLUSIONS

Fuel cell technology has matured to the point where a commercially competitive system can be produced, provided there is government support. With such support domestic fuel cell installation ranges from 10,300 to 39,800 MW in 1985 and 38,200 in 1990. When compared with the best alternative, that is, the most economical advanced system available in the decade of the 1980s, associated energy savings range from 95,000 to 279,000 bbl of oil per day in 1985 and from 228,300 to 561,500 in 1990 with base estimates at 174,000 bbl/day in 1985 and 380,500 in 1990. Compared to present technology, savings range as high as 400,000 bbl/day in 1985 and 722,000 in 1990.

There is also a large potential export market in which the USA will hold a monopoly. Estimated total exports range from $9.4 to $29.7 billion. Considering the base case, and net exports being half of total exports, then by 1990 federal government revenues will increase by greater than $450 million (in 1975 dollars). Furthermore, discounting net exports and base estimate cost savings for the Northeast at 10%, the present value of fuel cells to the U.S. economy on a benefit basis exceeds $1.5 billion (in 1975 dollars).

The conclusion is that the fuel cell is a nearly mature technology offering substantial environmental benefits, sizeable energy savings, and facing a large export market. Furthermore, it appears that success of the more efficient second-generation units is dependent upon successful introduction of the first-generation units. Based upon these findings government support of fuel cell development is appropriate.

APPENDIXES

Appendix 1—Characteristics of Advanced Fuel Cell Types

Base Electrolyte Systems (Low Temperatures): These systems provide very high efficiency and great flexibility on specialized fuel (hydrogen); hence they are used for spacegoing applications. It may be possible to develop a very attractive and efficient generating system based on carbonaceous fuels. Exxon and its French partner, Alsthom, are pursuing this option, but the problems are formidable, and this will doubtless be a second-generation system available perhaps in the mid to late 1980s.

Molten Carbonate Systems, High Temperature (500-700°C): These systems offer very high efficiency, near 50%, and also high-quality, useful reject heat that could be used for space conditioning in buildings or for process steam. They are well suited to the use of carbonaceous fuels, and provide all the benefits of flexibility, desirable environmental characteristics, etc., that are associated with other fuel cell systems. High operating temperature cells have historically caused more development difficulties than lower temperature cells, but considerable progress has been made in the last few years and commercial cells of this type may reasonably be expected to become available in the 1985-1988 time period.

High Temperature Oxide Electrolyte Systems: These systems may yield highest efficiency of operation on a variety of fuels, perhaps in direct conjunction with coal gasification. Their very high quality reject heat would certainly be recovered, leading to overall efficiencies much greater than 50%. However, major technical problems remain to be solved, so it is exceedingly difficult to project a commercialization date with any reliability.

Appendix 2—Effect of Electric Capacity on Fuel Cell Deployment

The estimation of demand for fuel cells is very sensitive to the growth rate of electric generating capacity. If the capacity additions are 6.3%/yr, which is in line with forecasts made in early 1975, then capacity in 1985 is 935×10^3 MW and in 1990, $1,272 \times 10^3$ MW. These are about 25 and 30% greater than the estimates used in this study. However, the difference in new capacity additions during the period is startling.

If the growth rate is 6.3%, then between 1981 and 1985 2.0 times the projection used will be required and 1.75 times as many megawatts for the full period. The reason behind this is that 1980 capacity is already committed, based upon growth rates of close to 7% per annum. Thus, there will be considerable excess capacity with the low projected growth rate. During the 1981-1985 period, the excess is worked off and, consequently, additions become somewhat below what they would be otherwise. Hence, there is a great difference in new capacity additions resulting from changes in the growth rate.

The possibility of the higher growth rate is not remote, for there is one factor that could compensate for others that lead to lower growth; the unavailability of natural gas. In many cases electricity is the best substitute for natural gas and, therefore, the demand for electricity is sensitive to the extent of future natural gas shortfalls, and/or the price levels that pertain when synthetic gas becomes available.

Appendix 3—Regional Fuel Prices 1985

The following prices are in 1975 dollars/10^6 Btu. They assume that unit train costs are $\frac{2}{3}$ that for a nonunit train.

Northeast

Fuel Cell Fuel	2.95
Distillate Fuel Oil, Low Sulfur	2.75
Coal	1.65

Midwest

Fuel Cell Fuel	2.80
Distillate Fuel Oil, Low Sulfur	2.52
Coal	1.22

Southwest

Fuel Cell Fuel	2.74
Distillate Fuel Oil, Low Sulfur	2.56
Coal	1.15

West
(Including Southern California)

Fuel Cell Fuel	2.80
Distillate Fuel Oil, Low Sulfur	2.52
Coal	1.35 (Florida 1.65)

Source: Reference (14)

Appendix 4—Regional Capacity Shares

The regions used are defined as follows: The Northeast: New England, New Jersey, Delaware, Maryland, and those portions of New York and Pennsylvania that face similar severe environmental restrictions; the Midwest: the East North Central and West North Central regions, Kentucky, and those portions of New York and Pennsylvania not severely impacted environmentally; the South: those states east of the Mississippi not included above; the Southwest: Texas, Arkansas,,Oklahoma, and Louisiana; and the West: the remaining states except Southern California, which is a region of its own.

With these definitions, the share of the U.S. private electric generating capacity in 1972 is given below. Total exceeds 100% due to inclusion of publicly owned utilities in Southern California.

```
                    Regional Generation Shares (in %)
        Northeast                              13.1
        Midwest                                34.2
        South                                  25.4
        Southwest                              15.6
        West                                    8.6
        Southern California                     4.3
```

Source: References (15) (16)

The regional shares for the replacement market (including internal combustion units as a separate entry) are:

```
                Replacement Market Regional Shares (in %)
        Northeast                              25.4
        Midwest                                28.8
        South                                  22.5
        Southwest                               3.5
        West                                    6.8
        Southern California                     1.8
        Internal Combustion                    11.3
```

Source: Reference (3)

Appendix 5—Annual Generating Cost per kW Based on 1987 Installation

Southern California: Southern California, or more properly the Los Angeles basin, faces a unique electric generating condition. Because of the severe environmental problems, electric generating stations are NO_x dispatched. Thus, if a unit with low NO_x is placed on line, it must be brought up to full load as fast as practicable. Fuel cells in this market would be competing primarily in the intermediate market, since they would have to be either all on or all off. The competition would be combined-cycle units.

These units must be remotely located at a 3% energy penalty and about a $15 transmission line capital cost (17). Estimated load factor is about 45%, with annual hours of operation at 4,000 hours. Because they are run from a cold start straight up to full load, fuel cell heat rates would probably be about 9,300 Btu/kWh, the end-of-life design conditions.

The situation in Southern California is quite unstable. There is no coal-fired electric station in the state. California has been able to build coal plants in neighboring states, but this policy is being viewed with increasing disfavor by the host states. If nuclear power plants are not developed in Southern California, the only options are geothermal, solar, and oil-powered generation. At least, initially, the bulk of the generation would be oil based, and the fuel cell because of its sitability would probably dominate this market. In the following cost comparison, the superior reliability of fuel cells is accounted for. (See Table 7.2. This is done by requiring some gas turbines to operate to equalize electricity production.)

Annual Cost per kW, Assuming 1987 Installation*

Southern California

Fuel Cells

Capital Cost	= $230/kw x 0.17 capital charge rate	=	$ 39.10
Fuel Cost	= 9300 Btu/kw-hr x 4000 kw-hr x 3.08/10^6 Btu	=	114.58
Operating and Maintenance Cost	= 3.0 mills/kw-hr x 4000 kw-hr + $0.26	=	12.26
		Total Annual Cost	$165.94

Alternative Power Generation

Combined Cycle Plant

Capital Cost	= $210/kw + $15/kw transmission x 1.03 x 0.17	=	$ 39.40
Gas Turbine Peaking	= $140/kw x 1.03 x .17 x 0.1 (Units required to equalize system reliability)	=	2.46

Fuel Cost

Combined Cycle	= 9000 Btu/kw-hr x 1.03 x 3160 kw-hr x $2.74/$10^6$ Btu	=	80.21
Gas Turbine	= 13000 Btu/kw-hr x 1.03 x 840 kw-hr x $2.74/$10^6$ Btu (operation required due to differences in reliability)	=	38.82

Operating and Maintenance Cost

Combined Cycle	= $0.90 + 3160 kw-hr x 3.7 mills/kw-hr	=	$ 12.59
Gas Turbine	= $0.26 + 840 kw-hr x 5.0 mills/kw-hr	=	4.46
		Total Alternative Cost =	$177.94
Fuel Cell Saving	= $ 12.00		
Break-even Capital Cost	= $299.00/kw		

---

*The fuel cost used is levelized for the 30-year period. Base case fuel cell cost is $230/kw.

Northeast:

Fuel Cells

Capital Cost	= $230/kw x 0.17 capital charge rate	= $ 39.10
Fuel Cost	= 9000 Btu/kw-hr x 2061 kw-hr x $3.30/$10^6$ Btu	= 61.20
Operating and Maintenance Cost	= 3.0 mills/kw-hr x 2061 kw-hr + $0.26	= 6.44

                                         Total Annual Cost = $106.74

Alternative Power Generation

Because of environmental constraints the baseload alternative
is an oil-fired unit.

Capital Cost

Oil Baseload Unit	= $330 x 1.03 x 0.15 kw x 0.17	= $ 8.67
Gas Turbine Peaking	= $140 x 1.03 x 0.95 kw x 0.17	= 23.27

Fuel Cost

Oil Baseload Unit	= 9800 Btu/kw-hr x 1.03 x 1060 kw-hr x $3.08/$10^6$ Btu	= 32.96
Gas Turbine	= 12000 Btu/kw-hr x 1.03 x 1001 kw-hr x $3.08/$10^6$ Btu	= 41.00

Operating and Maintenance Cost

Oil Baseload	= $3.60 + .35 mills/kw-hr x 1060	= 3.97
Gas Turbine	= $0.26 + 5.0 mills/kw-hr x 1001	= 5.27

                                         Total Annual Cost = $115.14

Fuel Cell Saving	= $ 8.40	
Break-even Capital Cost	= $286.00/kw	

Midwest:

Fuel Cell

Capital Cost (see Northeast)		= $ 39.10
Fuel Cost	= 9000 Btu/kw-hr x 2061 kw-hr x 3.14/10^6 Btu	= 58.24
Operating and Maintenance Cost (see Northeast)		= 6.44

                                         Total Annual Cost = $103.78

(continued)

Alternative Power Generation

Capital Cost

Coal Baseload = $500/kw x 1.03 x .15 kw x 0.17	=	$ 13.13
Gas Turbine (see Northeast)	=	23.27

Fuel Cost

Coal Baseload = 10200 Btu/kw-hr x 1.03 x 1060 kw-hr
x $1.35/$10^6$ Btu   =   15.03

Gas Turbine = 12000 Btu/kw-hr x 1.03 x 1060 kw-hr
x $2.82/$10^6$ Btu   =   34.98

Operating and Maintenance Cost

Coal Baseload = $4.0 + 1.0 mills/kw-hr x 1060/kw-hr   =   5.06

Gas Turbine   (see Northeast)   =   5.27

Total Annual Cost   = $ 96.74

Fuel Cell
Saving   = $ 7.04

Break-even
Capital Cost   = $189.00/kw

Southwest:

Fuel Cell

Capital Cost (see Northeast)   = $ 39.10

Fuel Cost = 9000 Btu/kw-hr x 2061 kw-hr x $3.07/$10^6$ Btu   =   56.95

Operating and
Maintenance Cost (see Northeast)   =   6.44

Total Annual Cost   = $102.49

Alternative Power Generation

Capital Cost

Coal Baseload (see Midwest)   = $ 13.13

Gas Turbine   (see Northeast)   =   23.27

Fuel Cost

Coal Baseload   = 10,200 Btu/kw-hr x 1.03
x 1060 x $1.27/$10^6$ Btu   =   14.14

Summer Peak Gas Turbine*   = 13,000 Btu/kw-hr x 1.03
x 1001 x $2.87/$10^6$ Btu   =   38.46

Winter Peak Gas Turbine   = 12,000 Btu/kw-hr x 1.03
x 1001 x $2.87/$10^6$ Btu   =   35.50

(continued)

<u>Operating and Maintenance Cost</u>

Coal Baseload (see Midwest)   =   5.06

Gas Turbine   (see Northeast)   =   5.27

                                Winter Peak Total Annual Cost   = $ 96.37
                                Summer Peak Total Annual Cost   = $ 99.33

Winter Peak Fuel Cell Saving   = $ -6.12
Summer Peak Fuel Cell Saving   = $ -3.16
Winter Peak Break-even Capital Cost   = $194.00/kw
Summer Peak Break-even Capital Cost   = $212.00/kw

---

[*]Texas, Louisiana, and Oklahoma, based upon *Electricity Market Fact Sheets by States, 1970* (refers to time-of-system peak), J. G. Asbury and R. F. Talkie, Argonne National Laboratory (Jan 1976).

West:

<u>Fuel Cell</u>

Capital Cost (see Northeast)   = $ 39.10

Fuel Cost = 9000 Btu/kw-hr x 2061 kw-hr x $3.08/$10^6$ Btu   =   57.15

Operating and
Maintenance Cost (see Northeast)   =   6.44

                                Total Annual Cost   = $102.69

<u>Alternative Power Generation</u>

<u>Capital Cost</u>

Coal Baseload (see Midwest)   = $ 13.13
Gas Turbine   (see Northeast)   =   23.27

<u>Fuel Cost</u>

Coal Baseload   =   10,200 Btu/kw-hr x 1.03 x 1060
                             x $1.17/$10^6$ Btu   =   13.03

Winter Peak Gas Turbine =   12,000 Btu/kw-hr x 1.03 x 1001
                             x $2.64/$10^6$ Btu   =   32.66

Summer Peak Gas Turbine* =   13,000 Btu/kw-hr x 1.03 x 1000
                             x $2.69/$10^6$ Btu   =   35.88

<u>Operating and Maintenance Cost</u>

Coal Baseload (see Midwest)   =   5.06

Gas Turbine   (see Northeast)   =   5.27

                                Winter Peak Total Annual Cost   = $ 92.42
                                Summer Peak Total Annual Cost   = $ 95.64

(continued)

```
Winter Peak Fuel Cell Saving            = -$ 10.27
Summer Peak Fuel Cell Saving            = -$  7.05
Winter Peak Break-even Capital Cost     = $176/kw
Summer Peak Break-even Capital Cost     = $192/kw
```

---

*Arizona, based upon Asbury and Talkie, *op. cit.*

South:

Fuel Cell

Capital Cost (see Northeast)	=	$ 39.10
Fuel Cost = 9000 Btu/kw-hr x $314/10^6 Btu	=	58.27
Operating and Maintenance Cost (see Northeast)	=	6.44
Total Annual Cost	=	$103.81

Alternative Power Generation

Capital Cost

Coal Baseload (see Midwest)	=	$ 13.13
Gas Turbine (see Northeast)	=	23.27

Fuel Cost

Coal Baseload   = 10,200 Btu/kw-hr x 1.03 x 1060 x $1.51/10^6 Btu   = 16.81

Gas Turbine

Mid South Winter Peak   = 12,000 Btu/kw-hr x 1.03 x 1001 x $2.82/10^6 Btu   = 34.98

Deep South Summer Peak   = 13,000 Btu/kw-hr x 1.03 x 1001 x $2.82/10^6 Btu   = 37.88

Operating and Maintenance Cost

Coal Baseload (see Midwest)	=	5.06
Gas Turbine (see Northeast)	=	5.27
Total Annual Cost (Mid South)	=	$ 98.52
Total Annual Cost (Deep South)	=	$101.42

Fuel Cell Saving (Mid South)   = -$ 5.29

    (Deep South)   = -$ 2.39

Break-even Capital Cost (Mid South)   = $200/kw

    (Deep South)   = $217/kw

Due to higher coal prices in Florida ($1.85/10^6 Btu levelized), the fuel cell has about a $3.00 edge in that market, which is 4.5% of the national market, for the base case cost of $230/kw.

REFERENCES

(1) P.O. Steiner, "The Role of Alternative Cost in Project Design and Selection," *Quarterly J. of Econ., 79,* p 429 (1957).

(2) Temple, Barker, and Sloane, Inc., *The Economic Impact of EPA's Air and Water Regulations on the Electric Utility Industry,* preliminary draft (November 1975).

(3) Electrical World, *1975 Annual Statistical Report, 183* (March 15, 1975).

(4) Public Power, *1975 Directory Issue,* (January/February 1975).

(5) United Technologies based on raw data from the Federal Power Commission.

(6) R.A. Fernandes, *Optimum Peak Shaving for Electric Utilities,* paper presented at IEEE Power Engineering Society, winter meeting (January 1975).

(7) J.M. King, A.P. Grasso, J.V. Clausi, *Final Report, Study of Fuel Cell Power Plant with Heat Recovery,* NASA 14220, p 11 (April 24, 1975).

(8) G. Samuels and J.T. Meador, *Mius Technology Evaluation; Prime Movers,* ORNL (April 1974).

(9) G.M. Wolfer, *The Potential Benefit of an Advanced Integrated Utility System,* NASA (TR-X-5812) (November 1975).

(10) Division of Energy Bldg. Technology and Stds., HUD, *Final Environmental Statement, Application of Modular Integrated Utility Systems Technology* (October 1970).

(11) T. McCrory, Amana Corp., personal communication.

(12) Westinghouse Electric Corporation, *The Westinghouse Templifier...A New System for Producing Process Heat,* p 8 (January 1976).

(13) S.H. Nelson, *Utilization of Low Temperature Heat from Steam-Electric Power Plants: Techniques, Economics, and Institutional Issues,* unpublished doctoral dissertation, U. of Wisconsin (1975).

(14) *Assessment of Fuels for Power Generation by Electric Utility Fuel Cells,* Arthur D. Little Inc. for Electric Power Research Institute (October 1975).

(15) *Steam Electric Plant Factors 1973,* National Coal Assn. (1974).

(16) *Electric Power Statistics,* FPC (January 1972).

(17) Dr. I. Thierer, Southern California Edison, personal communications.

PROPRIETARY PROCESSES

This chapter lists, by company, U.S. Patents since 1970 that deal with fuel cells, fuel cell materials and related materials. Companies are arranged alphabetically and patents are arranged chronologically within each company. Each entry includes patent number, date, inventors' names, title and the patent abstract. Following the listing by company is a group of patents which have no company assignment and are arranged alphabetically by inventor.

Agence Nationale de Valorisation de la Recherche (Anvar), Puteaux, France

3,709,735; January 9, 1973; M.G. Bonnemay, G.R. Bronoel and D. Doniat "Electrochemical Generator with Disperse Carbon Electrode"

> The electrochemical generator has at least one electrode whose active material is constituted by active carbon powder suspension in a liquid electrolyte. The suspension is circulated in contact with conducting grid elements. Under the effect of a charging voltage applied to the grid elements the electrolyte provides, by electrolytic decomposition, a fuel gas or an oxidizing gas according to the sign of the rechargeable electrode. The compact electrode may be covered at least in part with a coating of porous polyvinyl chloride, whose pores are permeable to the electrolyte but not to the powdered material. The electrode plates may be of cadmium and the electrolyte of aqueous potassium hydroxide. A branch circuit may be provided for the circulating electrolyte and a reservoir of the powder material and filtration device for the electrolyte so that the disperse electrode may be replenished with fresh material.

Allis-Chalmers Manufacturing Co., Milwaukee, Wis.

3,575,719; April 20, 1971; C.R. Nelson and H.P. Bruss "Compact Cell Construction"

> A separable sealing construction for fuel cells and other electrochemical and membrane process cells which permits the nondestructive separation and restacking of the several unitary cells in a plurality of cells, or the disassembly and reassembly of elemental members comprising a single cell. The sealing construction comprises gasket means lying in the same geometric plane or planes as the fluid distribution matrix or matrices of the cell or cell assembly. A further feature is the provision of a symmetrical inlet and an outlet port arrangement in the cell frame which permits reverse edge orientation and reverse surface orientation of the cell frame, while still maintaining a proper fluid flow relation of the inlet and outlet conduits to the inlet and outlet ports.

3,595,699; July 27, 1971; J. Baude
"Fuel Cell Temperature Control"

The invention comprises a fuel cell temperature control which anticipates heating by internal losses generated as a result of fuel cell load current variation and initiates cooling in advance of the actual time the temperature rise is sensed by the temperature detecting means. A differential amplifier receives a first input signal which is a function of the rate of change of fuel cell current and time and a second input signal which is a function of fuel cell temperature, and means responsive to the output from the differential amplifier regulate the flow of coolant through a heat exchanger in heat transfer relation with the fuel cell.

3,615,844; October 26, 1971; H.H. Spengler
"Method of Operating a Battery Having Consumable Anode Material"

A method is disclosed for enhancing the energy capacity and extending the period of useful life of a battery or cell of a type, such as a zinc-air cell, wherein an anode comprises a stable substrate coated or plated with a deposit material which during discharge is at least partially soluble in an electrolyte contained in the cell. The method includes the step of completely discharging the cell to strip the deposit (e.g., the zinc, in a zinc-air cell) from its support structure after about five charge-discharge cycles, to provide clean support structure for accepting the deposit during charging with minimized formation of elongated dendritic forms of the deposit which if permitted to grow and regrow on unclean surfaces of the substrate, tend to span space between electrodes and short circuit such a cell.

3,682,709; August 8, 1972; D. Pouli and T.L. Larson
"Process for Improving Nickel Base, Nonnoble Metal Fuel Cell Catalysts"

A heat treating process for improving the performance and life characteristics of nickel base, nonnoble metal fuel cell anode catalysts wherein the catalyst after synthesis is stored under water. The wet material is then heated in an inlet atmosphere containing at least 5 mol % water vapor for a period of about one hour, and at a temperature of from $250°$ to $700°C$.

3,758,339; September 11, 1973; J.P. Manion
"Method of Operating Fuel Cell with Hydrogen Peroxide Oxidant"

A method of operating a fuel cell with a liquid oxidant, hydrogen peroxide, wherein the hydrogen peroxide is sprayed against the surface of the fuel cell oxidant electrode. Excess hydrogen peroxide is permitted to run off the oxidant electrode surface to be collected and reused.

Allmanna Svenska Electriska AB, Vasteras, Sweden

3,575,716; April 20, 1971; O. Lindstrom
"Method of Operating a Fuel Cell Battery Having Sloping Inlet and Outlet Channels"

In a fuel cell battery which comprises electrodes and spaces therebetween and also outer spaces outside the electrodes for a combustible substance or oxidant respectively, there are two channels, one inlet and one outlet channel. At least the outlet channel is arranged at an angle to the horizontal plane and with its uppermost part connected to a device for feeding back electrolyte to the inlet channel.

3,578,578; May 11, 1971; O. Von Krusenstierna
"Measuring Means for Measuring the Oxygen Content in Liquid and Gaseous Media"

Measuring means for measuring the oxygen content in a liquid or gaseous medium, comprising a solid oxygen ion conducting electrolyte arranged in contact with a reference system having known oxygen potential and consisting of a metal and its oxide which is arranged in contact with a current collector. The current collector is designed as a carrying element for the reference system and the electrolyte which are both arranged as layers on the current collector with the electrolyte outside the reference system.

3,634,140; January 11, 1972; O. Von Krusenstierna
"Fuel Cell Device Utilizing Air as Oxidant"

A fuel cell arrangement includes an electrolyte chamber with a fuel electrode arranged in contact with electrolyte and the fuel of the cell and having an air electrode arranged in contact with the electrolyte and with the atmosphere. The air electrode is formed of two porous layers having different porosity, the layer located nearest the electrolyte side having the finer pores. The electrolyte chamber is connected to a large storage vessel by electrolyte circulating means. The top of the storage vessel and the top of the electrolyte chamber are connected by a valved conduit to an evacuated container, which is also connected by a valved conduit to a suction pump capable of being operated by a battery charged by the output of the fuel cell.

American Cyanamid Co., Stamford, Conn.

3,704,171; November 28, 1972; H.P. Landi
"Catalytic Membrane Air Electrodes for Fuel Cells and Fuel Cells Containing Same"

Diffusion electrodes for fuel cells, and particularly suitable for metal-air fuel cells comprise two layers of porous sheet material laminated with a conductor screen. Both porous layers comprise polytetrafluoroethylene (PTFE) binder with partial filler of a low-melting resin incorporated to effect lamination at temperature below the sintering temperature of PTFE. One layer is a conductive porous electrode layer with fillers of carbon and silver. The other is a porous backing layer with filler of polyolefin fibers. The electrodes are especially suited as air electrodes for cells requiring low cost materials and long service life.

American Gas Association, Inc., Arlington, Va.

3,592,941; July 13, 1971; E.B. Shultz, Jr. and L.G. Marianowski
"Method for Producing Electron Flow in Carbonate Electrolyte Fuel Cell"

A fuel cell which utilizes hydrogen as fuel to produce electrical energy having an anode only permeable by hydrogen, a cathode permeable by an oxygen containing oxidant, and an alkaline carbonate electrolyte. The internal electric circuit utilizes only the carbonate ions from the alkaline carbonate for transporting electrons, internal of the cell, from the cathode to the anode where they are released as electrical energy. The method of producing electricity using the apparatus described comprises (1) heating the fuel cell apparatus to above the melting point of the alkaline carbonate, (2) allowing hydrogen fuel to permeate the anode and react with carbonate ions in the electrolyte thereby releasing electrons as electrical energy, and (3) providing an oxygen containing oxidant at the cathode in sufficient excess to provide oxygen atoms which combine with the carbon dioxide created at the anode and thereby produce carbonate ions at the cathode. The resultant reactions provide a carbon dioxide-carbonate ion balance internal of the fuel cell.

4,009,321; February 22, 1977; B.S. Baker and L.G. Marianowski
"High Temperature Fuel Cell Electrolyte"

The specification discloses an improved paste electrolyte composition for use in a high-temperature molten carbonate fuel cell, using substantially pure alkali metal aluminate formed from reactive alumina as the only inert material in the electrolyte. Finely divided reactive alumina substantially completely free of silica is admixed with alkali metal carbonates and fired to re-

move carbon dioxide to form as final inert carrier material substantially pure
inert alkali metal aluminate. The initial composition of the starting mixture
is selected so that the final composition is between 40 and 70% by weight
alkali metal carbonates. In contrast to magnesia supported fuel cells which
show a 40% drop-off in power output, the aluminate paste electrolyte-util-
izing fuel cells of the invention show essentially no deterioration.

American Hydrocarbon Co., Salt Lake City, Utah

3,679,486; July 25, 1972; W.A. Proell
"Fuel Cell and Method of Operating Said Cell"

A fuel cell with its anode comprising a carbon electrode and fossil fuel, its
cathode a porous carbon electrode through which molecular oxygen is introduced into the fuel cell, and employing nitric acid as an electrolyte for the
simultaneous production of electricity and humates. There is also disclosed
an electrochemical process for the conversion of the chemical energy into
electrical energy while simultaneously preparing a soil enriching agent from
a fossil fuel.

Amicon Corp., Lexington, Mass.

3,651,030; March 21, 1972; C.W. Desaulniers, C.A. Ford and R.W. Mayo
"Process for Making a Membrane"

A process for forming highly wettable and rewettable polysulfone membrane
useful in fuel cells as moisture vapor transmissive, yet hydrogen retentive,
barriers. This process comprises the steps of subjecting a microporous membrane to treatment in an acidic bath to enable the pores to imbibe water
more readily and more completely. In the preferred and most advantageous
aspects of the invention, an anisotropic polysulfone membrane having a microporous barrier layer about 0.1 to 5 microns thick which layer comprises about
1% by volume of pores having an average diameter of from 20 to 100 A is
treated with the acid to form a novel, high performance membrane.

Atlantic Richfield Co., Philadelphia, Pa.

3,544,374; December 1, 1970; A.F. D'Alessandro and H. Shalit
"Fuel Cell with D.C. Potential Means and Method of Operating Same"

This invention deals with fuel cells and in particular with a system for reducing the deterioration of anodes containing metallic hydrogen transfer membranes during periods of start-up and shut-down. During such operation hydrogen is removed from the vicinity of said membrane and a direct current
potential is supplied to the membrane in a sufficient amount to maintain the
membrane negative with respect to the cathode and chemically inactive with
respect to the electrolyte.

3,544,376; December 1, 1970; J.E. Connor, Jr., A.F. D'Alessandro and
H. Shalit
"Method and Apparatus for Monitoring Fuel Cell Feed"

This invention relates to monitoring the methane concentration in a fuel cell
effluent and adjusting the rate of feed to the fuel cell responsive to said monitoring.

3,577,329; May 4, 1971; H. Shalit
"Process for the Production of High Purity Hydrogen"

An improved process and apparatus for the production of hydrogen of high
purity whereby a carbonaceous fuel is reformed to produce hydrogen and
said hydrogen is recovered economically from said reforming operation
through the use of an electrolytic process using chemical energy of the reforming process to reduce electrical energy needed and wherein a moist gas

is added to the electrolyte to facilitate the electrolytic recovery of hydrogen at the cathode.

SA Automobiles Citroen, Paris, France

3,861,959; January 21, 1975; J. Cadiou
"Batteries Composed of Fuel Cells"

> A battery comprises a plurality of fuel cells which convert into electrical energy, chemical energy provided by the reaction of a gaseous fuel with a combustible gas. The cells are arranged in the form of a stack on a rod which extends through a central aperture in each cell. The axis of the rod constitutes the axis of symmetry of the battery.

SA Automobiles Citroen, Paris, France (with Battelle Memorial Institute)

3,887,400; June 3, 1975; D. Doniat, K. Beccu and A. Porta
"Method and Apparatus for Electrochemically Producing an Electric Current"

> Electric power is produced by electrochemical oxidation of zinc in suspension in an alkaline electrolyte. In order to prevent passivation of the zinc particles, the zinc-electrolyte suspension is continuously recycled during discharge so as to provide repeated passage of each zinc particle in an electric power producing manner past a negative current collector, with continual intermediate regeneration of each zinc particle by removal thereof from the collector between successive passages, for diffusion of the discharge byproducts into the electrolyte before the particle is returned to the collector for the next passage and for renewed participation in said discharge.

Bailey Meter Co., Alliance, Ohio

3,598,711; August 10, 1971; L.R. Flais
"Electrochemical Oxygen Analyzer"

> An electrochemical cell having a zirconium-oxide tubular electrolyte with a porous platinum electrode bonded to the inner surface of the electrolyte tube and a similar platinum electrode bonded to the outer surface of the electrolyte tube. A platinum lead extending from each electrode transmits the cell EMF generated when the inner and outer surfaces of the electrolyte are in contact with gases having unequal concentrations of oxygen.

Batelle Institut, Frankfurt, Germany

3,708,342; January 2, 1973; H. Binder, W.H. Kuhn, W. Lindner and G. Sandstede
"Hydrogen Electrodes for Fuel Cells"

> Porous electrodes for fuel cells which comprise tungsten carbide, electrically conductive activated carbon and thermoplastic polymer. A process for preparing such electrodes is also disclosed. The electrodes are particularly suitable as porous anodes in low-temperature fuel cells containing acidic electrolytes and utilizing hydrogen fuel.

3,756,860; September 4, 1973; H. Binder, W.H. Kuhn and G. Sandstede
"Electrodes with Mixed Catalysts of Metal Carbide for Hydrogen Fuel Cells"

> The invention relates to hydrogen electrodes for fuel cells whose high efficiency results from the use of supplemental treated tungsten-carbide catalysts and a process for the manufacture of such treated catalysts.

Battelle Memorial Institute, Switzerland (with SA Automobiles Citroen)

3,887,400; June 3, 1975; D. Doniat, K. Beccu and A. Porta
"Method and Apparatus for Electrochemically Producing an Electric Current"

Bayer AG, Leverkusen, Germany

3,857,760; December 31, 1974; W. Breuer, J. Deprez and B. Sturm
"Process of Measuring the Concentration of a Dissociatable Component in a
Gas and Apparatus Therefor"

> In the measuring of a component of a gas stream in a measuring cell similar
> to a Nernst concentration cell with the gas stream being conducted past and
> in contact with the electrolyte and one of the electrodes, the electrolyte es-
> sentially comprises a solid, organic material of low vapor pressure, such as a
> mixture of benzoic acid and propylene carbonate.

Beckman Instruments, Inc., Irvine, Cal.

3,558,528; January 26, 1971; R.P. Buck and R.W. Nolan
"Alkali Metal Ion Sensitive Glass"

> A soda-silicate glass containing about 6 to 9 mol % of an admixture of ZrO_2
> and TiO_2. The glass has exceptionally low electrical resistance and is sensitive
> to alkali metal ions, such as potassium and sodium ions. The glass is particu-
> larly suited for formation into ion sensitive bulbs of glass electrodes utilized
> for measuring alkali metal ion concentration of solutions.

Robert Bosch GmbH, Stuttgart, Germany

3,573,993; April 6, 1971; W. Pabst, G. Sandstede and G. Walter
"Oxygen Electrode Composed of Mixed Oxides of Praseodymium, Chromium,
Nickel and Cobalt"

> Oxygen electrode suitable as cathode in fuel cells. Mixed oxides of praseody-
> mium, chromium, nickel and/or cobalt are lodged in spaces formed on the
> surface of the solid electrode material and the electrical conductor of the
> electrode is made of heat resistant steel.

3,607,418; September 21, 1971; A. Ortlieb and G. Schnepf
"Fuel Cell"

> A fuel cell wherein an electrically insulating annular holder of natural or syn-
> thetic rubber comprises inwardly extending collars provided with sealing lips
> which engage marginal portions of disk-shaped oxygen and fuel electrodes
> which define between themselves oxygen and fuel chambers. The oxygen
> chamber receives oxygen and discharges residues through partly radial and
> partly axially parallel inlets and outlets of the holder. Similar inlets and
> outlets are provided in the holder for admission of fuel and electrolyte into
> and for evacuation of combustion products from the combustion chamber.
> The outlets and inlets are surrounded by annular sealing lips provided on
> one end face of the holder and arranged to abut against the other end face
> of an adjoining holder so that the inlets and outlets of one holder commu-
> nicate with the corresponding inlets and outlets of the adjoining holder but
> are sealed from the atmosphere.

3,650,840; March 21, 1972; H. Dietz
"Fuel Cell with Electrode Holder"

> A method of connecting an electrode member to an electrode holder of a
> fuel cell. An electrically conductive adhesive substance is applied to one or
> both of the electrode member and the electrode holder, and thereupon the
> two are placed into contact with one another so as to adhere via the adhe-
> sive substance, whereby the electrode is electrically conductive but in
> fluid-tight relationship connected to the electrode holder.

3,778,311; December 11, 1973; H. Metzger, K. Brill and F. Hornung
"Foil for Gas Diffusion Electrodes for Electrochemical Cells and Process of
Making the Same

An air-permeable, electrolyte-impermeable composite foil for self-breathing
gas diffusion electrodes for use in electrochemical cells and in particular fuel
cells, comprises a rigid sintered carrier foil of a fluorine-containing polymeric
compound and, firmly bonded thereto, a nonsintered microporous layer of a
fluorine containing polymeric compound. The foil is of light weight, thin
and mechanically stable and thus forms an excellent support for a catalyst
coating.

**3,821,028; June 28, 1974; H. Ziener, L. Weber, H. Jahnke, H. Magenau and
G. Zimmermann
"Fuel Cell with Metal Chelate Electrode Catalyst"**

A catalyst for fuel cell electrodes wherein a gaseous fuel or a fuel dissolved
in the electrolyte of the cell is subjected to anodic oxidation, comprises a
metal chelate of 5,14-dihydro-dibenzo(5,9,14,18)tetraaza(14)annulene or an
electrochemical oxidation product thereof. The catalyst permits formation
of electrodes which can be operated with an acidic electrolyte and compara-
tively inexpensive fuel, such as methanol, formaldehyde, formic acid, carbon
monoxide, oxalic acid, etc.

**3,850,695; November 26, 1974; H. Keller, J. Bregler, H. Rhein and H. Jahnke
"Voltage Regulator System for Use with Fuel Cell Battery"**

Intermittent operation of a pump feeding fuel to the battery is controlled by
circuits which produce pump operation cycles at intervals determined by energy used in the load circuit as well as additional cycles if the voltage falls
below a fixed limit. Compensation for the drop in concentration of the fuel
as the result of recirculation of partly spent fuel is provided by lengthening
the pump operation cycles or making them more frequent with respect to
power energy consumption in the load, or both.

**3,902,917; September 2, 1975; D. Baresel, W. Gellert and P. Scharner
"Process for the Production of Tungsten Carbide Catalyst Adapted for Use
in Fuel Cells"**

Processes of preparing tungsten carbide catalysts adapted for use in making
electrodes for use in electrochemical fuel cells for the direct production of
electrical energy by oxidation of hydrogen, formaldehyde, or formic acid, in
which a precipitated tungstic acid hydrate is freeze-dried, heated to expel the
water of crystallization and the water formed by dehydration of the tungstic
acid and the resulting tungsten trioxide is then reacted with carbon monoxide or a hydrocarbon at a temperature between 600° and 700°C to convert
it to an active tungsten carbide catalyst.

**3,930,884; January 6, 1976; G. Zimmermann, M. Schonborn, H. Magenau,
H. Jahnke and B. Becker
"Activated Stable Oxygen Electrode"**

A stable oxygen electrode which is in an active state and retains said activity
for a long service life. The electrode comprises an electrode carrier and an
activated transition metal chelate of 5,14-dihydrodibenzo(5,9,14,18)tetraaza-
(14)annulene. Said metal chelate is activated by heating in an inner gas atmosphere at a temperature of between about 600° and 1000°C for at least 5
minutes.

Brown, Boveri & Cie AG, Mannheim, Germany

**3,554,808; January 12, 1971; W. Fischer and W. Baukal
"High-Temperature Fuel-Cell Battery"**

A high-temperature fuel-cell battery consists of a column of fuel cells connected electrically in series, and each cell is composed of a plurality of superimposed layers consisting of a porous electron-conducting layer through
which fuel gas is fed transversely, a gas-tight oxygen-ion-conducting layer of

cubically stabilized zirconium oxide such as ZrO_2 with an additive of Y_2O_3 forming the electrolyte, and a porous electron-conducting layer such as Mn_2O_3 through which an oxidant gas is passed transversely. A gas-tight electron-conducting layer separates each cell from an adjacent cell in the column, a similar gas-tight electron-conducting layer is provided at the upper and lower ends of the column and all of the layers of the fuel cells and the separation layers between adjacent cells are sintered together to establish a compact columnar structure.

3,554,811; January 12, 1971; A. Isenberg, W. Pabst and G. Sandstede
"Oxide Cathode Material for Primary Fuel Cells for High Temperatures"

An oxide cathode material for primary fuel cells with oxygen-ion-conducting mixed-oxide solid electrolytes working at high temperatures with gaseous oxidants. The material consists of indium oxide doped with gallium oxide and/or zinc oxide, and the quantity of these latter two oxides in the indium oxide is between 0 and 30% by weight.

3,615,851; October 26, 1971; H.H. Eysel
"Battery with Fuel Cells of Solid Electrolyte"

A battery made up of fuel cells each having a solid electrolyte for direct conversion of chemical energy liberated in the oxidation of a combustible gas into electrical energy. The fuel cells are arranged in series each forming a subchamber and being combined into a chamber for the combustible gas. Each fuel cell has one electrode on an inner wall of the subchamber and its other electrode on an outer wall of the cell. The subchamber is formed by each of two successive cells with a temperature-resistant packing between cells so as to exert a sealing effect obtained by a force compressing the cells and packings. Each fuel cell is in the form of a portion of a cylinder made in one or two pieces having a cylindrical periphery and faces perpendicular to its axis, at least one face being recessed and the subchamber formed by putting two successive cells together.

3,718,506; February 27, 1973; W. Fischer and F.-J. Rohr
"Fuel Cell System for Reacting Hydrocarbons"

A fuel cell system of the high temperature type includes one or more fuel cells operating with an oxygen ion conductive solid or a carbonate melt to which fresh reaction gases containing hydrocarbon fuel and oxygen respectively are fed after preheating in heat exchangers. The system may also include a separate reformer for converting the hydrocarbons into hydrogen and other gases. A portion of the exhaust gases from the fuel side of the cell is recycled through the fuel cell by entrainment with fresh fuel gas by means of the aspirating effect of a nozzle through which fresh fuel gas is injected into the cell and the remainder of the exhaust gas from the fuel side together with oxygen impoverished gas discharged from the cell are led to an after-combustion catalyst in the form of a tubular structure surrounding the fuel cells and also the reformer component if one is utilized for burning of the remaining hydrocarbon content of the fuel gas. The heat generated during the after-burning phase in the catalytic zone is mostly transferred over a very short path to the fuel cells to maintain the fuel cells at their proper operating temperature and to supply the power requirements of the reformer if the latter is utilized. The remaining heat in the exhaust gases is conducted to the exchangers for heat transfer to the incoming fresh reaction gases.

3,754,995; August 28, 1973; H. Kleinschmager
"High Temperature Fuel Cell"

A high temperature fuel cell is disclosed. The fuel cell is of the kind wherein a porous electrolyte body of compressed zirconium dioxide (ZrO_2) is admixed with additives of alkaline earth metal oxide or rare earth metal oxide. An anode layer of metallic nickel is deposited on a first surface of the electrolyte body which has a rough surface configuration. In accordance with the invention, a cathode layer is deposited on a second surface of the electrolyte body, the cathode layer being formed from a mixture of tin oxide (SnO_2) and antimony oxide (Sb_2O_3), the amount of antimony oxide in the mixture ranging between about 7 and 30 mol %.

3,776,777; December 4, 1973; B. Houpert and A. Reich
"Unitary Assembly of Serially Connected Fuel Cells"

A unitary assembly of serially connected fuel cells is provided comprising a fuel cell electrolyte of solid material formed in two mating semicylindrical bodies, with the electrodes of the fuel cells arranged in a spiral pattern in opposed relationships on opposite sides of the electrolyte bodies. The semicylindrical bodies are joined together at abutment surfaces in a gas-tight relationship to form a cylinder through which a gaseous fuel flows, with the exterior of the cylinder being exposed to a gaseous oxidant. The electrodes extend across the abutment surfaces of the semicylinders, and when the semicylinders are assembled, selected electrodes are placed in electrical contact internally of the fuel cell assembly to effect series interconnection between the individual fuel cells of the assembly.

Brunswick Corp., Skokie, Ill.

3,623,912; November 30, 1971; B.W. Wessling
"Fuel Cell with Fluid Circulating Means"

This disclosure describes a new fuel cell and a three dimensional fuel cell electrode. The fuel cell comprises an electrolyte and a pair of three dimensional electrodes made of catalyzed micron-size metal filaments fabricated into skeletons. The hydrogen and oxygen gases are bubbled up through their respective electrodes in a direction closely parallel to the electrode's axes. By bubbling the gases through the electrodes in this manner a large number of three phase reaction sites are obtained for a given volume of electrode material. These reaction sites lead to higher energy and power densities while requiring less catalyst material and lower initial cost. This fuel cell utilizes a three dimensional electrode whereas conventional fuel cells are limited to utilizing effectively only a two dimensional electrode surface.

3,826,686; July 30, 1974; S.L. Epstein and B.W. Wessling
"Electrochemical Electrodes"

The specification describes a molten carbonate fuel cell electric energy source incorporating a novel pair of electrodes comprising a pair of permeable thin metallic films painted on a magnesia electrolyte matrix and a plurality of fiber metal wicks flocked onto the films. During operation, the molten carbonate electrolyte permeates the wicks, which extend into the gas manifolds, to provide very large surface areas for 3-phase fuel cell reactions covered by thin films of electrolyte supplied by large reservoirs in the wicks to minimize cell overvoltage. In addition to the preferred embodiment, a process for fabricating fuel cells and electrochemical electrodes is described. Also, applications to other types of fuel cells, batteries and electrochemical systems are described.

3,905,831; September 16, 1975; P.H. Brown and M.H. Tremblay
"Electrochemical Electrodes"

An electrochemical electrode of a metallic velvet-like material comprising a woven textile pile fabric wherein at least a portion of the woven base fabric and/or the velvet surface-forming pile yarns is metallic is described. The metallic yarn may comprise a blended yarn formed of staple metal fibers and conventional nonmetallic textile fibers, or may be formed of continuous metal filament material. The metal fibers, or filaments, are preferably formed with rough, unmachined, unburnished, fracture-free outer surfaces for improved retention in the velvet pile fabric. The electrode has particular application in fuel cells, batteries and other electrochemical systems where high usable surface areas (per unit volume) are needed to promote three-phase reactions.

3,953,237; April 27, 1976; S.L. Epstein and B.W. Wessling
"Electric Energy Sources Such as Fuel Cells and Batteries"
> Abstract identical to 3,826,686 shown on p 232.

Canadian Minister of Defence, Ottawa, Canada

4,020,239; April 26, 1977; W.A. Armstrong
"Cathode for Hydrazine/Air Cell"
> A cathode for use in a hydrazine/air fuel cell which comprises a sintered nickel substrate having a silver electrocatalyst deposited thereon and having, on one side, a semipermeable hydrophobic membrane, and on the other side thereof a hydrophilic microporous separator membrane. The silver-sintered nickel substrate may be wet-proofed by dipping in a solution of PTFE and in order to lower polarization of the cathode the silver-sintered nickel substrate may be amalgamated by treatment with a mercuric salt.

The Carborundum Co., Niagara Falls, N.Y.

3,563,802; February 16, 1971; R.L. Ogden
"Fuel Cell Construction"
> This invention pertains to a fuel cell which comprises a case with porous positive and negative electrodes inserted therein dividing the interior of the case into three sections. Electrolyte fills the inner section or space. Means for introducing an oxidizing gas into the space behind one porous electrode and a fuel behind the other are provided. The inner section is divided by a porous membrane substantially parallel to the electrodes comprising synthetic fibers prepared from resin condensation products of phenols and aldehydes which can be fiberized and cured. This membrane permits communication of electrolyte between electrodes but prevents passage of small bubbles of oxidizing or fuel gases.

Catalytic Technology Corp., Manchester, Conn.

3,637,437; January 25, 1972; M. Goldberger
"Raney Metal Sheet Material"
> A Raney metal catalytic sheet material is prepared by spraying molten particles of a Raney alloy, such as a silver, nickel, iron, platinum or cobalt alloy with aluminum on a substrate, and leaching aluminum from the alloy. The sheet catalyst is claimed as is the process of manufacture.

3,716,414; February 13, 1973; M. Goldberger
"100-Watt Fuel Cell"
> A commercially useful and marketable fuel cell is described. The cell is capable of operating at ambient conditions of temperatures, pressure and the like for long periods of time. The complete cell includes one or more banks of individual cells, each cell including an anode, an air breathing catalytic cathode, and means for controlled feed of fuel and an aqueous electrolyte. The catalytic electrodes may be catalytic silver. A 100-watt commercial fuel cell is described.

Communications Satellite Corp., Washington, D.C.

3,850,694; November 26, 1974; J.D. Dunlop, M.W. Earl and G. Van Ommering

"Low Pressure Nickel Hydrogen Cell"

A new sealed rechargeable electric fuel cell of the type using a reoxidizable nickel hydroxide as the positive electrode reactant and hydrogen as the negative electrode reactant. The cell uses a hexagonal nickel-rare earth metal hydride compound for storage of hydrogen gas which causes the cell to be maintained at a low gas pressure during charging and discharging.

Compagnie Francaise de Raffinage, Paris, France

3,578,502; May 11, 1971; H. Tannenberger and P. Kovacs
"Stabilized Zirconium Oxide Electrode for Solid Electrolyte Fuel Cell"

An electrode has been provided for a solid electrolyte containing fuel cell suitable for high temperature operations, comprising a first porous layer and a second porous layer, said first porous layer comprising at least one layer of granules of a ceramic material capable of conducting oxygen ions and electrons generated in said fuel cell, said granules being in contact with the electrolyte with at least a part of surfaces of said granules, at least a part of other surfaces of said granules being in contact with a material possessing good electron conductivity, said material being in turn in contact with a second porous layer, said second porous layer being a good conductor of electrons and in an electrical contact with said granules, said layers constituting an electrode which in the cell allows the ready access to said granules of fuel or comburant gas.

3,834,943; September 10, 1974; P. Van den Berghe and H. Tannenberger
"Electrolyte-Electrode Unit for Solid-Electrolyte Fuel Cell and Process for the Manufacture Thereof"

Method and apparatus to improve a fuel cell operating in the range of $800°$ to $1100°C$, the gas-impermeable solid electrolyte preferably being formed of stabilized zirconia with a suitable metal anode and a cathode formed of at least one layer of doped indium oxide affixed to the electrolyte. The improvement consisting of having the doped indium oxide layer formed with dendrites whose principal axes are at least approximately perpendicular to the surface of the electrolyte with the axes forming trunks having a diameter of between 500 and 10,000 A, so that said doped indium oxide layer is permeable to gas and the thickness is at most equal to 0.02 cm. The improvement also includes the method of obtaining such a doped indium oxide layer with dendrites by utilizing an electron beam to vaporize a target of the doped indium oxide and condense the latter on the electrolyte which has been heated to at least $300°C$ under total gas pressure of between 10^{-6} and 10^{-2} mm Hg, containing an oxygen partial pressure of at least 10^{-6} mm Hg. A further improvement of the foregoing includes the use of a chromite layer between the electrolyte and the indium oxide layer to prevent the diffusion of the one into the other.

3,981,746; September 21, 1976; J. Bezaudun and J.E. Weisang
"Electrode for Fuel Cell"

Fuel cell has been provided having an electrolyte of solid, fritted ceramic suitable for operation at high temperatures and for operation on a hydrocarbon or hydrogen as a fuel and oxygen or oxygen containing gas as a comburant comprised of: as an electrolyte, a cubic phase-stabilized zirconia; a cathode; and as an anode, an electron collector and an oxidation catalyst. A method for preparing the anode therefor has been provided as well as a method directed to the operation of said fuel cell.

Compagnie Generale d'Electricite, Paris, France

3,547,699; December 15, 1970; J.-C. Charbonnier and B. Goue

"Fuel Cell with Aqueous Ethylene Diamine Electrolyte Solution"

An electrolyte for fuel cells which essentially comprises an aqueous ethylene diamine solution. Soluble compounds may be added to improve the conductivity of the electrolytic ethylene diamine solution. Exemplary soluble compounds are potassium or sodium salts. Preferred concentrations of the aqueous ethylene diamine solution range from about 15 to about 40 grams of ethylene per liter. The electrolyte finds particular application for use in a cell wherein the fuel contains carbon dioxide or wherein the fuel cell is capable of yielding carbon dioxide by reaction at the electrodes.

3,547,702; December 15, 1970; C. Hespel
"Electrode Arrangement for a Fuel Cell"

Fuel cell having a plurality of thin electrodes, similarly shaped and being either hollow or laterally spread and method of manufacture thereof.

3,547,704; December 15, 1970; J.-P. Pompon
"Automatic Means for Regulating the Concentration of Electrolyte of a Fuel Cell"

Automatic regulation of electrolyte concentration within a fuel cell by siphoning various strength portions to and from a gauge chamber in response to change in chamber level.

3,668,010; June 6, 1972; J. Fally, Y. Lazennec and C. Lasne
"Fuel Cells and Fuel Cell Batteries Operating at High Temperature and Process of Manufacture Thereof"

A solid electrolyte fuel cell operating at a high temperature, comprising several elements electrically connected in series. These elements are supported by one electrolyte tube, and one end of the external electrode of an element protrudes slightly above the internal electrode of the following element. This enables simplified assembly and a reduction in the weight and the cost price of the fuel cell.

3,703,416; November 21, 1972; J. Jacquelin
"Basic Cell Operating with Hydrocarbon Fuel"

Basic cell with hydrocarbon fuel, in which a decarbonization of the electrolyte is carried out. It comprises two diaphragms respectively anodic and cathodic which insulate a central compartment wherein there locally appears a pH value sufficiently low to give rise to a carbon dioxide emission. The invention may be used for fuel cells having a high overall energy yield.

3,849,202; November 19, 1974; J.-P. Pompon
"Method of Manufacturing a Metal-Air Battery"

Electrochemical storage battery of the forced flow type comprising several reaction elements of tubular shape inside which the electrolyte containing zinc in suspension flows, such elements comprising, more particularly inside a copper negative collector in contact with an insulating separator, and outside a positive collector formed by a grid of unfolded metal surrounding the active layer and pressed tight against that layer subsequent to axial traction having caused plastic deformation.

3,902,918; September 2, 1975; J.-P. Pompon
"Method and Device for Feeding Cells of an Electrochemical Battery of Forced Circulation Type"

The invention, which makes it possible to feed in series with electrolyte at a predetermined flow rate and near pressure specified electrochemical cells grouped into identical modules, consists in placing between each module, a pump suitable for compensating the pressure drop in the module immediately upstream, at the same time placing downstream from the assembly of modules a pump suitable for developing a pressure equal to the difference between the said predetermined pressure and half the pressure drop in the module arranged upstream. A pump suitable for creating a pressure drop equal to the sum of the said average pressure and of half pressure drop in the module arranged downstream is also placed upstream from the assembly of the cells.

Consolidated Natural Gas Co., Pittsburgh, Pa. (with Southern California Gas Co. and Southern Counties Gas Co. of California)

3,589,942; June 29, 1971; F.B. Leitz, Jr. and D.K. Fleming
"Bipolar Collector Plates"

A bipolar collector plate having vertically oriented ridges and grooves that are hemispherical in cross-section. The adjoining ridges are connected by cross-grooves of a lesser depth than the vertical ridges and grooves. The range of radius of curvature of the ridges is critical and varies between 0.25" and 0.01". The plates may be of a single material, preferably columbium, tantalum, stainless steel or nickel, or may be of a composite material such as particles of carbon, carbides, nitrides and borides impregnated with a binder such as poly-tetrafluoroethylene, poly-chlorotrifluoroethylene, copolymers of poly-chlorotrifluoroethylene with vinylidene fluoride, polyolefins, or an epoxy, polyester or phenolic resin. The plate may also be a laminate having a heat conductive core. An embodiment has extended peripheral edges forming a heat conductive flange for contact with a heat conductive fluid to control the operating temperature of the cell.

Dead Sea Works Ltd., Beersheva, Israel (with Israeli Prime Minister's Office)

3,578,503; May 11, 1971; M.R. Bloch, C. Forgacs and R.V. Bobker
"Devices for the Electrochemical Generation of Electrical Current"

A fuel cell comprising a cathode system and an anode system disposed in a chamber and separated by a permselective membrane. A liquid catholyte is continuously introduced into and circulated through the catholyte compartment, and an anolyte is continuously introduced into and circulated through the anolyte compartment. The anode system preferably includes a slurry of zinc particles dispersed in a zinc-bromide solution (the anolyte), and the cathode preferably includes a carbon electrode, bromine, and a zinc-bromide solution (the catholyte).

Decktronic, Inc., Washington, D.C.

3,705,056; December 5, 1972; H.P. Doble, Jr. and H.C. Langelan
"Method for Producing and Operating a Fuel Cell"

The invention contemplates a fuel cell having a porous electrode forming the septum between the reducing and oxidizing gas chambers, the electrode comprising a grid pasted with a simple metallic compound such as ferrous oxalate which has been electrolytically reduced while under controlled mechanical pressure to produce a relatively specific, high porosity of extreme surface area.

Delta F Corp., Woburn, Mass.

3,895,102; July 15, 1975; J.P. Gallagher
"Solid Fuel for the Generation of Hydrogen and Method of Preparing Same"

A solid porous rigid fuel composition for the generation of hydrogen gas by the reaction of a liquid hydroxide solution with the solid fuel, the fuel composition comprising silicon-containing metal particles, such as ferrosilicon, and a powdered salt compound, such as an alkali metal halide like sodium chloride, the metal silicon particles bonded together into a rigid porous mass,

and the salt compound present within the porous mass to inhibit the formation
of slow-dissolving silicate cement on the surface of the silicon metal particles
on repeated reaction of the silicon particles with a liquid hydroxide solution.

The Dow Chemical Co., Midland, Mich.

3,591,419; July 6, 1971; C.E. Hamilton
"Process for Generating Electrical Energy Using Manganese Dioxide with Oxidizing Gas"

A fuel cell and method of its operation wherein organic materials are contacted
with a suitable oxidizing system in the presence of an electron-collecting system to produce electrical current.

E.I. du Pont de Nemours and Co., Wilmington, Del.

3,560,568; February 2, 1971; P.R. Resnick
"Preparation of Sulfonic Acid Containing Fluorocarbon Vinyl Ethers"

Vinyl ethers of the formula $CF_2{=}CFOCF_2CFXSO_3Me$, wherein X is F or CF_3
and Me is alkali metal, are made by contacting and reacting an alkali metal
alkoxide with a compound of the formula

$$CF_2CF{-}SO_2{-}CFX$$
$$\lfloor O{-}CF_2 \rfloor$$

in an inert organic solvent. The vinyl ether reaction products can be copoly-
merized with fluorocarbon monomers, e.g., tetrafluoroethylene, to form ionically conductive material for use as fuel cell membranes.

3,607,420; September 21, 1971; L.H. Cutler
"Process of Operating Fuel-Cell Half Cell with Cupric-Cuprous Redox Couple"

A redox cathode system of an electrical conductor immersed in an acidic solution containing cupric chloride and excess chloride ions is disclosed for use in
fuel cells. The system is isolated from the anode system by an ion-exchange
membrane, and an anode having catalytic material associated therewith and an
acid anolyte are employed. An organic hydrogen-containing fuel, e.g., formaldehyde, and an oxygen-containing oxidant, e.g., air and hydrogen peroxide, are
preferred.

3,615,840; October 26, 1971; W.R. Wolfe, Jr.
"Fuel Cell and Fuel Cell Electrode Comprising a Sulfurated Compound of Tungsten and Oxygen"

A relatively inexpensive catalyst for a fuel cell electrode, particularly useful
with acid electrolytes, is an acid-insoluble solid material composed of at least
one compound of tungsten and oxygen wherein the valence of tungsten ranges
from four to six and at least one sulfurated compound of tungsten, wherein
tungsten has a valence of four, e.g., tungsten disulfide, at least the exposed
regions of said solid material containing said sulfurated compound of tungsten,
the ratio of oxygen-to-sulfur in said solid material being 80:1–1:80.

3,702,267; November 7, 1972; W.G. Grot
"Electrochemical Cell Containing a Water-Wettable Polytetrafluoroethylene Separator"

Polytetrafluoroethylene is provided containing from 20 to 95% by weight of
inorganic particulate solids which are characterized by a particle diameter of
no greater than 0.2 micron, a bulk density no greater than of 0.25 g/cm^3, and
a specific surface area of at least 20 m^2/g. The particulate solids impart a high
degree of microporosity and wettability which makes the polymer useful as a
filter or as an ion permeable separator in electrochemical cells.

3,822,149; July 2, 1974; H.J. Hale
"Rechargeable Zinc Electrochemical Energy Conversion Device"

> A rechargeable zinc cell and battery, preferably a zinc-air cell and battery, are disclosed in which the cell comprises (1) a circular casing having a circular reservoir with electrolyte contained therein; (2) a rotatable electrode, having at least one planar zinc surface, enclosed coaxially in the reservoir; (3) a stationary planar counterelectrode spaced from the rotatable electrode; (4) means for preventing the electrochemical reaction in the axial area of the zinc surface; (5) wiper means disposed between the electrodes for lightly abrading the zinc surface; and (6) means for agitating the electrolyte to maintain particulate matter in suspension.

Electrocell Limited, Toronto, Canada

3,682,704; August 8, 1972; R.M. Keefer
"Redox Fuel Cell Regenerated with Sugar"

> An electrochemical cell has an anodic half-cell which contains an oxidizable organic fuel, such as a sugar, and a redox couple, such as silver-argentous, the cell generating a current on interaction of the fuel and the redox couple whereby the fuel alone is consumed.

Electromedia, Inc., Summit, N.J.

3,847,671; November 12, 1974; A.F. Leparulo and C. Grun
"Hydraulically Refuelable Metal-Gas Depolarized Battery System"

> A metal-gas depolarized battery system is so constructed that both solid and liquid contents can be drained from the battery after discharge. The negative grid and the gas depolarized electrode in each cell define a compartment which can be refilled with a slurry of electrolyte and active metal powder, thereby recharging the battery in the time it takes to drain same and refill with said slurry. A system for collecting the discharge products of such a battery and reducing metal ion to metal which can then be slurried with electrolyte and returned to the cells of the battery is also described. The preferred system utilizes zinc, KOH and air and is especially suitable for propulsion of vehicles and the like.

Energy Conversion Ltd., London, England

3,565,693; February 23, 1971; P. Tapsell
"Gas Depolarized Cell Having a Support Assembly for Electrodes"

> A support member for the cathode of a gas-depolarized cell incorporates on one side thereof a plurality of spaced projections for engaging an adjoining support member of an adjacent cell when mounted in a battery. The support member may comprise an intercell separator member with a mount for the cathode on one side thereof and a plurality of spaced projections on the other side, said projections being so arranged as to cooperate with those of the next cell when assembled in the battery. Preferably the cathode is mounted within a recess in one face of the member. Further projections and/or recesses are provided to enable adjoining support members to be registered with respect to each other when brought face-to-face. A molding technique is preferred for the manufacture of the support member. The invention lends itself particularly to the production of a bicathode member comprising two support members as described above arranged face-to-face with the projections extending outwardly and the cathode-bearing faces being spaced to receive an anode member for the cell.

3,762,956; October 2, 1973; P.J. Gillespie
"Electrochemical Cells"

Gas depolarized cell which includes a vent to allow the pressures interior and exterior of the cell to equalize. The vent comprises an aperture or volume closed by electrolyte-phobic, gas permeable material. This material is preferably a portion of the cathode of the cell. End cap crimping arrangements are also provided for sealing the ends of the cell.

3,840,404; October 8, 1974; D.F. Porter and D.P. Redfearn
"Gas Depolarization Cell"

An electrochemical cell or battery of cells which is depolarized by a gas at the cathode. A support structure or outer case for the cell or battery is so formed as to restrict the amount of depolarizing gas reaching the cathode to reduce electrolyte evaporation from the cell or battery.

3,864,171; February 4, 1975; T. Mills, M. Wiacek, P.J. Gillespie and C.D. Hatcher
"Electrochemical Cells"

An electrochemical cell including a current collector which extends between end caps of the cell so as to impart structural rigidity to the cell. The current collector may extend through an apertured anode of the cell and engage the anode so as to serve as its current collector.

Energy Development Associates, Madison Heights, Mich.

3,881,958; May 6, 1975; P. Carr and C.J. Amato
"Mass Flow Rate Control of Chlorine Content of Electrolyte for High Energy Density Battery"

A high-energy density battery system of the metal-chlorine electrode type, in which chlorine dissolved in or dissolved and entrained in aqueous metal chloride electrolyte is utilized as a feed to an electrode compartment, includes a conduit for withdrawing of electrolyte from the electrode compartment, means for adding chlorine to the electrolyte circulating through said conduit and a return conduit, a pump to maintain the electrolyte in circulation, a sensor to measure chlorine:electrolyte ratio in a conduit by measuring the mass flow rate of material being pumped compared to said mass flow rate when a desired ratio of chlorine to aqueous electrolyte is pumped, means for controlling the addition of chlorine to the circulating electrolyte in response to the sensed mass flow rate and conduit means for returning the rejuvenated electrolyte to the electrode compartment. Sensors utilized are responsive to the mass flow rate and can distinguish between flows which are high or low in chlorine content, the former being of lower mass flow rates. The sensors may be connected to means for charging chlorine to the circulating electrolyte and thereby can automatically control the chlorine content of the electrolyte, maintaining it at about a desirable level.

4,020,238; April 26, 1977; P.C. Symons
"Control of Generation of Chlorine Feed from Chlorine Hydrate for Use in a Metal Chlorine Electric Energy Storage Device"

A simple, inexpensive and trouble-free method of controlling the chlorine feed to a metal-chlorine battery system from a source of chlorine hydrate includes circulating aqueous electrolyte from an electrode compartment through a pump and recycling back to the electrode compartment, taking off a portion of the electrolyte, passing through a container of chlorine hydrate from which it releases chlorine and water, the chlorine generation increasing the pressure in the container, controlling the flow of electrolyte to the container, allowing the chlorine generated to flow from the container to the recycle line and back to the electrode compartment, which diminishes the pressure in the container, at which time electrolyte is again admitted to it, and the steps are continually repeated during the discharging of a battery system.

An apparatus for carrying out the process of the invention includes a check valve between the pump and the container of hydrate which closes, preventing admission of the electrolyte to the container when the pressure increases enough to threaten to force the contents back through the line in which the check valve is located.

Energy Research Corp., Danbury, Conn.

3,935,029; January 27, 1976; B. Baker and M. Klein
"Method of Fabricating a Carbon-Polytetrafluoroethylene Electrode-Support"

An electrode substrate is described which is suitable for supporting noble metal electrodes in fuel cells and similar electrochemical cells. It is composed of fine graphite particles enmeshed in a web of fibers of polytetrafluoroethylene (PTFE) produced by forming a homogeneous mixture of graphite particles and PTFE powder in an organic liquid, removing the liquid and milling the mixture in a series of steps to form a thin, porous, electrically conductive sheet of graphite-PTFE which is an excellent support for thin film metallic electrodes and which is resistant to corrosive electrolytes.

Engelhard Minerals & Chemicals Corp., Newark, N.J.

3,575,718; April 20, 1971; O.J. Adlhart and V.V. Tanna
"Composite Electrolyte Member for Fuel Cell"

Thin flexible fuel cell electrolyte members, which also serve as effective barriers between the cathode and anode compartments in a fuel cell and which are useful for fuel cells operating in the temperature range of $80°$ to $250°C$ are composite structures having two dissimilar layers, each consisting of a matrix and a concentrated liquid acid immobilized in such matrix. The matrix of each layer is comprised of fluorocarbon gel and inert filler particles. The matrix of one layer has carbon powder as an essential component. The second, which is a thinner layer contains no carbon powder, forms a dielectric barrier between the cathode and the layer containing a carbon-powder.

3,623,913; November 30, 1971; O.J. Adlhart and P.L. Terry
"Fuel Cell System"

In an air-breathing fuel cell stack, the air sweep simultaneously supplies oxidant and maintains the temperature and water balance in the cell. The individual cells of the stack contain an immobilized stable acid system as the electrolyte. A bipolar thermally conductive plate having a cooling fin on one side is disposed between and in contact with electrodes of adjacent cells. Gas flow channels on each of the electrode-contacting surfaces of the plates intercommunicate on the active area of the electrodes for distribution of the fuel and air over the electrodes. In a compact system, especially adapted for use in remote areas, the fuel cell assembly is used in combination with a hydrogen generator.

3,709,736; January 9, 1973; O.J. Adlhart and P.L. Terry
"Fuel Cell System Comprising Noncirculating, Counter-Current Gas Flow Means"

In an air-breathing fuel cell stack utilizing a stable acid electrolyte system and electrically and thermally conductive plates between the individual cells, the air sweep simultaneously supplies oxidant and maintains the temperature and water balance in the cell. The triple function control by air flow is made possible by virtue of a combination of the particular type of electrolyte and by an improved arrangement of air feed to the cells.

3,716,416; February 13, 1973; O.J. Adlhart and P.L. Terry
"Fuel Metering Device for Fuel Cell"

A fuel cell system designed for self-regulated power output at a predetermined level includes a capillary flow control device for throttling the fuel flow in response to the power output of the fuel cell.

ESB Inc., Philadelphia, Pa.

3,741,810; June 26, 1973; J.R. Dafler and R.P. Niederberger
"Battery Construction"

> An air depolarizable galvanic battery construction is disclosed. The battery
> includes a vessel structure having a portion thereof comprised of at least one
> air electrode. A closure means overlies the air electrode for preventing direct
> contact of foreign objects with the air electrode to thereby protect the air
> electrode from being damaged. The closure means is provided with air access
> openings and is disposed in spaced relation to the air electrode to permit cir-
> culation of air between it and the air electrode.

3,746,580; July 17, 1973; W.E. Aker and R.J. McCormick
"Gas Depolarizable Galvanic Cell"

> A gas-depolarizable galvanic cell having an insert molded cathode subassembly
> is disclosed. The cathode subassembly has portions thereof formed of elec-
> trically nonconductive material and a portion thereof comprised of a gas-
> depolarizable electrode.

3,791,896; February 12, 1974; J.C. Sklarchuk
"Bi-Functional Gas Electrode"

> A bi-functional gas electrode is described suitable for use in a rechargeable
> gas-metal cell or gas-metal oxide cell. A grid structure such as nickel screen
> is coated with carbonized nickel powder bonded in place by an organic poly-
> mer. The electrode is waterproofed by adhering a sheet of microporous
> polyfluorohydrocarbon on one of the surfaces of the electrode. The elec-
> trode is suitable for use with such gases as oxygen, hydrogen, carbon monox-
> ide, and numerous carbon-hydrogen compounds.

Ethyl Corp., New York, N.Y.

3,607,438; September 21, 1971; T.H. Coffield
"Pentacyanocobaltate Electrolytes for Fuel Cells"

> Alkali metal, alkaline earth metal, and ammonium pentacyanocobaltate solu-
> tions are efficacious electrolytes for fuel cells.

Exxon Research and Engineering Co., Linden, N.J.

3,544,378; December 1, 1970; B. Broyde
"Fuel Cell Comprising a Metal Tungstate Anode"

> The invention concerns the use of certain tungsten bronzes as catalysts for
> the anodic oxidation of a fuel in a fuel cell.

3,551,207; December 29, 1970; W.A. Herbst
"Fuel Cell"

> This invention relates to a novel fuel cell wherein fuel and oxidant compart-
> ments containing an electrolyte slurry of catalyst comprising particulate
> solids are separated by a central zone containing a catalyst-free electrolyte
> positioned between opposing electrodes.

3,615,836; October 26, 1971; J.S. Batzold
"Fuel Cell Containing and a Process of Making an Activated Fuel Cell Catalyst"

> A catalyst selected from the group consisting of noble metals, transition metals,
> lanthanide series metals, mixtures and alloys thereof, which have limited activ-
> ity because of absorbed halide ions, may be employed in electrodes of a fuel
> cell containing a halide ion-free acid electrolyte provided that, prior to its oper-
> ation in the fuel cell, the catalyst has been treated with a basic solution having
> a pH in the range of 8 to 14 for a period of time sufficient to desorb the halide
> ions.

3,615,837; October 26, 1971; M. Beltzer
"Alkali Metal Dihydrogen Phosphate Melt Electrolytes"

Due to their scarcity and expense as compared with their activity, noble metals
have become less desirable as fuel cell catalysts. Therefore, nonnoble metals have
found increasing use in fuel cells. Satisfactory performance, however, can only be
obtained when they are operated at relatively high temperatures of $200°$ to $300°C$.
One problem encountered at these high temperatures is that art-known electrolytes
become too corrosive, necessitating the use of other electrolyte systems. Satisfactory
electrolytes can be made from several alkali metal dihydrogen phosphates as the initial components. A particularly desirable electrolyte consists of a melt of potassium
dihydrogen phosphate and sodium dihydrogen phosphate.

3,644,147; February 22, 1972; A.R. Young, II
"Mixed Valency Transition Metal Perovskites as Cathodic Reduction Catalysts"

Ionic crystals having a perovskite structure and comprising a solid solution of
oxides of the formulas ABO_3 and $A'BO_3$ where A and A' are nontransition
metals and B is a transition metal present in two different valency states are
effective cathodic electrocatalysts for use in electrochemical cells such as fuel
cells.

3,658,595; April 25, 1972; J.S. Batzold
"Fuel Cell with Anode Containing a Catalyst Composition of Iridium and
  Ruthenium"

In a fuel cell comprising an anode, a cathode and an electrolyte positioned
between and communicating with said anode and cathode, the improvement
wherein said anode comprises a catalyst composition consisting essentially of
coreduced iridium and ruthenium where the atom ratio of each component
to the total catalyst is in the range of from 25 to 75%.

3,728,159; April 17, 1973; G.M. Varga, Jr.
"Electrochemical Cell with Modified Crystalline Disulfide Electrocatalysts"

An electrochemical cell is improved when at least catalytic amounts of a
modified crystalline disulfide of a Group VI-B metal is used as an electrocatalyst. The metal disulfide is modified with from 1 to 10 mol % of a
lower altervalent atom. The electrocatalyst can be used as an anode or cathode catalyst.

3,793,081; February 19, 1974; G.M. Varga, Jr.
"Sulfided Transition Metal Fuel Cell Cathode Catalysts"

An improved electrochemical cell has an acid electrolyte and a catalytic
amount of at least one acid insoluble sulfide of a transition metal in the
cathode. The preferred cathode catalyst is a mixture of a transition metal
disulfide and thio spinel.

3,957,535; May 18, 1976; W.J. Asher
"Fuel Cell Heat and Mass Plate"

A novel method and apparatus is provided for controlling heat and mass in-
ventory in a fuel cell. Heat and mass, e.g., water, generated in the cell are
removed by heat transfer and capillary action.

Ford Motor Co., Dearborn, Mich.

3,547,701; December 15, 1970; P.R. Basford and M.H. Boyer
"Fuel Cell Having an Electrolyte Comprising a Fused Vanadate Salt"

A high temperature fuel cell in which a gaseous fuel and an oxidizer are
maintained in contact with a molten electrolyte composed of a vanadate salt
or a mixture of vanadate salts.

3,951,689; April 20, 1976; F.A. Ludwig
"Alkali Metal/Sulfur Cell with Gas Fuel Cell Electrode"

An improved alkali metal/sulfur cell of the type including: (A) at least one anodic reaction zone containing a molten alkali metal reactant-anode in electrical contact with an external circuit; (B) at least one cathodic reaction zone including an electrode of porous conductive material and in which, during discharge of the cell, cations of the alkali metal combine with polysulfide or sulfur-saturated polysulfide ions to form reduced alkali metal polysulfide salts which at least partially fill said porous conductive material; and (C) a cation-permeable barrier to mass liquid transfer interposed between and in contact with said anodic and cathodic reaction zones, said porous conductive material being in electrical contact with both said cation-permeable barrier and said external circuit. The improvement comprises adapting the cathodic reaction zone to operate as a gas fuel cell electrode by employing a sulfur storage chamber containing molten sulfur connected with said cathodic reaction zone so as to allow sulfur vapors to pass therebetween, the storage chamber being adapted to be maintained at a temperature (i) above the temperature of said cathodic reaction zone when said cell is being discharged such that sulfur distills into said cathodic reaction zone and (ii) below the temperature of said cathodic reaction zone when the cell is being charged such that sulfur vapor condenses in said storage chamber. The application also relates to a process for reducing alkali metal polysulfide salts to form elemental alkali metal using said cell.

3,985,576; October 12, 1976; J.N. Lingscheit and T.J. Whalen "Seal for Energy Conversion Devices"

An improved energy conversion device of the type comprising: (A) an anodic reaction zone, (i) which contains a molten alkali metal anode-reactant in electrical contact with an external circuit, and (ii) which is disposed interiorly of a tubular cation-permeable barrier to mass liquid transfer; (B) a cathodic reaction zone (i) which is disposed exteriorly of said tubular cation-permeable barrier, and (ii) which contains an electrode which is in electrical contact with both said tubular cation-permeable barrier and said external circuit; (C) a reservoir for said molten alkali metal which is adapted to supply said anode-reactant to said anodic reaction zone; and (D) a tubular ceramic header (i) which connects said reservoir with said anodic reaction zone so as to allow molten alkali metal to flow from said reservoir to said anodic reaction zone, (ii) which is sealed to said tubular cation-permeable barrier, and (iii) which is impervious and nonconductive so as to preclude both ionic and electronic current leakage between the alkali metal reservoir and the cathodic reaction zone. The improvement of the invention comprises a lap joint seal between the tubular ceramic header and said tubular cation-permeable barrier which is formed by (1) disposing the end portion of a first one of said tubes, which has been sintered to final density, inside the end portion of the second of said tubes which (i) is not sintered to final density, (ii) has an inner diameter in the unsintered state greater than the outer diameter of said first tube, and (iii) upon being sintered to final density is adapted to shrink to the extent that the inner diameter thereof is at least 0.002" less than the outer diameter of said first tube; and (2) sintering said second tube to shrink the same and effect a seal between said first and second tubes.

Fuji Electric Co. Ltd., Kawasaki, Japan

3,956,013; May 11, 1976; N. Miyoshi, Y. Watabe and M. Tsuruoka "Fuel Cell"

In a fuel cell comprising (1) a generator cell having a fuel gas (hydrogen gas) chamber, an oxidizing gas (oxygen gas) chamber and an electrolyzing solution chamber; (2) a fuel gas circulating circuit for supplying fuel gas to the fuel gas chamber of said generator cell; (3) an oxidizing gas circulating circuit for supplying the oxidizing gas to the oxidizing gas chamber of said generator cell; and (4) an electrolyte circulating circuit for supplying an electrolytic solution to said electrolyzing chamber, the improvement characterized in that a means for regulating the concentration of the electrolytic solution at a constant value is provided on said electrolyzing chamber and a means for removing the formed water from each of said gas chambers, a means for regulating simultaneously the gas pressures in both of said gas chambers, and a means for keeping constant the temperature of said electrolytic solution are provided on each of said gas chambers.

3,956,016; May 11, 1976; N. Miyoshi, Y. Watabe and H. Hoshikawa
"Fuel Cell"

Abstract identical to 3,956,013 shown on p 243.

3,957,536; May 18, 1976; N. Miyoshi and M. Tsuruoka
"Fuel Cell"

Abstract identical to 3,956,013 shown on p 243.

3,959,019; May 25, 1976; N. Miyoshi and Y. Tachibana
"Fuel Cell"

Abstract identical to 3,956,013 shown on p 243.

Fuji Electrochemical Co., Ltd., Tokyo, Japan

3,881,959; May 6, 1975; T. Tsuchida, K. Shinoda, K. Yamamoto and T. Murata
"Air Cell"

In an air cell comprising an outer metal casing constituting an anode terminal of the cell, and an anode inner casing containing anode active mass therein, there is provided an alkaline resistance insulative thin film covering over the outer vertical side surface of the anode casing, and a conductive metal penetrating through the vertical center portion of the insulative thin film to electrically and liquid-tightly connect the anode inner casing with the outer metal casing.

Furukawa Denki KK, Yokohama, Japan

3,563,803; February 16, 1971; M. Katoh
"Aluminum-Air Battery"

An aluminum-air battery having an anode of aluminum or an alloy thereof and an alkaline electrolyte containing as an additive plumbite, plumbate or stannate, the additive being present in a concentration lower than 0.2 M. The additive inhibits self-corrosion of the aluminum electrode and raises the current efficiency, while limiting the electrolyte temperature with passage of time.

General Electric Co., Schenectady, N.Y.

3,554,809; January 12, 1971; D.W. Craft
"Process and Apparatus for Distributing Fluid Inerts with Respect to the Electrodes of a Fuel Battery"

A fuel battery is provided having an electrolyte element, such as a matrix immobilized electrolyte or ion exchange structure, carrying first and second segmented electrodes on opposite faces. The electrode segments are electrically serially related. Fluid reactant is distributed to the first electrode segments through one or more channels which direct the fluid reactant to flow serially across the electrode segments along laterally spread interconnected courses which are free of interconnection mediate their ends. This causes fluid inerts present in the fluid reactant to be collected in the terminal courses so that the fluid inerts are distributed between the electrode segments.

3,558,361; January 26, 1971; C.C. Christianson
"System and Process for Selectively Diverting an Electrochemically Consumable and Regenerable Fluid"

> A fluid stream is pumped to and from a receiving zone. A cell bridges the fluid stream upstream and downstream of the receiving zone and has one electrode in contact with the fluid stream upstream of the receiving zone and a second electrode in contact with the fluid stream downstream of the receiving zone. The electrodes are ionically communicated by an electrolyte.

3,565,692; February 23, 1971; J.L. Weininger
"Rechargeable Non-Aqueous Alkali Metal-Halogen Electrochemical Cells"

> A rechargeable nonaqueous alkali metal-halogen electrochemical cell is described which includes an alkali metal anode, a halogen cathode, a nonaqueous electrolyte, and an ion permeable barrier between the electrodes separating the electrolyte into anolyte and catholyte reservoirs. The barrier has finely divided particles having through pores and a binder joining the particles into a unitary structure.

3,573,104; March 30, 1971; C.W. Snyder, Jr. and A.D. Thumim
"Fuel Cell Unit with Novel Fluid Confining and Directing Features"

> A fuel battery is disclosed provided with apertured reactant and electrolyte gaskets. The electrolyte frames are provided with notched baffle fingers to uniformly distribute fluid while the reactant frames are provided with stacked screens having offset apertures. The frames making up the fuel cell unit are held assembled by rigid frames compressively urging the peripheral portions of the gaskets and frames into engagement while not compressing the central portion.

3,573,988; April 6, 1971; D.W. McKee and M.S. Pak
"Electrode Comprising Non-Noble Metal Disulfides or Phosphides and Electrochemical Cell Utilizing Same"

> An electrode is composed of particles of a metallic disulfide, a metallic phosphide, or such particles coated with a nonnoble metal bonded together with a binder, and an electrical lead in electrical contact with the bonded particles. The metallic component of the disulfide or the phosphide is selected from the class consisting of molybdenum, tungsten, zirconium, niobium, hafnium, and tantalum. This electrode, which does not contain any noble metal catalyst, is particularly useful in electrochemical cells employing alkaline electrolytes.

3,576,730; April 27, 1971; H.S. Spacil
"Nickel/Nickel Oxide Reference Electrodes for Oxygen Partial Pressure Measurements"

> A high temperature oxygen sensing cell with means for producing a partial pressure of oxygen as a reference. A mixture composed of nickel, nickel oxide and an antisintering material is compacted in a tube formed of a solid oxygen-ion electrolyte. When the resulting cell is sealed to prevent substantial quantities of environmental oxygen from reaching the mixture, a stable partial pressure of oxygen, dependent upon temperature, is produced within the tube.

3,580,741; May 25, 1971; T.C. Hovious and R.A. Torkildsen
"Universal Automatic Sensing and Actuating Control System for Fluid Consuming Power Sources and the Like"

> A combined sensing and output power actuating control system for electrically operable devices such as electric power generation assemblies of the fluid consuming battery type, such as fuel batteries. The control system is comprised by a ratio detector circuit which is coupled, for example, across the output terminals of a fuel battery and also coupled across the last fuel

cell unit in the fuel battery for comparing the ratio of the proportional voltage appearing across the last fuel cell to the output terminal voltage of the fuel battery to a preset value and deriving an output trigger signal in response to the ratio varying from the preset value. Timing circuit means are coupled to the output of the ratio detector circuit and are responsive to the output trigger signal from the ratio detector. An output power actuating circuit which is comprised by a purge valve actuator is energized for a predetermined time period by the timing circuit in response to the output from the ratio detector for purging the fuel battery of undesired accumulated inert gaseous matter thereby restoring the fuel battery to full operating capacity.

3,589,943; June 29, 1971; W.T. Grubb and H.A. Liebhafsky
"Electrochemical Battery"

An electrochemical battery has a plurality of individual, fluid-tight fuel-electrolyte chambers in each of which an anode and a gas diffusion cathode mounted therein are connected electrically to the respective gas diffusion cathode and anode of adjacent cells. The electrolyte containing dissolved fuel in each cell is separated from adjacent cells thereby eliminating load paths across cells. Such a battery provides a high voltage output from a compact device.

3,589,945; June 29, 1971; H.A. Christopher and P.J. Moran
"Stacked Metal Gas-Cells"

An electrochemical cell has a first electrode and a second electrode, an aqueous electrolyte positioned between and in contact with the electrodes, and at least one of the electrodes is a gas diffusion electrode immersed partially in the electrolyte and the remaining portion of the electrode exposed to a gaseous medium. A modified electrochemical cell has a casing with an aqueous electrolyte therein, a first electrode in contact with the electrolyte, a second electrode in contact with the electrolyte and spaced from the first electrode, and at least one of the electrodes is a gas diffusion electrode immersed partially in the electrolyte and the remaining portion of the electrode exposed to a gaseous medium. For example, such cells are operated as metal-air cells, metal-fuel cells, and fuel cells.

3,592,695; July 13, 1971; P.J. Moran
"Metal-Air Cell Including a Composite Laminar Gas Diffusion Cathode"

A composite gas diffusion electrode has an electrically conductive porous substrate, nonnoble metal catalytic material impregnated into the substrate, at least one chemically inert porous separator positioned adjacent one surface of the catalytically impregnated substrate, and a porous, electrically conductive sheet positioned adjacent the opposite surface of the separator and in electrical contact with the substrate, the porous sheet having a lower oxygen overvoltage than the substrate. During the charging of a cell employing the above electrode, the porous sheet provides isolation of oxygen gassing at the porous sheet while the porous separator provides physical spacing of the oxygen gassing at the porous sheet from the catalytically impregnated substrate.

3,594,235; July 20, 1971; P.J. Moran
"Method of Using a Metal-Air Cell with a Tetra-Alkyl-Substituted Ammonium Electrolyte"

A method of generating electrical energy from a metal-air cell includes providing at least one gas diffusion cathode, supplying an oxygen oxidant to the cathode, providing at least one anode, and providing for the cathode and anode an aqueous electrolyte selected from the class consisting of tetraalkyl-substituted ammonium hydroxides and tetraalkyl-substituted ammonium halides. The electrolyte eliminates or substantially reduces the accumulation of carbonate reaction products which are formed from the reaction of carbon dioxide from the air with some other alkaline electrolytes.

3,598,655; August 10, 1971; R.P. Hamlen and E.C. Jerabek
"Metal-Air Cell"

> A metal-air cell has at least one cathode electrode, at least one precipitate-producing anode electrode spaced from the cathode electrode, and an absorbent electrolyte matrix positioned between and in contact with the electrodes, the matrix extending beyond one of the edges of the respective electrodes and adapted to contact a separate electrolyte source. This cell provides a structure in which a self-cleaning anode is produced by the employment of the extended matrix and separate electrolyte source.

3,600,228; August 17, 1971; A.B. La Conti
"Multiple Electrolyte High Voltage Cell"

> The potential of a fuel cell is increased by placing an alkaline electrolyte adjacent the anode and an acid electrolyte adjacent the cathode. The alkaline and acid electrolytes are separated by a cation exchange membrane. Hydrogen and oxygen may be obtained by electrolysis at a lower impressed potential than with cells having an acid or alkaline electrolyte only. The hydrogen evolving electrode is placed in contact with the acid electrolyte and the oxygen evolving electrode contacts the alkaline electrolyte with the cation exchange membrane separating the electrolytes.

3,600,229; August 17, 1971; R.A. Torkildsen
"Fuel Cell Automatic Purge System Providing Reliable Initiation and Termination of Purging"

> A combined sensing and output power actuating control system for power generation assemblies of the fluid consuming cell type such as fuel cells and fuel batteries. The control system is designed to sense or detect build-up of undesired inert gaseous matter in the consumable and regenerable fluid supply system of such assemblies by sensing the need for purging of the consumable and regenerable fluid supply system. After some predetermined safe period of purging, the assembly will indicate the need for termination of the purging operation. A signal is then generated in response to the reduction in the inert concentration to actuate the purge control circuitry and terminate the purging operation.

3,607,323; September 21, 1971; C.S. Tedmon, Jr. and S.P. Mitoff
"Sintered Stabilized Zirconia Containing Discontinuous Phase of Cobalt-Containing Oxide"

> The resistance of a zirconia mass or layer to gas penetration is maximized by sintering the zirconia body with a cobalt-containing sintering agent present therein as a separate phase. In the case of stabilized zirconia the cobalt-containing sintering agent may be introduced either during or after stabilization.

3,607,422; September 21, 1971; P.J. Moran
"Metal-Air Cell"

> A metal-air cell has a cathode, an aqueous electrolyte, a magnesium anode, and an electronically conductive porous grid positioned in contact with the electrolyte and between the cathode and the anode, and connected electrically to the cathode. A modified metal-air cell has such a porous grid connected electrically to a separate reference cathode which is in contact with both the air supply and the electrolyte.

3,607,425; September 21, 1971; W.A. Titterington, J.P. Gallagher and R.W. Milgate, Jr.
"Fuel Battery Including Apparatus Innovations Relating to Heat Transfer, Reactant Distribution, and Electrical Output"

> An internally tapped fuel battery which serially feeds a first group of cells between one terminal and an internal tap and a second group of cells be-

tween the internal tap and a second terminal. In addition, the external wall of the battery includes a deformable thermally conductive member to facilitate heat transfer from the battery.

3,607,426; September 21, 1971; L.W. Niedrach
"Fuel Cell Electrode"

Fuel cell electrodes are composed of a mixture of catalytic and gas-adsorbing materials of a chromium-tungsten oxide and a metal selected from the class consisting of noble metals and alloys of noble metals, a current collector, and a binder bonding the materials together and to the current collector in electronically conductive relationship. An adhesive binder is employed which is not chemically attacked by the electrolyte or the reactant fluid of the cell in which the electrode is used. Such an electrode is particularly useful in a fuel cell employing a fuel containing carbon monoxide, such as reformer gas, an acid electrolyte, and under various operating conditions.

3,607,427; September 21, 1971; D.W. White
"Electrically Coupled Fuel Cell and Hydrogen Generator"

A solid oxygen-ion electrolyte fuel cell electrically coupled directly to a solid oxygen-ion electrolyte water dissociation cell (for the generation of hydrogen) is described. Hydrocarbon fuel is reacted with air and/or steam to produce a reducing gas mixture, which is admitted to the coupled cells to depolarize the anode of the dissociation cell and serve as fuel for the fuel cell. The fuel cell produces low voltage direct current useable, as produced, by the dissociation cell and, when these cells are directly coupled electrically, either cell will automatically adjust to altered operating conditions or cell characteristics of the other cell.

3,615,842; October 26, 1971; D.W. Craft and F.J. Porter, Jr.
"Method and Apparatus Comprising an Electrochemical Ion Exchange Membrane Purge Pump in Combination and Fuel Cell Combination"

A purge pump is arranged for removing inert gases and impurities from the reactant gases in fuel cells or fuel cell systems. Normal operation of fuel cells results in the accumulation of inert gases in the reactant gases after a period of operation, ultimately tending to blanket an electrode, causing the performance of the fuel cell to deteriorate. The purge pump, which is arranged to remove such materials at intervals, includes a housing incorporating an ion exchange membrane with electrodes on opposite sides thereof. Current collectors engaging the electrodes are formed of corrugated shape, the corrugations of one collector being transverse to those of the other collector. A chamber is provided adjacent one electrode for accumulating inert gases received from the fuel cell system. The housing of the purge pump has incorporated therewith an inlet valve at one end of the chamber for conducting inert gases from the system to the chamber and a second valve at the other end of the chamber for effecting discharge of inert gases from the chamber at intervals. When the inert gases accumulate in the chamber to an extent affecting performance by a predetermined amount an associated electrical circuit causes reversal of current through the purge pump. This results in an increase in pressure in the chamber causing the inlet valve to close and the discharge valve to open, thereby effecting a purging of the inert gases from the chamber. The circuit is arranged to return the system to normal operation after a time which insures complete purging without excessive loss of reactant gases.

3,615,843; October 26, 1971; P.J. Moran
"Method of Charging a Metal-Air Cell"

A method of charging a secondary metal-air cell includes providing a composite laminar gas diffusion cathode which has a porous, electrically conductive sheet in electrical contact with the substrate of the cathode, the sheet

having a lower oxygen overvoltage than the cathode substrate, positioning
the cathode with the sheet facing the anode, and applying a charging current
across the electrodes whereby oxygen evolution during charging occurs substantially only at the porous sheet.

3,615,846; October 26, 1971; R.E. Plank
"Fluid Expansion Bladder"

An expandable bladder for permitting expansion of a fluid within the bladder
due to temperature, pressure or chemical changes without damage to fragile
members adjacent the fluid. The thin, flexible, fluid-tight bladder is disposed
about a pool of liquid having gas dissolved therein and a fragile member located adjacent thereto. When fluid expansion occurs the bladder member
is displaced by the gas or liquid to provide accommodation thereof, thus preventing any substantial fluid pressure force from acting on the fragile member.

3,615,850; October 26, 1971; P. Chludzinski and J.W. Harrison
"System and Process Employing a Reformable Fuel to Generate Electrical
Energy"

A system for generating electrical energy reacts fuel and water to form hydrogen which is delivered to a fuel cell in an amount exceeding its requirements, the surplus hydrogen being burned to maintain the endothermic fuel-
water reaction. Product water from the burning and fuel cell operations is
recovered to continue the fuel-water reaction, which water is preheated by
the electrolyte, which advantageously may be phosphoric acid.

3,616,335; October 26, 1971; W.N. Carson, Jr. and J.L. Manganaro
"Electrochemical Method of Generating Hydrogen"

Method of producing hydrogen which involves a magnesium reactant and an
aqueous electrolyte with a dewatering agent, and an electrical energy source
in contact with the magnesium reactant and the cathode. The dewatering
agent reduces substantially the volume of magnesium hydroxide precipitation
within the cell during its operation by freeing water from the precipitate
thereby increasing its density.

3,619,296; November 9, 1971; L.W. Niedrach and G.J. Haworth
"Fuel Cell Electrode"

A fuel cell electrode is composed of catalytic and gas adsorbing materials
which consist of a chromium-tungsten oxide as a support, a metal dispersed
on the support, the metal selected from the class consisting of noble metals
and alloys of noble metals, a current collector, and a binder bonding the materials together and to the current collector in electrically conductive relationship. Further, the catalytic and gas adsorbing materials comprise a chromium-tungsten oxide as a support, a support of carbon with boron dissolved
therein, a mixture of carbon with boron dissolved therein and boron carbide,
or carbon, and a metal dispersed on both of the supports, the metal selected
from the class consisting of noble metals and alloys of noble metals. An adhesive binder is employed which is not chemically attacked by the electrolyte
or the reactant fluid of the cell in which the electrode is used. Such an electrode is particularly useful in a fuel cell employing a fuel containing carbon
monoxide, such as reformer gas, an acid electrolyte, and under various operating conditions.

3,619,297; November 9, 1971; P.J. Moran
"Method of Charging Secondary Metal-Air Cell"

A composite gas diffusion electrode has an electrically conductive porous
substrate, a noble metal or nonnoble metal catalytic material impregnated
into the substrate, and a porous, electrically conductive sheet adjacent to
and spaced from one surface of the substrate and electrically insulated there-

from, the porous sheet having a lower oxygen overvoltage than the substrate, and electrical insulation means disposed between the sheet and the substrate. A modified gas diffusion electrode has electrical insulation means of at least one chemically inert, porous separator positioned between the substrate and the porous sheet. During the charging of a cell employing the above types of electrodes, the porous sheet provides isolation of oxygen gasing at the porous sheet while the porous separator provides additional physical spacing of the oxygen gasing at the porous sheet from the catalytically impregnated substrate.

3,629,008; December 21, 1971; W.T. Grubb and C.E. Cliche "Composite Electrode and Electrochemical Cell with at Least One Gas Diffusion Electrode"

A composite gas diffusion electrode has an electrode body, at least one water-soluble polymeric thickening or gelling agent swellable in an aqueous electrolyte, and a layer of the thickening agent adhering to and covering at least a portion of one of the surfaces of the electrode body. Upon subsequent use in contact with an aqueous electrolyte, the thickening agent swells on the covered surface of the electrode body preventing electrolyte leakage therethrough, preventing drowning of the electrode, reducing loss of any dissolved fuel in the electrolyte, and immobilizing at least partially the electrolyte. Additionally, an electrochemical cell with at least one gas diffusion electrode has the above type of thickening agent covering various portions of the interior structure of the cell. Upon subsequent use in contact with an aqueous electrolyte, the thickening agent swells on covered portions of the interior structure preventing electrolyte leakage therethrough, and immobilizing at least partially the electrolyte.

3,635,763; January 18, 1972; L.W. Niedrach and W.T. Grubb "Fuel Cell Electrode"

Fuel cell electrodes are composed of a mixture of catalytic and gas adsorbing materials which consist of a chromium-tungsten oxide and a metal dispersed on a support, the metal selected from the class consisting of noble metals and alloys of noble metals, with a current collector, and a binder bonding the materials together and to the current collector in electronically conductive relationship. An adhesive binder is employed which is not chemically attacked by the electrolyte or the reactant fluid of the cell in which the electrode is used. Such an electrode is particularly useful in a fuel cell employing a fuel containing carbon monoxide, such as reformer gas, an acid electrolyte, and under various operating conditions.

3,635,764; January 18, 1972; J.L. Setser and E.P. Schneider, Jr. "Combined Wastewater Treatment and Power Generation"

Both apparatus and a method for the treatment of domestic wastewater are described. A substantial reduction in refractory nutrient content of such waters is effected by the addition of an electrolyte, if needed, and passage of the wastewater through battery-cell structure having as one electrode thereof a metal the oxide of which, when formed in water, acts as a flocculent. Air introduced into the cell prevents suspended sludge from settling in the cell and provides (or supplements) the depolarizing oxidant. Conduct of the process simultaneously produces direct current electrical power, as well, which may be employed to release chlorine from an aqueous chloride solution for combating pathogenic organisms in the wastewater.

3,649,361; March 14, 1972; J. Paynter and J.R. Morgan "Electrochemical Cell with Manganese Oxide Catalyst and Method of Generating Electrical Energy"

An electrochemical cell is described which includes a gas diffusion cathode, an alkaline electrolyte and operates at room temperature. The electrochemical cell has an improved gas diffusion cathode which comprises a finely divided, electrically conductive carbon powder and a base metal catalyst of a manganese oxide, or a manganese oxide with one or more other base metal catalysts. Such a cell provides performance approaching that of a cell including a gas diffusion cathode with a platinum catalyst. A method of generating electrical energy from such a cell is also described.

3,650,836; March 21, 1972; W.T. Grubb and R.A. Macur
"Electrochemical Cell with at Least One Gas Diffusion Electrode"

An electrochemical cell has at least one positive wet-proofed gas diffusion electrode with its major portion within the casing covered with a fuel and oxidant impervious mask, and with its remaining portion within the casing covered with at least one water-soluble polymeric thickening agent swellable in the aqueous alkaline electrolyte of a fuel-electrolyte solution and an alkali metal silicate dissolved in the solution. This cell eliminates or reduces substantially the conventional problems of carbonate crust formation, electrolyte leakage, drowning of the gas diffusion electrode, and loss of fuel.

3,653,969; April 4, 1972; W.A. Titterington, R.W. Milgate, Jr. and J.P. Gallagher
"Fuel Cell System with Plural Fuel Cells"

A system for purging a main fuel cell of impurities comprises an auxiliary fuel cell which utilizes at least one of the streams of gas provided to the main fuel cell, which stream carries gaseous impurities downstream from the main fuel cell to the auxiliary fuel cell whereat they are trapped.

3,676,220; July 11, 1972; W.J. Ward III
"Facilitated Transport Liquid Membrane"

A confined liquid membrane having an electrode in contact with each surface thereof is provided with a concentration of a nonvolatile transporting specie in solution therein, which transporting specie can be oxidized or reduced from a first to a second valence state. The transporting specie must be reactive with a specific gas in at least one of the multiple valence states. If the transporting specie is reactive with the gas in both valence states, such reactions must have different equilibrium constants. By bringing the specific gas into contact with one surface of such a liquid membrane and applying an electrical potential across the electrodes on each side thereof, the specific gas may be pumped through the membrane against a concentration gradient. Similarly, while a concentration gradient in the specific gas exists across the structure described, an electrical potential difference develops across the film and the electrodes.

3,701,687; October 31, 1972; W.T. Grubb and R.A. Macur
"Electrode Containing a Catalytic Coating of Platinum, Palladium, and a Third Metal and Fuel Cell Utilizing Same"

An electrode is described which has a catalytic coating of platinum, palladium, and a third less noble metal on a substrate. Such an electrode provides a high performance anode for electrochemical cells.

3,703,358; November 21, 1972; W.N. Carson, Jr. and J.L. Manganaro
"Method of Generating Hydrogen with Magnesium Reactant"

A method of generating hydrogen is described which has a magnesium reactant and an aqueous electrolyte with a dewatering agent. A method is also described for generating hydrogen. The dewatering agent reduces substantially the volume of magnesium hydroxide precipitation within the cell during its operation by freeing water from the precipitate thereby increasing its density.

3,713,892; January 30, 1973; P.J. Moran
"Method of Charging Secondary Metal-Air Cell"

A composite gas diffusion electrode has an electrically conductive porous substrate, nonnoble metal catalytic material impregnated into the substrate, at

least one chemically inert porous separator positioned adjacent one surface
of the catalytically impregnated substrate, and a porous, electrically conductive sheet positioned adjacent the opposite surface of the separator and in
electrical contact with the substrate, the porous sheet having a lower oxygen
overvoltage than the substrate. During the charging of a cell employing the
above electrode, the porous sheet provides isolation of oxygen gassing at the
porous sheet while the porous separator provides physical spacing of the oxygen gassing at the porous sheet from the catalytically impregnated substrate.

3,730,774; May 1, 1973; D.W. McKee and A.J. Scarpellino, Jr.
"Ammonia Fuel Cell with Iridium Catalyst and Method of Using the Cell"

A fuel cell has a catalyst material of iridium or an iridium alloy in the anode
chamber and heating means associated with the chamber whereby ammonia
vapor, which is fed to the chamber, is dissociated into its constituents, nitrogen and hydrogen. The dissociated gases are supplied to the anode as fuel.
A hydrogen permeable membrane can also be employed between the catalyst
material and the anode whereby hydrogen is supplied to the anode as fuel.
A wide variety of electrodes and electrolyte are suitable in the fuel cell.

3,819,416; June 25, 1974; W.T. Grubb and C.E. Cliche
"Electrochemical Cell with Inside Layer of Water-Swellable Polymer"

A composite gas diffusion electrode has an electrode body, at least one water-
soluble polymeric thickening or gelling agent swellable in an aqueous electrolyte, and a layer of the thickening agent adhering to and covering at least a
portion of one of the surfaces of the electrode body. Upon subsequent use
in contact with an aqueous electrolyte, the thickening agent swells on the
covered surface of the electrode body preventing electrolyte leakage therethrough, preventing drowning of the electrode, reducing loss of any dissolved
fuel in the electrolyte, and immobilizing at least partially the electrolyte.
Additionally, an electrochemical cell with at least one gas diffusion electrode
has the above type of thickening agent covering various portions of the interior structure of the cell. Upon subsequent use in contact with an aqueous electrolyte, the thickening agent swells on covered portions of the interior structure preventing electrolyte leakage therethrough, and immobilizing
at least partially the electrolyte.

3,833,420; September 3, 1974; F.G. Will
"Battery Casing and Regenerative Metal-Water Battery"

A battery casing is disclosed which includes a reservoir in sealed communication with an inner vessel of a solid crystalline ion-conductive material, an
outer vessel surrounds the inner vessel and defining therewith a narrow fluid
passageway, and a fluid storage and circulating system for providing fluid
flow through the passage. A regenerative metal-water battery has the above
type of casing with a sodium or potassium anode in the inner vessel and water
as the oxidant in the outer vessel.

3,833,422; September 3, 1974; F.G. Will and S.P. Mitoff
"Regenerative Metal-Water Battery"

A regenerative metal-water battery is disclosed which includes a metallic sodium or potassium reservoir in sealed communication with an inner vessel of
a solid metal ion-conductive material, the metallic sodium or potassium in
the reservoir, means for heating the metal, the metal adapted to provide an
anode in the inner vessel of the battery, and water as the oxidant in the
outer vessel of the battery and in contact with the outer wall of the inner
vessel.

General Engineering Laboratories, Inc., Westford, Mass.

3,881,956; May 6, 1975; T.S. Williams

"Fuel Cell Construction"

A stacked array of unit cells, fed and/or drained by a conductive liquid, is subdivided into cell groups which are further subdivided into blocks, which in turn have subblock subdivisions. The blocks have integral elongated passages for carrying liquid, consistent with limiting intercell I^2R losses through the liquid distribution lines to tolerable levels. The blocks also have a flexible modular construction of components thereof which accommodates model variations and also enhances uniformity of said liquid flow in all such variations.

3,920,474; November 18, 1975; R. Zito, Jr. and L.J. Kunz, Jr.
"Method of Operating a Fuel Cell Using Sulfide Fuel"

A fuel cell with all-liquid feed and effluent is made feasible through the use of fuels comprising a liquid source of sulfide ion. The fuel cell comprises anolyte and catholyte compartments, and anode and cathode electrodes. The compartments are separated by an ion-transfer or mechanically porous membrane or the like. Gas producing reactions are suppressed through competing liquid product reactions.

General Motors Corp., Detroit, Mich.

3,544,373; December 1, 1970; D.A.J. Swinkels
"Electrochemical Cell with Layered Electrode of Ceramic and Carbon and Method"

An electrode for an electrochemical cell which electrode comprises at least two porous materials, one of which is electrically conductive and not wettable by the cell's electrolyte, and the other is electrically nonconductive and wettable by the cell's electrolyte. The wettable, nonconductive material faces the electrolyte. The nonwettable, conductive portion of the electrode faces the cell reactant.

3,546,021; December 8, 1970; G.M. Craig
"Galvanic Cell and Method of Operation"

A reversible electrode assembly for a liquid reactant of a regenerative fuel cell in which the assembly contains a reservoir for receiving and dispensing the liquid reactant, means for buoying the reactant up on cell discharge and means for effecting a differential pressure in the direction of the reservoir during cell recharge.

3,551,206; December 29, 1970; P.T. Ross
"Electrical Insulation in Lithium-Chlorine Electrochemical Cell"

A method and means for electrically insulating a lithium half-cell structure from a chlorine half-cell structure in the presence of an Li-LiCl solution. A pocket of protective gas is provided between the Li-LiCl solution and a ceramic oxide to preclude loss of the ceramic's insulator characteristics.

3,560,263; February 2, 1971; D.A.J. Swinkels
"Electrolysis"

Improvement to process for electrolytically dissociating fused alkali metal halides by adding an alkali metal to the fused halide to cause increased wetting of the carbon or graphite anodes by the fused halides. A preferred current reversal pulsing technique for introducing the alkali metal into the halide is described.

3,560,265; February 2, 1971; G.M. Craig
"Galvanic Cell with a Matrix Electrode"

A matrix electrode for the molten lithium of a Li|LiCl|Cl$_2$ fuel cell, which electrode is a porous bundle of specifically oriented filaments of a material which is preferentially wetted by the lithium. A particular filament bundle has spacer filaments intermittently disposed between layers of woven or meshed filaments. A simply scaled battery of cells having matrix electrodes is described.

3,575,720; April 20, 1971; G.M. Craig
"Insulator Means for Lithium-Chlorine High Temperature Battery"

Substantially environmentally inert electrical insulators and insulated conductors for use in the Li |LiCl |Cl$_2$ electrochemical system. The insulators and conductors are composites having Cl$_2$-inert and Li-inert barriers which shield vulnerable composites from chlorine or lithium attack. Specific materials disclosed include combinations of aluminum nitride, carbon, graphite, molybdenum, Kovar and stainless steel.

3,575,722; April 20, 1971; G.M. Craig
"Level Indicator for Conductive Liquids"

A liquid level indicator measures the level of a conductive liquid contained in a metallic reservoir. A sidewall portion of the reservoir is included in a current conductive path supplied by a source of current applied to one end of the reservoir. A measuring system monitors both the voltage drop produced by current flow through a portion of the reservoir sidewall and also the value of current which produces the voltage drop. The ratio of the amount of voltage drop to the value of current flow is measured to determine the resistance of the sidewall portion not in contact with conductive liquid to give an indication of the level of the conductive liquid.

3,586,540; June 22, 1971; J.J. Petraits and G.T. Sinnet
"Attitude and Gravity Insensitive Galvanic Cell"

An alkali metal |alkali metal halide |halogen galvanic cell including an excess electrolyte reservoir, an electrochemically active region filled with electrolyte and containing the cell's electrodes, and a porous, carbon or graphite, non-electrolyte-wettable, electrolyte flow restrictor located between the reservoir and the region. The restrictor keeps the region filled and the electrodes covered with electrolyte regardless of the cell's orientation and acts as an electrolyte relief means to prevent excessive electrolyte pressure buildup in the region. The restrictor's pores are larger than the pores of the carbon or graphite halogen electrode such that the flooding pressure of the restrictor is less than the flooding pressure of the electrode.

3,607,421; September 21, 1971; J.J. Werth
"Fuel Cell Gas Supply System and Method"

Chlorine is stored in a low vapor pressure condition by adsorption on activated charcoal and is continuously released from the charcoal by means of a closed loop recirculation system between the charcoal and the fuel cell. The closed loop system recirculates a quantity of chlorine in excess of that required for electrochemical consumption in the cell. The excess chlorine acts as an efficient heat transfer medium for conducting heat away from the cell and into the charcoal whereby the temperature of the charcoal is increased, more chlorine is released and the pressure in the closed loop maintained. The rate of heat transfer from the cell to the charcoal is controlled by the recirculation rate of the excess chlorine, which is, in turn, controlled by the system's pressure.

3,674,567; July 4, 1972; T.G. Bradley and D.A.J. Swinkels
"Electrolysis Cell and Process Using a Wick Electrode"

A combined electrode and liquid reactant transporting means for at least one liquid reactant of an electrolysis cell. A wick extends from within the cell into a liquid reactant collection reservoir. The wick is preferentially wet by the liquid reactant providing a plurality of capillary flow paths between the storage and cell regions of the system. An electrolysis process using the wick electrode is described.

3,719,529; March 6, 1973; D.P. Lake
"Voltaic Cell and Method Using Dilute Fuel Gases to Generate Electrical Power"

A voltaic cell operating on oxidizing gaseous mixtures containing a dilute
fuel gas. Functionally similar catalytic gas electrodes are used with a phosphoric acid electrolyte and an ion exchange resin gas diffusion barrier in between. The electrodes are saturated with electrolyte. Means are provided to
reduce current variation due to water content variation in the phosphoric
acid. The cell can operate on fuel gas mixtures of air containing low concentrations of carbon monoxide.

3,772,086; November 13, 1973; E.J. Zeitner, Jr., M.E. Wheatley, R.R. Witherspoon and S.G. Meibuhr
"Method of Making Anodes for Hydrazine Fuel Cells"

A method of making nickel boride catalyzed anodes for hydrazine fuel cells
including the step of codepositing electrolytic nickel and carbonyl nickel
particles onto an appropriate conductive substrate to form a catalyst-receiv-
ing surface. Sufficient electrolytic nickel is deposited to tack the carbonyl
nickel particles to the substrate and to themselves, yet not so much as to appreciably degrade the characteristic rough and jagged surfaces of the carbonyl
nickel particles.

3,844,842; October 29, 1974; R.A. Sharma and J.F. Rhodes
"Insulating Seal for Molten Salt Battery"

An interelectrode insulating seal comprising lithium fluoride and a low thermal expansion filler for use in high temperature batteries having lithium
fluoride-saturated molten salt electrolytes. The filler mitigates the effects of
lithium fluoride's thermal expansion.

3,859,181; January 7, 1975; S.G. Meibuhr
"Stabilization of Nickel Boride Catalyst in Potassium Hydroxide Electrolytes"

This invention relates to hydrazine fuel cell electrodes employing nickel
boride as a catalyst and used with aqueous potassium hydroxide electrolytes.
More particularly, this invention relates to contacting the nickel boride with
a hydrogen atmosphere for a time and temperature sufficient to retard degradation of its effectiveness upon continued exposure to aqueous potassium
hydroxide and to thereby extend the useful life of electrodes made therefrom.

3,985,578; October 12, 1976; R.R. Witherspoon and R.L. Adams
"Tailored-Carbon Substrate for Fuel Cell Electrodes"

In accordance with a preferred embodiment of this invention, the polarization or overvoltage of a carbon black-based air cathode is significantly reduced by adding from 1 to 5%, by weight based on the total carbon composition, of carbon derived from poly(vinylidene chloride) (hereinafter
PVdC) to the carbon black, prior to the formation of the electrode. To accomplish this, the porous and particulate carbon black is initially blended
with a solution of PVdC; the resulting blend is dried, and then dehydrochlo-
rinated and baked at a temperature in the range of from $900°$ to $1100°C$.
The electrocatalytic layer of the cathode is then prepared from this tailored
carbon substrate.

4,007,059; February 8, 1977; R.R. Witherspoon, E.M. Domanski and J.A. Davis
"Electrochemical Cell Electrode Separator and Method of Making it and Fuel
Cell Containing Same"

An ion permeable material particularly useful for electrode separators. The
material is a uniform mixture of 30 to 70% by weight colloidal size asbestos
fibers and 20 to 40% by weight unsintered fine polytetrafluoroethylene fibers,
bonded together with uniformly dispersed particles of a fluorinated ethylene-
propylene copolymer comprising about 5 to 15% by weight of the material.

Colloidal size particles of an alkaline earth metal oxide are dispersed throughout the material. The material is initially prepared in a modified papermak-
ing process and then heated to a temperature high enough to fuse the fluorinated ethylene-propylene copolymer but not high enough to sinter the
polytetrafluoroethylene fibers. An improved zinc-air hybrid fuel cell and an
improved silver oxide-zinc secondary battery are made possible with electrode
separators such as described herein.

Globe Union Inc., Milwaukee, Wis.

3,645,795; February 29, 1972; W.E. Elliott and J.R. Huff
"Electrochemical Cells"

Electrochemical primary and secondary cells which utilize alkali or alkaline
earth metal anodes such as lithium, potassium, sodium, calcium and magnesium, and catalytic air electrodes which utilize components of the atmosphere, namely, oxygen and water as a source of electrochemical power.
The electrolyte system in the electrochemical cells comprises a nonaqueous
electrolyte solvent, namely, an organic solvent of sufficient polarity to dissolve an electrolyte salt.

Gould Inc., Mendota Heights, Minn.

3,801,375; April 2, 1974; A.M. Jaggard
"Assemblying Electrodes in Electrochemical Cells"

A process for forming electrode leads through a cell housing by inserting of
a unitary cathode pin through the housing so that the cathode pin simultaneously functions as a low resistance electrical contact with a collector screen
and as an external cell terminal. Another process of the invention is the inserting of a unitary anode pin through the cell housing so that the anode pin
simultaneously functions as an external cell terminal and an internal current
collector.

GTE Sylvania Inc., Seneca Falls, N.Y.

3,730,706; May 1, 1973; W.E. Buescher and D.R. Kerstetter
"Strengthened Cathode Material and Method of Making"

A cathode material containing discrete islands of thoria enriched nickel particles is disclosed, as well as a method for fabricating the same. Powder
metal techniques are utilized and the material comprises a major amount of
substantially pure nickel and a minor amount of a composite material. The
composite material comprises composite particles of cotreated nickel and
thoria. The thoria is homogeneously dispersed with respect to its cotreated
nickel but nonhomogeneously dispersed with respect to the remainder of the
nickel.

Gulf Oil Corp., Houston, Tex.

3,576,679; April 27, 1971; P.R. Shipps
"Electrode Assembly"

The disclosure shows a thin electrode assembly for a storage battery or fuel
cell using a gaseous reactant and a liquid electrolyte. A storage battery cell
stack employs several electrode assemblies each constructed of a thin plate
porous to the gaseous reactant which plate is connected mechanically and
electrically at a plurality of points across its surface to a very thin metal
plate that is deformed to provide a gas plenum chamber between the two
plates. The exposed surface of the deformed plate is used as a substrate to
support a deposit of an electropositive material, like zinc, for the anode of
the next cell in the cell stack.

3,708,345; January 2, 1973; J.F. Loos and T.F. Unkle, Jr.
"Electrochemical Energy Conversion System"

> An energy conversion system utilizing a plurality of electrochemical cells employing the zinc-oxygen electrochemical couple and a circulating alkaline electrolyte which is pumped through the cells to remove the zinc oxide reaction products therefrom. A high electrolyte flow rate is achieved through each of the cells, which scours the zinc oxide reaction products from the zinc anodes, by providing a jet pump in association with each electrochemical cell unit. The individual jet pumps transform a relatively low flow rate of high pressure liquid electrolyte from a high pressure pump into a substantially higher circulation flow of low pressure electrolyte within each of the individual electrochemical cells.

3,717,505; February 20, 1973; T.F. Unkle, Jr. and J.F. Loos
"Electrochemical Cell Stack"

> A cell stack for use in a battery or fuel cell using a liquid electrolyte. A plurality of individual frames made from electrically insulating material, each having a cavity extending transversely therethrough and separate elongated supply and return electrolyte passageways leading to and from the cavity. An electrode assembly is disposed in a recess provided in one surface of one of the frames and sandwiched between it and another frame. Each frame is bonded to the next adjacent frame to provide a complete seal surrounding the periphery of the cavity, each group of two adjacent electrode assemblies and three frames providing an electrochemical cell. The frames may be injection-molded from polysulfone and solvent-bonded by a chlorinated hydrocarbon.

Gulton Industries, Inc., Metuchen, N.J.

3,625,769; December 7, 1971; A.E. Lyall
"Fuel Cell"

> A room-temperature-operated fuel cell comprising an oxygen electrode, a lithium metal-containing electrode, and an electrolyte comprising an inert, aprotic organic solvent, exemplified by dimethylsulfoxide, which contains an inorganic or organic ionizable salt sufficient in amount to attain saturation. The salt may, or may not, be an oxygen carrier. In those instances where a salt which is not an oxygen carrier is employed, lithium oxide initially is incorporated into the electrolyte.

Hitachi Ltd., Tokyo, Japan

3,607,424; September 21, 1971; Y. Maki, M. Matsuda and T. Kudo
"Solid Electrolyte Fuel Cell"

> A solid electrolyte which consists of a mixture of cerium oxide and gadolinium oxide or said mixture with magnesium oxide added thereto, and which has an excellent oxygen-ion-conductive characteristic at a relatively low temperature as compared with conventional solid electrolytes.

3,698,956; October 17, 1972; M. Emoto and Y. Maki
"Alkaline Electrolyte-Zinc Anode Air-Depolarized Battery"

> An air-depolarized battery of the alkaline electrolyte-zinc anode type, in which an alkaline electrolyte-zinc anode composed of amalgamated zinc powder, alkali metal electrolyte and one or more kinds of alkaline earth metal hydroxide in combination at a given proportion is utilized.

3,861,961; January 21, 1975; T. Kudo, M. Yoshida, H. Obayashi and T. Gejyo
"Gas Electrode"

> An oxide having the perovskite type crystal structure and represented by the general formula of $Ln_{1-x}A_xCo_{1-y}Ni_yO_{3-\delta}$, (where $0 \leqslant x \leqslant 0.6$, $0.01 \leqslant y \leqslant 1.0$

and $0 \leqslant \delta \leqslant (x/2)$, and Ln denotes at least one element selected from the group consisting of La, Pr, Nd, Sm, Gd and Y, while A indicates at least one element selected from the group consisting of Ca, Ba, and Sr) is sintered into a porous sintered body. The porous sintered body is subjected to a water repellency treatment, to form a gas electrode. The gas electrode thus formed possesses far more excellent characteristics as compared with a prior art gas electrode of the carbon series, of the nickel series, or formed by kneading a perovskite type oxide.

Hooker Chemical Corp., Niagara Falls, N.Y.

3,544,379; December 1, 1970; G.T. Miller
"Fuel Cell Electrode and Process"

A fuel cell, utilizing liquid carbonaceous fuels insoluble in the aqueous electrolyte, is provided with a fuel electrode having catalytic points or projections and means for flowing a film of liquid fuel over the electrode face in contact with both such projections and the electrolyte.

3,544,380; December 1, 1970; A.N. Dey
"Method of Activating Fuel Cell Electrode by Direct Current"

The activity of the fuel electrode in a fuel cell is increased by periodic application for a short time, of a superimposed direct current in the same direction as the normal current.

3,544,381; December 1, 1970; G.T. Miller
"Fuel Cell with Wicking-Type Electrode"

A fuel cell with wicking type fuel electrode is provided by wrapping or pressing an electrically conductive base or support with a porous, hydrophobic sheet of fiber or paper upon which a catalyst is deposited. A liquid fuel, substantially insoluble in the electrolyte wicks up between the electrode base and the hydrophobic porous sheet of material to contact the catalyst at the electrolyte, catalyst, fuel interface, generating electrical current.

3,706,602; December 19, 1972; G.T. Miller
"Fuel Cell Electrode Structure Utilizing Capillary Action"

A fuel electrode is provided by wrapping or pressing a hydrophobic, porous sheet of fiber or paper, upon which a catalyst is deposited, into contact with an electrically conductive base or support. In use, a liquid fuel, substantially insoluble in the electrolyte moves by capillary action or wicks between the electrode base and the hydrophobic, porous sheet of material to contact the catalyst at the electrolyte, catalyst, fuel interface, generating electrical current.

Hughes Aircraft Co., Culver City, Cal.

3,727,058; April 10, 1973; A.F. Schrey
"Hydrogen Ion Device"

A dihydrogen orthophosphate salt crystal body, either KH_2PO_4 or $NH_4H_2PO_4$, is provided with a hydrogen transparent metal coating on opposing surfaces, and is protected on the remaining surfaces with a protective insulator coating. When the metal coated surfaces are in contact with gas mixtures having different hydrogen partial pressures, current flow between the metal coatings and hydrogen ion flow through the salt body are directly related. Thus, when connected to a power supply, the salt body acts as a hydrogen pump.

Hydronautics, Inc., Laurel, Md.

3,573,103; March 30, 1971; C.E. Brown
"Gill or Sea Water Fuel Cell"

A fuel cell utilizing sea water as the cell's supply of oxidant reactant and as the electrolyte for the cell.

Institut Francais du Petrole, Hauts-de-Seine, France

3,775,186; November 27, 1973; J. Cheron
"Fuel Cell"

> A fuel cell apparatus comprising a cell block and an oxygen-containing gas
> purifier placed by the side of cell block, a fuel recirculation circuit, an elec-
> trolyte recirculation circuit, a static device for oxygen-containing gas draft,
> fed through a fuel leakage circuit and a heat exchanger exchanging heat be-
> tween the hot electrolyte issuing from the cell block and the cold oxygen-
> containing gas purifying liquid issuing from the oxygen-containing gas pur-
> ifier, thereby causing the respective motions of the electrolyte and the pur-
> ifying liquid by double thermosiphon effect.

3,855,001; December 17, 1974; J. Cheron
"Fuel Cell with Air Purifier"

> A fuel cell comprising a cell block, fed with a comburent, a fuel and electro-
> lyte, and a comburent purifier, wherein the cell block forms a hot part of the
> fuel cell and is placed above the comburent purifier which forms the cold
> part of the fuel cell, said hot part and said cold part are connected to each
> other through a duct for the conveyance of the purified comburent to the
> cell block by natural convection and static suction means are provided for
> the exhaust of the comburent and the hot combustion products from the
> cell block.

3,861,958; January 21, 1975; J. Cheron
"Process and Devices for Recovering Electrolyte Leaks in a Fuel Cell"

> A process for the complete recovery of possible electrolyte leaks of a fuel
> cell fed with oxidizing and reducing gases comprising separately recovering
> the leaks from the respective circuits of said two gases, collecting said leaks
> while preventing any mixture of said gases and recycling to the cell the whole
> of said collected electrolyte leaks.

4,015,052; March 29, 1977; J. Cheron
"Fuel Cell"

> This fuel cell comprises circuits for collecting electrolyte leaks in the fuel
> and comburent chambers, said circuits being connected to the inlet orifice of
> the electrolyte pump. The electrolyte tank is located above the electrolyte
> pump and adjustment means provide for a pressure drop in the electrolyte
> flow between the tank and the pump inlet.

Institute of Gas Technology, Chicago, Ill.

3,718,418; February 27, 1973; D.K. Fleming, S.S. Randhava and E.H. Camara
"CO Shift Process"

> Process of conversion of CO_2 to CO in the presence of hydrogen, using cata-
> lysts of Rh and Ru metals, and alloys thereof with Pt, the Rh metals and al-
> loys being beneficiated with admixtures of reduced, amorphous tungsten or
> molybdenum oxides, and the Ru metals and alloys being beneficiated by the
> molybdenum oxide only. Precise control of the product gas composition is
> obtained by predetermined control of process parameters of input gas flow
> rate and temperature conditions in the shift reactor.

3,723,186; March 27, 1973; A.Z. Borucka, L.G. Marianowski and B.S. Baker
"Electrolyte Seal Means"

> An electrolyte mass for high temperature electrochemical cells particularly
> for use in multiple cell batteries wherein the mass includes an integral seal.
> The electrolyte mass is the molten carbonate electrolyte type and is sup-
> ported by an inert matrix or filler material. The improvement comprises a
> substantially rigid and substantially inert load supporting outer electrolyte
> periphery which is capable of withstanding the compressive forces necessary

to hold each of the electrolyte masses in place in a multiple cell battery; the central portion of the electrolyte mass is integrally formed interior of the rigid outer periphery and is the electrochemically active portion of the cell.

3,774,243; November 27, 1973; D.Y.C. Ng, S.K. Wolfson, Jr. and A.J. Appelby "Implantable Power System for an Artificial Heart"

The invention involves a hybrid power system in which the electrical energy producing device (fuel cell) is combined with an electrical energy storage device. In the hybrid power system, a fuel cell is combined with a storage battery, preferably of the silver-cadmium or nickel-cadmium type. Thus, the hybrid system contains two independent power supplies, each of which is selected to be capable of powering the blood pump alone. This important feature gives the fuel cell-storage battery hybrid system an outstanding redundancy capability. In the case of failure of the main fuel cell unit, the storage battery will provide sufficient power to operate the blood pump until the patient can be safely transferred to a hospital for emergency replacement of the fuel cell. While the safety time depends in part upon the battery capacity and the activity of the patient, the capacity of presently available batteries will not be exceeded by a safety time of up to 10 hours. Conversely, in the event of the failure of the battery, the patient can survive indefinitely on the fuel cell alone by refraining from only those activities which require peak power. The hybrid system also provides an improved peak power capability and efficiency of the system. The fuel cell reliability and life is increased since in a hybrid system the fuel cell can be permitted to operate within a narrow band of power output thus more nearly approximating steady state conditions. During normal demand, the fuel cell powers the blood pump and continuously recharges the storage battery. At peak demand almost all of the extra power is supplied by the battery while the fuel cell continues to operate at nearly a steady state. Still other benefits of the hybrid system include increased efficiency and reduction of the fuel cell size both important in view of the limited volume available for implantation, particularly in small adults. Another advantage of the fuel cell-storage battery hybrid power source system is the fact that the hybrid is silent except for a normal type of "breathing" noise. Thus the hybrid does not have the psychological and physiological effects on the patient as in the case with vibration of a "prime mover" type of system such as a steam engine.

3,823,038; July 9, 1974; D. Gidaspow, R.W. Lyczkowski and B.S. Baker "Continuous Bleed Fuel Cells"

Apparatus and method of operation of fuel cells by means of a controlled bleed at the "dead-end" of electrode gas compartments to continuously remove impurities, inevitably present in even ultrapure fuels, which accumulate and cause a drop in voltage and current during the lifetime of the cells. The bleed at such locations establishes a steady state fuel-impurity concentration distribution in the cell resulting in improved electrical characteristics. The preferred rate of bleed may be only slightly greater than the percentage of input of impurities or inerts coming in with the reactant or fuel gas. The complicated and heavy valving apparatus, undesirable voltage transients, power interruption, and permanent loss of electrolyte of prior art periodic or recycle purge systems are avoided by the present construction and method. Apparatus and method of providing supplemental thrust for spacecraft propulsion, navigation and altitude control are also disclosed.

3,837,922; September 24, 1974; D.Y.C. Ng, S.K. Wolfson, Jr. and A.J. Appelby "Implantable Fuel Cell"

An implantable fuel cell power source for an artificial heart or pacemaker device which utilizes blood carbohydrates as the anode fuel. The cathode of the implantable fuel cell is an oxygen utilizing cathode, and may be air breathing, for example, following being ventilated through a percutaneous airway by a balloon system. The fuel cell may be integrated as part of a heart pump prosthetic mechanism, or placed at some remote location where a portion of the venous blood can be diverted to the fuel cell on its way to the heart. The anode chamber of the fuel cell may form a portion of the atria or the blood collecting plenum of the heart pump, but preferably is placed at the outflow passage of the prosthetic heart pump right ventricle so that the pumping force provides enough power to overcome the resistance of both the anode and the pulmonary vascular bed. The anode is separated from the whole venous blood by a thin, porous membrane capable of passing a blood ultrafiltrate containing the oxidizable organics. The organics contact the catalyst anode which is preferably of a "felt"-type construction, with the ultrafiltrate at approximately neutral pH acting also as the electrolyte. The anode is backed by a similar membrane, on the obverse side of which is coated the cathode catalyst. The implanted fuel cell may be connected in parallel with an implanted storage battery to form a hybrid power system which gives a fuel cell-storage battery hybrid system an outstanding redundancy capability.

3,865,924; February 11, 1975; D. Gidaspow and M. Onischak
"Process for Regenerative Sorption of CO_2"

A process and apparatus for the removal of carbon dioxide from a gaseous stream, wherein the CO_2 content ranges from trace amounts up to 50 mol %, and for subsequent transfer to another gas stream by thermal regeneration. A special composition comprising a finely ground mixture of potassium carbonate and alumina, which has synergistic properties, is used as the absorbent. In a preferred apparatus embodiment, the mixture is incorporated into a rotary regenerative wheel.

Ionics, Inc., Watertown, Mass.

3,607,417; September 21, 1971; W.A. McRae, and J.O'M. Bockris
"Battery Cell"

A primary and/or secondary battery cell is disclosed in which the high electrochemical capacity and high voltage obtainable from the alkali metal-oxygen system is achieved without the inherent risk of the explosively violent reaction of the alkali metal with the aqueous catholyte of the cell by interposing an ion-permeable but water-impermeable substantially insoluble laminar membrane-diaphragm between said metal and said aqueous electrolyte. The laminar membrane-diaphragm comprises a supporting microporous material such as, for example, a ceramic material which is fused or otherwise attached to a thin nonporous substantially catholyte insoluble alkali metal composition which is permeable to the alkali metal ions.

Isotopes, Inc., Westwood, N.J.

3,615,848; October 26, 1971; D.D. Sibenhorn
"Thermal Control for Fuel Cell Module"

An assembly of fuel cells or like source of electrical power having mounted on an outer surface of the fuel cell module interleaved sets of high and low thermal emissivity surfaces, with one set of surfaces being stationary and the other set of surfaces being movable. Means are provided for sensing the thermal condition of the fuel cell module and for moving one set of surfaces relative to the other set of surfaces to vary the thermal radiation from the fuel cell module and thus to control the temperature of the fuel cell module.

3,697,325; October 10, 1972; J. Baude
"Purge Control for Fuel Cell"

Impurities accumulated within a fuel cell from the reactants are flushed from the cell by periodic operation of valves which permit reactant flow at a sufficiently high rate to purge, i.e., clean the cell plates. Operation of the valves is governed by the ampere-hour output of the fuel cell. The voltage developed by the flow of fuel cell load current through a shunt controls the frequency of output pulses from a voltage-to-frequency converter, and the pulses are supplied to frequency dividing means including a binary type counter which controls the purge valves. The submultiple of the input pulse frequency to the counter is selectively variable to adjust the number of ampere-hours between purges, and the frequency dividing means includes a monostable multivibrator whose period is selectively variable to adjust the duration of each purge.

Israeli Prime Minister's Office, Jerusalem, Israel (with Dead Sea Works Ltd.)

3,578,503; May 11, 1971; M.R. Bloch, C. Forgacs and R.V. Bobker
"Devices for the Electrochemical Generation of Electrical Current"

Leesona Corp., Warwick, R.I.

3,553,022; January 5, 1971; D.P. Gregory
"Electrochemical Cell"

> An improved lightweight electrode comprising an inert polymer in a major amount and a conductive material in a minor amount formed as a conductive matrix and having a catalytic layer on at least one surface of the conductive matrix and the disposition of the electrode in an electrochemical cell are described.

3,553,024; January 5, 1971; J.H. Fishman
"Electrochemical Device Utilizing an Electrode Comprising a Continuous PTFE Film Which is Gas Permeable and Free from Liquid Electrolyte Seepage"

> An improved electrode for use in electrochemical cells and electrochemical cells utilizing the improved electrodes are described. The improvement resides in the use of an unsintered polytetrafluoroethylene film in contradistinction to a sintered polytetrafluoroethylene film in an electrode construction. The unsintered polytetrafluoroethylene film is adjacent a catalyst layer. The electrodes provide enhanced electrochemical performance and stability in the electrochemical cell.

3,553,027; January 5, 1971; H.G. Oswin, J.E. Oxley and C.W. Fleischmann
"Electrochemical Cell with Lead-Containing Electrolyte and Method of Generating Electricity"

> An improved rechargeable cell for generating electrical energy is described. The cell comprises a zinc anode, an alkaline electrolyte and a cathode. The electrolyte contains minor amounts of lead ions. The lead ions suppress dendrite formation upon charging the cell greatly increasing the cell's life.

3,556,849; January 19, 1971; H.G. Oswin and K.F. Blurton
"Recharging Alkaline/Zinc Cells"

> An improved method of recharging a secondary battery utilizing a zinc anode is described comprising maintaining the current density during charging between about 1 and 15 ma/cm^2 and pulse charging, i.e., charging with a periodic cut-off of current, with the on time being no greater than about 150 milliseconds and the off period being at least 0.5 of the on period.

3,592,693; July 13, 1971; M.G. Rosansky
"Consumable Metal Anode with Dry Electrolyte Enclosed in Envelope"

> A consumable metal electrode for an electrochemical cell containing an alkaline metal hydroxide electrolyte in dry form is described. Accordingly, a cell employing the electrode can be activated by the addition of water to the cell forming a conductive aqueous electrolyte solution.

3,592,696; July 13, 1971; N.I. Palmer
"Direct Hydrocarbon Fuel Cell Containing Electrolyte Comprising a Halogenated Acid and a Lewis Acid"

A fuel cell system is described employing a super acid electrolyte. The fuel electrochemically oxidized at the fuel electrode is a carbonaceous material, preferably being a C_1-C_{16} alkane. The preferred super acid electrolyte is antimony pentafluoride-fluorosulfonic acid. This fuel cell system provides a rapid and efficient electrochemical conversion of alkanes and the like to their oxidation products.

3,623,911; November 30, 1971; H.G. Oswin
"High-Rate Consumable Metal Electrodes"

Zinc electrodes containing a minor amount of a metal which is capable of creating cation vacancies in a zinc oxide lattice such as gallium or tin are described. The gallium or tin additive helps to prevent passivation of the zinc during rapid discharge of the electrodes in an electrochemical cell and reduces the amount of cell electrolyte needed for efficient performance. The gallium or tin is preferably introduced into the zinc by means of alloying, e.g., by electrolytic coreduction of zinc and gallium and/or tin compounds.

3,630,785; December 28, 1971; B. Jagid and H.P. Louie
"Anode with a Two Layered Separator"

An improved anode primarily for use in a metal/air or metal/oxygen cell is described. The anode comprises a porous metal body wrapped in a first layer of hydrophilic gas impermeable material, such as a cellulosic membrane, having a plurality of slits or holes therein. A second layer of material which is hydrophilic, dimensionally stable, and heat sealable, such as copolymers of vinylchloride and acrylonitrile is placed over or around the first layer of material.

3,650,837; March 21, 1972; N.I. Palmer
"Secondary Metal/Air Cell"

An electrically rechargeable metal/air or metal/oxygen depolarized electrochemical cell and a method of generating electricity therefrom are described. The electrochemical cell comprises an anode capable of being regenerated and a composite cathode having an oxygen-permeable, hydrophobic member layered with catalyst and a biporous unit. The coarse pore layer of the biporous unit is in contact with the catalyst. A bubble barrier which can be integral with either the anode or composite cathode is in contact with the fine pore layer of the biporous unit. During discharge of the cell, the biporous unit is flooded with electrolyte and merely acts as a current collector. When charging, oxygen evolution occurs at the fine pore layer and, due to the bubble barrier, the evolved oxygen is forced into the coarse pore layer displacing the electrolyte, thereby protecting the catalyst layer from damage due to oxygen evolution and the high charging potential.

3,660,165; May 2, 1972; N.I. Palmer
"Electrodes Comprising Fluorinated Phosphonitrile Polymer and Fuel Cell in Combination Therewith"

Lightweight electrodes for use in an electrochemical device are described. In one embodiment the lightweight electrodes comprise a catalyst layer including an admixture of catalyst and a fluorinated phosphonitrile polymer. In another embodiment the hydrophobic backing layer is a continuous film of fluorinated phosphonitrile polymer with a contiguous layer of catalyst thereon. The catalyst can also be admixed and bound with particles comprising the same fluorinated phosphonitrile polymer used as the backing layer, derivatives of the polymeric material or other hydrophobic polymers such as polytetrafluoroethylene. The electrode is adapted to be disposed in an electrochemical cell with the catalytic layer in contact with the cell electrolyte.

3,668,014; June 6, 1972; E.G. Katsoulis and W.S. Pryor
"Electrode and Method of Producing Same"

A mat of polyfluorocarbon fibers having controlled pore size and hydrophobicity
is impregnated with a catalytic substance to form a light-weight catalytic mass
having a low, uniform loading of the catalytic substance. This catalytic mass,
particularly when combined with an electrically-conducting element, and/or a
continuous hydrophobic polymer membrane is suitable for use as an electrode
in an electrochemical cell, e.g., as a fuel or oxidant electrode in a fuel cell or
as the cathode in a metal-air battery, wherein it provides high current densities
at relatively constant voltages over a long period of time.

3,759,748; September 18, 1973; N.I. Palmer
"Electrically Recharged Metal/Air Cell"

A two-electrode rechargeable metal/air or metal/oxygen depolarized electrochemical cell and method of generating electricity therefrom are described.
The electrochemical cell comprises an anode capable of being regenerated,
a composite cathode having an oxygen permeable hydrophobic member
layered with catalyst, and a current collector in contact with the layered
catalyst on the member. The cathode is constructed and arranged in order
that the current collector is in removable electrical contact with the catalyst.
Thus, for cell discharge, the current collector contacts the catalyst and trans-
ports electrons to the external circuit. For cell recharge, the current collector is electrically separated from the catalyst whereby numerous discharge-
recharge cycles of the cell may be effected without damage to the cathode.

Licentia Patent-Verwaltungs GmbH, Frankfurt, Germany

3,645,796; February 29, 1972; Harold Bohm and J. Heffler
"Electrochemical Cell Containing Electrical Contact and Method of Using"

To prevent corrosion at the contact to a gas-diffusion electrode of an electrolytic cell, the electrode is provided on its gas-side with a hydrophobic
porous coating of, for example, polytetrafluoroethylene and with a layer of
graphite, tungsten carbide, or niobium silicide in conductive contact with
the electrode.

3,799,809; March 26, 1974; H. Bohm and G. Louis
"Fuel Cell"

A fuel cell having an anode and a cathode each in the form of a particulate
filling, the anode and cathode being separately contained. The anode and
cathode are soaked by an electrolyte continuous between the anode and
cathode, while fuel is interspersed in the anode and an oxidant is interspersed in the cathode.

3,833,423; September 3, 1974; F.A. Pohl and H. Bohm
"Fuel Cell with WC Catalyst-Material Protection"

Autoxidizable fuel cell catalyst material is covered after its production
with a substance protecting it against autoxidation. The substance is
then removed before placing the fuel cell in operation.

3,833,424; September 3, 1974; G. Louis and H. Bohm
"Gas Fuel Cell Battery Having Bipolar Graphite Foam Electrodes"

A gas fuel cell battery is formed of a series of alternating bipolar electrodes
and cell frames. Each bipolar electrode comprises a flexible graphite foam
foil and active (anode and cathode) layers secured to opposite foil faces.
Each face of a cell frame is in engagement with a bipolar electrode along
a marginal zone of the foil that is outside the perimeter of the active layer.
Gas is contacting that face of each active layer that is oriented towards the
foil to which it is secured, while electrolyte is contacting that face of each
active layer that is oriented away from the foil to which it is secured. The

active layers are provided with openings for the gas supply, while the cell
frames have through-going bores and channels for the electrolyte supply.

3,857,735; December 31, 1974; G. Louis and H. Bohm
"Fuel Cell System"

Apparatus for regulating the quantity of water present in a reaction chamber
of a fuel cell system, which has a circulating aqueous solution of electrolyte
and air operated cathodes. The reaction chamber includes an anode and a
cathode. The regulating apparatus includes two portions, the first of which
acts to control the temperature of the electrolyte solution and the second
of which controls the air throughput in the system in dependence upon the
battery current. The temperature control device raises or lowers the operating temperature of the electrolyte solution in response to an increase or
decrease, respectively, in the concentration of the solution with respect to
the predetermined value range. The output of the air pump which supplies
air to the cathodes is regulated in dependence upon the current flowing
between the anode and cathode.

3,871,922; March 18, 1975; H. Bohm and J. Heffler
"Gas Diffusion Electrode for Electrochemical Cells"

A gas diffusion electrode for an electrochemical cell having an acidic electrolyte, including a metal net, a corrosion-resistant, electrolyte- and gas-
impervious, graphite-foam coating on the net, sunken canals on the graphite-
foam coating for conducting electrolyte, a porous catalyzing layer conform-
ing with the net and in contact with the graphite-foam coating and an
electrically insulating coating on the graphite-foam coating.

3,874,930; April 1, 1975; F.A. Pohl, H. Bohm and J. Heffler
"Fuel Cell System Including Source of Fuel Containing Hydrogen Sulfide and Method of Using"

By including tungsten disulfide and/or molybdenum disulfide in the fuel
electrode of a fuel cell, poisoning effects previously experienced with fuels
containing hydrogen sulfide are avoided.

3,907,600; September 23, 1975; F.A. Pohl and H. Bohm
"Electrode of Selenides or Tellurides"

A fuel electrode whose catalytic component is chosen from the group consisting of the tellurides and selenides of the elements molybdenum, tungsten
and chromium.

Lithium Corporation of America, New York, N.Y.

3,578,501; May 11, 1971; S.C. Honeycutt
"Fuel Cell System and Method Using Lithium and Lithium Hypochlorite to Produce Hydrogen and Oxygen"

Regenerative hydrogen-oxygen fuel cell system in which gaseous hydrogen
and oxygen are separately generated and conveyed to a hydrogen-oxygen
fuel cell associated with the system. Hydrogen is generated by the action
of water on lithium metal. Oxygen is generated by heating an aqueous
solution containing lithium hypochlorite. The system is regenerated by
converting the lithium chloride resulting from the thermal decomposition
of lithium hypochlorite, to lithium metal, and by converting the lithium
hydroxide monohydrate, obtained from the action of water on the lithium
metal, to lithium hypochlorite.

Joseph Lucas (Industries) Ltd., Birmingham, England

3,607,423; September 21, 1971; M.M. Bertioli
"Rechargeable Hybrid Fuel Cells and Method of Recharging"

The process relates to a hybrid fuel cell having battery plates, oxygen electrode, counterelectrode and electrolyte. Such a cell is charged by connecting the oxygen electrodes and counterelectrodes to a DC source while isolating the oxygen electrodes and counterelectrodes electrically, and the feature of the process is that the required isolation is effected by supplying gas under pressure to the spaces between the counterelectrodes and oxygen electrodes.

Matsushita Electric Industrial Company, Ltd., Osaka, Japan

3,709,834; January 9, 1973; M. Fukuda, T. Miura and K. Takahashi
"Method of Making a Uranium Containing Catalyst for a Metal Electrode"

An electrochemical electrode adapted for use in fuel cell batteries, metal-air batteries and electrochemical oxygen evolution instruments, and comprising an active catalyst which contains uranium and which in a small amount exhibits a catalytic activity equal to that obtainable from a larger amount of the conventional catalyst.

3,793,085; February 19, 1974; T. Hino and M. Fukuda
"Gas Diffusion Electrode for Cells"

A gas diffusion electrode for cells comprising a layer participating in the electrode reaction, the layer consisting essentially of carbon powder and a binder therefor, and a preformed porous fluorocarbon resin film integrally attached to the carbonaceous layer.

4,001,040; January 4, 1977; M. Fukuda, T. Iwaki and Y. Kobayashi
"Fuel Cell"

A fuel cell is constructed in such a manner that a porous film which is electrolyte-resistant and inactive to fuel is allowed to adhere to one surface of a fuel electrode and an electrolyte containing hydrazine fuel supplied to the cell and passing through the porous film and fuel electrode reaches the side of an oxidation electrode. In such fuel cell, self-decomposition of fuel due to direct contact with the fuel electrode occurs in a small degree and utilization efficiency of fuel is high. Furthermore, such fuel cell can be operated with an electrolyte containing fuel in a high concentration.

McDonnell Douglas Corporation, Santa Monica, California

3,546,020; December 8, 1970; C. Berger
"Regenerable Fuel Cell"

A rechargeable battery-fuel cell combination in which the fuel cell comprises wick means in contact with the electrolyte of the battery and functions as a sealing and pressure relief means.

3,560,260; February 2, 1971; C. Berger
"Method of Eliminating Gas Pressure in Batteries by Using Gas in Fuel Cell"

Combination of a fuel cell with a battery arranged with a conduit means between the battery and the fuel cell so that gas generated in the battery, e.g., hydrogen, is conducted to the fuel cell for reaction therein with another gas, e.g., oxygen, to generate an electric current and thereby eliminate gas pressure developed in the battery and also sealing the battery.

3,565,691; February 23, 1971; M.P. Strier and H.A. Frank
"High Energy Density Silver Oxide-Hydrogen Battery"

A solid-fluid cell or battery having a silver oxide cathode, a hydrogen catalyst anode, a separator comprising an inorganic material selected from the group consisting of (a) a solid solution consisting essentially of magnesium silicate and iron silicate, (b) zirconia and (c) alumina, positioned between the electrodes, and electrolyte material, e.g., a 30% aqueous solution of KOH, retained in the separator, to provide an improved high energy density battery of this type.

3,607,403; September 21, 1971; F.C. Arrance
"Self-Charging Battery Incorporating a Solid-Gas Battery and Storage Battery
Within a Honeycomb Matrix"

> A self-charging battery unit incorporating solid-fluid battery cell sections and
> high energy density storage cell sections formed within a honeycomb matrix
> which acts to support and separate the electrodes, the solid-fluid cell sections
> being actuable to charge the storage cell sections when the latter are discharged.

3,647,542; March 7, 1972; C. Berger
"Solid-Fluid Battery"

> An electrode separator unit in the form of a nonmetallic honeycomb matrix,
> e.g., of porous organic or inorganic material, and having first catalyst electrode
> e.g., platinum, and second active battery electrode material. e.g., zinc, separately
> contained in cells of the honeycomb matrix. A fluid operable battery incorporat-
> ing the above honeycomb electrode separator unit.

3,711,336; January 16, 1973; J.S. Smatko
"Ceramic Separator and Filter and Method of Production"

> Production of a ceramic-like porous potassium titanate member having high
> strength, fine substantially uniform pore size and resistance to alkali, suitable
> for use as a battery separator, fuel cell membrane or filter medium, prepared
> according to one embodiment, by adding an organic binder, particularly a wax
> such as a polyethylene glycol wax (Carbowax), to potassium titanate fibers,
> compressing the resulting mixture into blocks, breaking and granulating the
> blocks into particles, compressing the granules into a member or sheet, slowly
> heating the resulting member at temperature of about $400°$ to about $600°C$
> to decompose the organic binder, and firing the resulting member or sheet at
> temperature ranging from about $1000°$ to $1370°C$.

McGraw-Edison Company, Elgin, Illinois

3,649,362; March 14, 1972; E.V. Steffensen, J. Orshich and J. Steffensen
"Primary Battery"

> The process resides in providing in an alkaline-zinc-anode primary cell an
> agent which regenerates the electrolyte by causing the zinc reaction end
> products to crystallize out as zinc hydroxide when the electrolyte becomes
> saturated with zincate during discharge of the cell. Further, the process
> resides in providing a fibrous mass below the electrodes having a large surface
> area onto which the crystals of zinc hydroxide grow in preference to forming
> onto the electrodes. In this way the zinc reaction end products are kept from
> coating or clogging the electrodes and from interfering with the continuing
> operation of the cell. The crystallizing agent may be any compound or com-
> pounds which function to release sulfide or silicate ions to the electrolyte
> and may be selected from the group consisting of the silicates of potassium
> and sodium, sulfur in its precipitated or sublimed forms, thiourea
> (NH_2CSNH_2) and the sulfides of calcium, zinc (precipitated form), potassium,
> sodium, strontium, antimony, mercury, barium, aluminum, and phosphorus.
> The action of the sulfide ion crystallizing agents is improved by adding a
> small amount of zinc oxide especially as to air depolarized cells which are
> to have good life on open circuit. The fibrous mass may be of glass wool,
> cotton linters, shredded or loosely folded kraft paper, polyethylene fibers,
> asbestos wool, steel wool or nylon fibers.

3,788,900; January 29, 1974; J.E. Schmidt
"Air-Depolarized Primary Battery"

> An air-depolarized battery of the zinc-anode, carbon-cathode and alkaline
> electrolyte type has a series of cavities leading from the top face of the
> carbon cathode to nearly the bottom thereof and a chimney system which
> produces a rapid transfer of oxygen from the outside atmosphere to the

gas-electrolyte interface of the cathode resulting in a marked increase in current capacity of the battery. As a further feature, which may be employed independently or in combination with the chimney system above mentioned, wire coils are inserted in the cavities to obtain a low resistance connection of the cathode terminal to the lower active portions of the cathode whereby a still greater efficiency and current capacity is achieved.

3,843,413; October 22, 1974; J.E. Schmidt
"Low Resistance Terminal Connector for Air-Depolarized Cathode"

An air-depolarized battery of the type having a carbon cathode with one or more cavities leading through the top face to a lower portion thereof is provided with a wire coil in each cavity which is inserted under tension so that when released it springs out to engage the cavity wall under pressure. The coil provides a low resistance connection continuously throughout the height of the cathode from the lower portion closed to the active interface of the cathode with the electrolyte to the upper portion engaged by the cathode terminal. Preferably, the coil has also an upper tail extension connected to the cathode terminal for providing a direct connection of the terminal with each such lower portion of the cathode.

Minnesota Mining and Manufacturing Company, St. Paul, Minn.

4,007,058; February 8, 1977; A.D. Nelson and L.E. Espelien
"Matrix Construction for Fuel Cells"

This process pertains to a fuel cell which comprises a case, positive and negative electrodes inserted therein and electrolyte filling the inner space between the electrodes. Means for introducing an oxidizing gas into the space behind one electrode and a fuel behind the other are provided. A porous membrane substantially parallel to the electrodes comprising a self-supporting compressible porous fibrous mat of randomly dispersed and entangled microfibers of a high molecular weight thermoplastic polymer impregnated with the electrolyte occupies the inner space. This porous membrane permits communication of electrolyte between electrodes but prevents passage of small bubbles of oxidizing or fuel gases.

Monsanto Research Corporation, St. Louis, Mo.

3,553,028; January 5, 1971; S. Matsuda
"Method for Generating Electricity in Hydrazine-Oxygen Fuel Cell"

The performance characteristics of a hydrazine fuel cell anode can be selectively adjusted and/or controlled by introducing minor amounts of sulfur or selenium to the site of anodic activity.

3,570,260; March 16, 1971; J.O. Smith and B.A. Gruber
"Heat Pump"

A system and method for transferring heat between an enclosure and an exterior heat sink. The method involves operating a fuel cell within an enclosure to be cooled, on a fuel selected from hydrogen or hydrazine and an oxidant selected from nitric acid or nitrogen tetroxide and expending the electrical output of the cell in driving an electrical resistance located in an exterior heat sink.

3,617,388; November 2, 1971; S. Matsuda, J.J. O'Connell and M.C. Freerks
"Method of Activating Fuel Cell Anodes"

A fuel cell comprising a cathode, means to introduce an oxidizing agent to the cathode, an electrolyte separating the cathode from an anode comprising a conductive base material and a metal chelate as a catalytic component thereof and means to introduce a fuel to the anode.

3,676,222; July 11, 1972; M.C. Deibert
"Conductive Carbon Membrane Electrode"

> Membrane electrodes of particulate electrode material comprising conductive
> hydrophobic carbon and polytetrafluoroethylene with the electrode material
> particles coating the polytetrafluoroethylene particles, and substantially all of
> the surface of the electrode material being exposed.

Murgatroyd's Salt and Chemical Company, Ltd., Elworth, Sandbach, England

3,622,490; November 23, 1971; M.J. Lockett
"Electrical Circuit"

> The invention is an electrical circuit comprising a source of DC power, a plurality of
> electrolytic cells in which mercury is used as the cathode, and a plurality of fuel
> cells. The electrolytic cells and fuel cells are connected alternately and electrically
> in series and a shunt is provided from the cathode of each of the electrolytic cells to
> the anode of the next but one electrolytic cell. Amalgam produced in the elec-
> trolytic cells is used as fuel in the fuel cells to reduce total energy requirements.

National Research Development Corp., London, England

3,668,101; June 6, 1972; I. Bergman
"Membrane Electrodes and Cells"

> A three component membrane electrode for use in electrical cells having in
> sequence, a gas-permeable membrane, an electrically conductive layer and a
> protective layer acting as a diffusion barrier to gas dissolved in the electrolyte.
> The gas permeable membrane preferably is a plastic material and the electrically
> conductive layer is preferably laid down on the gas permeable membrane in the
> form of minute noncoalescing globules of gold and/or silver.

Occidental Energy Development Company, Madison Heights, Mich.

3,772,085; November 13, 1973; H.K. Bjorkman
"Method and Apparatus for Improving Efficiency of High Energy Density
Batteries of Metal-Metal Halide-Halogen Type By Boundary Layer"

> Efficiencies of high energy density batteries having metal and halogen elec-
> trodes and aqueous metal halide electrolyte are improved by creating a boundary
> layer of electrolyte low in halogen content adjacent to the metal electrode.
> Such boundary layer or stagnant electrolyte prevents contact of the metal
> electrode with halogen from a first flow of electrolyte containing dissolved
> halogen which passes through the base of the halogen electrode. When the
> cell is vertical, best formation of the boundary layer is obtained when the
> second flow of electrolyte enters the bottom of the cell and is directed up-
> wardly and toward the halogen electrode with a horizontal velocity com-
> ponent equal to that of the first electrolyte flow entering the reaction zone.
> In preferred embodiments of the process, the cell comprises bipolar electrodes
> of zinc and a porous carbon base for the halogen, the halogen is chlorine
> and the electrolyte is aqueous zinc chloride.

3,773,560; November 20, 1973; H.K. Bjorkman
"Circulation of Electrolyte Over a Metal Electrode of a Cell in a High Energy
Density Battery"

> A cell of a high energy density secondary battery, such as one having zinc and
> chlorine-on-carbon electrodes and a separator dividing the cell electrolyte portion
> into a smaller flow zone adjacent to the zinc electrolyte and a larger flow zone
> adjacent the chlorine-on-carbon electrolyte, includes a common outlet from
> both cell portions of such a structure that the flow of electrolyte through the
> larger cell zone helps to determine the rate of flow through the smaller zone.

In preferred embodiments of the apparatus the same electrolyte is circulated to both zones by a single pump or driving means, the flow in the smaller zone is such as to cause a desirable circulation of electrolyte therein at a low speed which is still sufficient to prevent polarization at the zinc electrode and pitting and dendritic formations on it, and the rate of flow through such lesser flow zone is regulated by having the outlet from such zone communicate with a restricted portion of a Venturi or equivalent orifice in the main flow zone outlet passage.

3,773,561; November 20, 1973; H.K. Bjorkman
"Isolation of Cells of a Battery Stack to Prevent Internal Short-Circuiting During Shutdown and Standby Periods"

Internal short circuiting of a plurality of electric cells of a cell stack is prevented during shutdown or standby by sealing off the cells from electrical contact with each other by closing off inlet and outlet ports to isolate electrolyte portions in the individual cells. A preferred apparatus for carrying out the process includes membranes which, in response to selective hydraulic pressure, open or close ports of the cell inlets and/or exits, to allow flow of electrolyte through the cells during charging or discharging and to prevent it during shutdown or standby.

3,877,990; April 15, 1975; A.F. Laethem and P. Carr
"Method of Slowing Escape of Chlorine from Metal-Chlorine Battery"

The escape of chlorine from a source thereof, such as the chlorine hydrate store of a metal-chlorine battery, is prevented or further escape and circulation of chlorine is slowed by blanketing escaped chlorine and the source thereof with a foam, such as one which is generated from an aqueous system which includes a foaming agent, e.g., an anionic surface active compound, and a gas. In preferred embodiments of the process, neutralizing agents are present in the foam to convert the chlorine to a less toxic state or condition. Also described are pressurized compositions for generating the foam, articles for containing and dispensing the compositions and apparatuses for applying the compositions to sources of escaping chlorine, some of which apparatuses operate automatically when chlorine accidentally escapes from the battery. Also described is an apparatus for confining the escaped chlorine near the source thereof, in case of a motor vehicle accident, and contacting such chlorine with a neutralizing medium.

Onan Corp., Minneapolis, Minn.

3,556,857; January 19, 1971; A.R. Poirier and J.A. Briese
"Fuel Cell"

A fuel cell in which there are porous fuel and oxygen electrodes and an interposed porous matrix containing an electrolyte and in which an excess of hydrocarbon fluid fuel is introduced into the fuel electrode compartment and is forced through the fuel electrode, the electrolyte matrix, and the oxygen electrode, and out through an exhaust passage in the oxygen electrode compartment to carry with it any reaction products occurring at the fuel and oxygen electrodes.

3,589,941; June 29, 1971; J.J. Eaton and D.W. Tschida;
"Fuel Cell with Internal Manifolds"

A fuel cell battery consisting of a plurality of associated duplex cells. Each duplex cell is mounted on a single frame which defines an integrally formed fuel input manifold, exhaust manifold, anode terminal manifold, and cathode terminal manifold. Valve means is provided to individually regulate the fuel input to each duplex cell. A plurality of duplex cells may be associated in a highly convenient and efficient manner due to the mating design of the manifolds.

Adjacent duplex cells may be electrically connected in series or in parallel
with a minimum of adaptation or conversion of individual duplex cells.

Philip Morris, Inc., New York, N.Y.

3,698,955; October 17, 1972; A.C. Lilly, Jr. and C.O. Tiller
"Oxygen-Responsive Electrical Current Supply"

> A supply providing electrical current in accordance with both load resistance
> and environmental oxygen concentration incorporates an electrochemical cell
> having an anode, a cathode and an electrolyte composed of a rare earth
> fluoride. A rugged miniature supply adapted for use in space-limited applica-
> tions is provided by thin film deposition of the cell elements.

Prototech, Co., Cambridge, Mass.

3,575,717; April 20, 1971; W. Juda, D.M. Moulton and H.L. Gruber
"Method of Generating Power in Molten Electrolyte Fuel Cell"

> Halogen-containing molten electrolyte fuel cell is operated so that the actual
> output voltage exceeds the standard output voltage. Hydrogen-iodide gaseous
> reaction product insoluble in the electrolyte is swept out of the electrolyte
> immediately as formed to keep the reaction product activity low. The reac-
> tion product is decomposed and its constituents recycled.

3,640,773; February 8, 1972; R.J. Allen and R.L. Novack
"Method of Operating Fuel Cell and Preventing Corrosion"

> The process relates in an important aspect to the prevention of electrolyte
> creeping out of fuel and/or oxidant inlets and the like in closed fuel cell
> housings, through the use of high temperature substantially nonwetting
> plugs disposed at appropriate locations along the inlets.

3,669,749; June 13, 1972; R.J. Allen and H.G. Petrow
"Method of Operating Electrochemical Cells with Increased Current Density
and Oxygen Efficiency"

> This disclosure is primarily concerned with the discovery of the improvement
> in the current density and oxygen efficiency of alkali media fuel cells and the
> like effected by the introduction into the cathode region thereof of a source of
> oxidizable boron.

3,669,750; June 13, 1972; W. Juda
"Fuel Cell System"

> Stacked fuel cells in which successive anode and cathode electrodes are porous
> supports juxtaposed with a thin impervious conductive layer. Each anode
> support is provided with a thin hydrogen-permeable layer at the side thereof
> facing the electrolyte. Hydrogen-containing fuel is reformed in situ by pro-
> viding reformation catalysts internally of the anode supports within the pores
> thereof. Fuel and oxidant are applied to one end of the anode and cathode
> electrodes, and vents are provided at the opposite end. Venting is assisted by
> electrode tapering or tilting.

3,669,752; June 13, 1972; R.L. Novack, D.M. Moulton and W. Juda
"Method of Operating Fuel Cell"

> Electrolytic cell operation in which the electrolyte is agitated near one of a pair
> of electrodes and a barrier is interposed between the electrodes and is physically
> combined with structure of one of the electrodes. The cell may have an alkali-
> metal hydroxide electrolyte maintained at a temperature of at least 300°C to
> render the electrolyte molten and anhydrous. Peroxide and superoxide may
> function as the electrochemical oxidant of the cell.

3,776,776; December 4, 1973; H.G. Petrow
"Gold-Coated Platinum-Metal Black Catalytic Structure and Method of Preparation"

This disclosure is concerned with a gold-coated platinum-metal black catalytic structure, such as a fuel cell electrode and the like, formed by growing colloidal gold oxide upon the black and reducing the same.

Research Corp., New York, N.Y.

3,553,025; January 5, 1971; B.P.J. Bockris, D. Sepa and A. Damjanovic
"Fuel Cell with an Electrode Comprising Barium Tantalum Bronze or Strontium Niobium Bronze"

Barium tantalum and strontium niobium bronzes are catalysts for oxygen reduction at the cathode in fuel cells.

Rhone-Poulenc S.A., Paris, France

3,741,945; June 26, 1973; G. Bourat and R. Margraff
"Vinyl Alcohol Copolymers Containing Hydroxy Sulfonyl Groups"

Vinyl alcohol copolymers partially etherified with hydroxy sulfonyl organic radicals and optionally partially crosslinked by ether groups are useful in making cation-exchange membranes.

Rockwell International Corp., El Segundo, Cal.

3,663,298; May 16, 1972; L.R. McCoy and L.A. Heredy
"Rotatable Electrode Structure with Conductive Particle Bed"

An electrode structure for use with an electrolyte wherein an electrically conductive flowing particle bed stream within a compartment of the electrode structure establishes a plurality of electrochemical reaction sites at the interface of the particles and of the electrolyte. The flowing particle bed electrode structure finds a preferred application as an electrode in an electrically regenerable system, particularly as a rotatable zinc electrode in zinc-nickel oxide and zinc-air secondary cells.

3,767,466; October 23, 1973; L.R. McCoy and L.A. Heredy
"Electrode Structure and Battery"

Abstract identical to 3,663,298.

Saft—Societe des Accumulateurs Fixes et de Traction, Romainville, France

3,679,488; July 25, 1972; F.J. Dalard, J.W. Augustinksy and J.-C. Sohm
"Silver Oxide-Magnesium Cell or Electrochemical Generator"

The process relates to an electrolyte for a cell having electrodes based respectively upon magnesium and preferably silver oxide. Other electrodes in lieu of silver oxide are also disclosed as useful, e.g., silver chloride, copper chloride or oxygen or air electrodes. In accordance with the process, the electrolyte comprises an aqueous solution of sodium or lithium metaborate and sodium or lithium perchlorate and has a pH in the neighborhood of 11. In certain conditions, additives to the electrolyte solution, e.g., sodium tartrate improve operation of the cell in rapid discharge. The process is applicable notably to rapidly discharging electrochemical generators.

3,697,326; October 10, 1972; J.F. Jammet
"Battery Including Pair of Air-Depolarized Cells"

Primary electric battery comprising an outer metallic cup coated externally except at its bottom with a casing of synthetic material, the bottom constituting one terminal and a metallic cap in the cover of the casing constituting the other terminal. A pair of flat air depolarized cells connected in parallel are housed in the cup and embody main trays whose openings face each other and whose bottoms are parallel to the axis of the cup, each of the trays containing a negative electrode, an immobilized electrolyte preferably carried by a separator, and a positive electrode arranged in layers in each tray. The trays are separated by a space in communication with the exterior of the bat-

tery during its operation. The space is positioned preferably between the
positive electrodes of the facing trays. The positive and negative electrodes
of both trays are parallelly connected electrically respectively to the metallic
cap and the metallic cup. Provision is made for expansion of the negative
electrodes when the batteries are in use.

3,855,000; December 17, 1974; J. Jammet
"Air Depolarization Primary Cell and Process for Production Thereof"

Air depolarization type electric cell comprising positive and negative electrodes
and an electrolyte within a container. The container is closed in fluid-tight
manner. A free space within the container permits the expansion of the negative electrode without movement of the electrolyte. The process of manufacture of such cells is also recited.

3,871,920; March 18, 1975; G. Grebier and P. Arlot
"Air Depolarized Cell"

An air depolarized electrochemical cell comprising a negative electrode, electrolyte and a positive electrode provided with a flue. Air is introduced into
the flue which has a partitioning element located internally of the flue. The
length of the partitioning element is approximately equal to the height of
the flue. The partitioning element delimits separate chambers in the flue of
different volumes that communicate at their lower portions and whose pro-
portionality coefficients at the air arrival surfaces of the positive electrode via
perforations of like numbers at catalytic sites of the positive electrode with
respect to corresponding chamber volumes are different.

3,928,072; December 23, 1975; G. Gerbier and P. Depoix
"Air Depolarized Electric Cell"

Air depolarized electric cell comprising a negative electrode occupying
a peripheral position, an electrolyte which is preferably gelled and a positive
electrode mass occupying a central position within the negative electrode.
The positive electrode is provided with a funnel in the center of the positive
mass and with a metallic current collector disposed in the funnel in contact
with walls of the funnel. The collector comprises a metallic wire in the form
of a helix at least partially incrusted or embedded in the wall of the funnel.

Shell Oil Co., New York, N.Y.

3,703,446; November 21, 1972; E.W. Haycock and R. Prober
**"Method of Carrying Out Electrochemical Processes in a Fluidized Bed Elec-
trolytic Cell"**

An electrolytic cell using an electrolyte and a particulate electrode in a fluidized
condition to achieve a large, continuously depolarized electrode surface area.

Siemens AG, Munich, Germany

3,554,812; January 12, 1971; F. von Sturm and H. Nischik
**"Fuel Cell Comprising Asbestos Diaphragms and Nickel Mesh Electrolyte
Supports"**

An asbestos diaphragm is positioned between and next adjacent the supporting
structure and each of a pair of spaced thin electrodes in an electrochemical
cell. The supporting structure is porous and is impregnated with nickel mesh
having large pores and regions adjacent the electrodes of nickel mesh having
fine pores.

3,574,560; April 13, 1971; F. von Sturm and H. Kohlmüller
**"Device for Producing Gaseous Reactants Particularly Hydrogen and Oxygen
for Fuel Cells"**

Method of producing gaseous reactants, particularly hydrogen and oxygen for
fuel cells by reaction or catalytic dissociation of liquid or liquid-dissolved substances, with solid substances or solid catalysts, whereby the development of
gas is automatically adjusted to the gas consumed. The liquid substance or
the liquid-dissolved substance is placed into a storage container which is connected with a sealed gas chamber and also connected, via at least one line, to
a reaction tube containing the solid substance or the solid catalyst. Thus,
when the pressure drops in the reaction tube, the liquid located in the storage
container is further compressed into the reaction tube and when the pressure
in the reaction tube rises, the liquid is either completely or partially returned
to the storage container.

3,585,079; June 15, 1971; G. Richter, H. Cnoblock and H.-J. Henkel
"Fuel Cell Electrodes Having a Polymeric Metal-Containing or Metal-Free Phthalocyanine Catalyst"

A fuel cell comprising an anodic and a cathodic electrode for converting hydrogen and oxygen into fuel elements. The electrodes are characterized in
that they are completely or partly comprised of polymeric metal-containing
and/or metal-free phthalocyanine, which may be substituted in the core.

3,597,514; August 3, 1971; P. Jager
"Method for Producing Porous Sheets, Particularly Diaphragms, of Asbestos Free of Bonding Agents"

Described is a method of producing porous asbestos sheets and asbestos
diaphragms free of bonding agents. The method is characterized in that
asbestos fibers are first pretreated with alkaline lyes and are subsequently
tempered at temperatures of $300°$ to $700°C$.

3,615,852; October 26, 1971; J. Gehring and K. Strasser
"Fuel Cell"

Fuel cell includes a pair of electrodes spaced from one another, a porous support skeleton having electron-nonconductive cover layers at opposite sides
thereof disposed in the space between the electrodes and containing fluid electrolyte, a metal profile frame carrying the support skeleton and provided with
at least one supply duct and one discharge duct for electrolyte, and an elastic
metal frame carrying each of the electrodes and located adjacent the frame for
the support skeleton and separated therefrom by the respective electrically nonconductive cover layer, the frame for the electrodes being provided with respective supply and discharge ducts for fuel cell reactants.

3,629,075; December 21, 1971; H. Gutbier
"Method and Apparatus for Eliminating Waste Heat and Reaction Water Together from Fuel Cells"

Method of eliminating waste heat and reaction water together from a fuel cell
or battery includes bringing the fuel cell electrolyte into contact with a porous
diaphragm through the pores of which water necessary for removing waste heat
diffuses only in vaporous form, passing separated water vapor diffusing through
the diaphragm to a cooled condensation surface lying opposite and closely
adjacent the diaphragm, and depositing thereon condensation water produced
from the water vapor, and alternatively returning the condensation water thus
formed to the fuel cell electrolyte and removing the condensation water from
the fuel cell through a pressure lock. Apparatus for carrying out the foregoing
method includes a chamber traversible by the electrolyte, the chamber being
defined on at least one side by a porous diaphragm, and a condensation surface communicating with the porous diaphragm through a closed gas chamber
therebetween and defined thereby.

3,644,148; February 22, 1972; H. Gutbier
"Method of Placing and Holding A Fuel-Cell Battery in Inactive Maintenance-Free Ready Condition"

Method of placing and holding in inactive maintenance-free ready condition a fuel-cell battery having electrode chambers, respective electrodes in the chambers and respective supply means for two reactants, namely fuel and oxidant, the supply means having respective inlet valves for controlling the supply of the reactants, includes the steps of displacing one of the reactants of the electrodes and electrode chambers by filling the electrodes and electrode chambers with nonreactant substance, closing the inlet valve for the one reactant and holding the valve closed during the inactive-and-ready state of the battery.

3,679,485; July 25, 1972; H. Kohlmuller and D. Kuhl
"Fuel Cell Battery for Reacting Gaseous Reactants in Fuel Cells Operated with Liquid Electrolyte"

Fuel cell battery for reacting gaseous reactants in fuel cells operated with liquid electrolyte includes a stack of alternatingly superimposed components A and B with a component C at the respective ends of the stack, the stack being embedded in a casing of molding resin, the component A comprising a pair of diaphragms with a support frame and sealing frame sandwiched therebetween, the component B comprising a pair of diaphragms with bipolar electrodes and sealing frames sandwiched therebetween, and the component C comprising a diaphragm and a contact plate with an electrode, a spacer grid and a sealing frame sandwiched therebetween, the sealing frames, diaphragms and contact plates being provided with fins and formed with bores for supplying the reactants to and discharging the same from the battery, and the sealing frames being firmly bonded to the diaphragms and contact plates of the respective components; and method of producing the fuel cell battery.

3,681,145; August 1, 1972; H. Kohlmuller and D. Kuhl
"Fuel Cell Battery for Reacting a Reactant, Such as Hydrazine Especially, Dissolved in Electrolyte"

Fuel cell battery for reacting a reactant, such as hydrazine especially, dissolved in electrolyte, the same includes a stack of superimposed components A with respective components B and C at opposite ends of the stack thereof, the stack being embedded in a casing of molding resin, the component A comprising a pair of diaphragms with a support frame, a spacer screen, a contact plate, a sealing frame and electrodes of different polarities sandwiched therebetween; the component B comprising a diaphragm and a contact plate with an electrode, a support frame and a sealing frame sandwiched therebetween, and the component C comprising a diaphragm and a contact plate with a spacer screen, a sealing frame and an electrode sandwiched therebetween, the sealing frames, the diaphragms and contact plates being provided with fins and formed with bores for supplying the reactants to and discharging the same from the battery, and the sealing frames being firmly bonded to the contact plates and diaphragms of the respective components; and method of producing the fuel cell battery.

3,711,333; January 16, 1973; H. Kohlmuller
"Fuel Cell Battery"

A fuel cell battery is divided into several electrically series-connected blocks of fuel cells. These blocks and the individual fuel cells within each block are connected with lines for an electrolyte-hydrazine fuel mixture and the gaseous oxidation agent, whereby during the operation of the fuel battery, the fuel cell elements within each block are traversed in parallel by the gaseous oxidation agent, while the individual blocks are traversed in series. All the fuel cells of the fuel cell battery are passed in parallel by the mixture of electrolyte-hydrazine. The number of the fuel cells decreases per block, in flow direction of the gaseous oxidation agent and the flow rate of the gas in the oxidation agent supply line may be regulated by a valve situated at the gas outlet of the fuel cell battery, the valve being controllable by electrical signals obtained from a voltage comparison between various blocks.

3,775,184; November 27, 1973; D. Kuhl and H. Poppa
"Component for a Fuel Element and a Battery Produced with These Components"

> A component for a fuel cell element which operates on gaseous reactants or with a fuel dissolved in electrolyte. The component comprises an electricity conducting sheet. On at least one side of the sheet is at least one metallic spacer net, an electrode and a diaphragm, whose edge is tightly connected with the separating sheet and is cast into a synthetic frame, which projects unilaterally beyond the area of the separating sheet and contains channels closed with synthetic foils for inlet and outlet of the electrolyte. The separating sheet, spacer net and diaphragm are provided with diametrically opposed lugs.

3,783,028; January 1, 1974; H. Cnobloch, H. Kohlmuller and M. Marchetto
"Fuel Cell and Apparatus for Water Removal by Evaporating the Water from the Electrolyte of Fuel Elements"

> A process for the removal of water by vaporizing the water from the electrolyte of fuel cell elements, wherein the hydrogen-containing fuels are reacted and at least one reactant is produced through a preceding exothermal reaction. The heat energy which occurs during the exothermal reaction is used for the vaporization of the water. Apparatus for carrying out the process is also described.

3,783,107; January 1, 1974; H. Kohlmuller
"Water Depletion Unit for Fuel Cells"

> A filter-press type water depletion unit for removing reaction water from the electrolyte cycle of a fuel cell battery forms a depletion cell having at least one gas space and an electrolyte space. The depletion unit has a coolable structure bordering the gas space and a diaphragm member separates the gas space from the electrolyte space, the diaphragm member having a rim portion that is gas-tight and impervious to electrolyte. The rim portion has a thickness greater than the remaining portion of the diaphragm. A holding arrangement presses the rim portion at its lateral surfaces for sealing the gas space and the electrolyte space and the rim portion is provided with respective passages communicating with the gas space and the electrolyte space.

3,791,872; February 12, 1974; F.V. Sturm, W. Naschwitz, W. Rummel and D. Groppel
"Method for Production of Electrode for Electrochemical Cells"

> An electrode for electrochemical cells constituting a porous cover layer of fibrous material and a layer of pulverulent catalyst. The catalyst particles are connected with each other as well as with the cover layer by means of a binder. The cover layer contains a binder. A butadiene-styrene-acrylonitrile. copolymer is present in the catalyst layer. Also described is a method for producing the electrode.

3,793,084; February 19, 1974; M. Marchetto
"Electrode for Electrochemical Cells with Graduated Catalyst Concentration"

> An electrode, for electrochemical cells, with pulverulent catalyst material and a porous cover layer of fibrous material. The electrode comprises a layer of fibrous material with two sequential semilayers. One of the semilayers comprises fibrous material only and the other of the semilayers comprises fibrous material and pulverulent catalyst. Also described is a method of forming the electrode.

3,798,197; March 19, 1974; W. Naschwitz and E. Knust
"Method for the Manufacture of Latex Bonded Asbestos Cover Layers and Covers Produced"

> A process for the production of latex bonded cover layers for electrochemical cells, particularly fuel cells, is described. Amphibole asbestos, particularly

anthophyllite asbestos, bundles are broken down by a cationic wetting agent.
The nonbroken-down particles are removed. A suspension is made from the
asbestos fibers and filler substances in the form of powders, with a grain
size of up to 100 microns, which are stable in acid electrolyte, have ionic conductivity and can be swelled to several times their volume. Bentonite and
cationic ion exchangers are particularly suitable as the filler substances. The
suspension is mixed with an acid resistant latex and formed into a finished
cover layer.

3,806,370; April 23, 1974; H. Nischik
"Fuel Cell Battery with Intermittent Flushing of the Electrolyte"

An electrolyte interrupter system for providing intermittent flushing of the
electrolyte in a fuel cell battery having several fuel cells in which the electrodes are held in plastic frames. The electrolyte interrupter system is constituted by an electrolyte distributor and an electrolyte manifold arrangement
in the frames of the individual fuel cells. Electrolyte supply ducts for the individual fuel cells open into the electrolyte distributor, and electrolyte discharge ducts for the individual fuel cells open into the electrolyte manifold.
The electrolyte distributor and the electrolyte manifold are each formed by
mutually aligned holes in the upper portion of the frames, with the bottom
of the holes forming the electrolyte distributor being located at least at the
same height as the openings of the electrolyte discharge ducts leading into
the electrolyte manifold.

3,846,176; November 5, 1974; D. Kuhl
"Fuel Cell Encased in Molded Plastic Material"

Method of producing fuel cell encased in molded plastic material and composed of a substantially planar support frame defining a chamber for receiving liquid electrolyte and, superposed on each side of the support frame, a
substantially planar diaphragm member and a substantially planar cover plate
member sandwiching therebetween a substantially planar electrode and a substantially planar spacer defining a chamber for receiving reactant gas includes
cementing the edges of each of the diaphragm members to another of the
members with a bonding agent, and sealing at least one of the chambers with
additional bonding agent prior to molding a casing of plastic material about
the superposed components, and fuel cell produced thereby.

3,856,573; December 24, 1974; D. Groppel
"Method of Producing a Structural Member Formed of Electrodes, Cover
Layers and a Support Frame for Fuel Elements with a Liquid Electrolyte,
and Structural Member Produced by the Method"

Method of producing a compact, storable structural member, formed of electrodes, cover layers and a support frame, for fuel elements includes welding
a first metal screen to one side of a framework formed of plastic material,
placing a support frame of at least one screen of plastic material adjacent
the first metal screen, welding a second metal screen to the other side of the
framework so as to sandwich the support frame between the first and second
metal screens, spraying on the metal screens at least one layer of fine-granular
ion exchanger material suspended in a dissolved binding agent, and thereafter
applying to the ion exchanger layers at least one layer of a catalyst suspended
in a dissolved binding agent; and structural member produced by the method.

3,861,397; January 21, 1975; R. Rao and G. Richter
"Implantable Fuel Cell"

An implantable fuel cell is used particularly for the operation of heart beat
actuators, artificial hearts or the like. As its operating means are used an
oxidizable body substance, preferably glucose, as well as oxygen from the
body fluids. The fuel cell is particularly characterized in that a cell fuel

electrode as well as one or several selective oxygen electrodes are spatially so
arranged with respect to each other that the operational mixture diffused in
operational condition from the body liquid into the cell, is guided substantially
initially to the corresponding oxygen electrodes and thereupon to the fuel electrode.

3,900,602; August 19, 1975; W. Rummel
"Method and Device for the Manufacture of Catalytic Layers for Electrodes in Electrochemical Cells, Particularly Fuel Cells"

Method and apparatus for the manufacture of catalytic layers by deposition of
a catalyst powder on a substrate comprises the acceleration of the catalyst powder
in a direction approximately parallel to the direction of motion of the substrate
and the deposition of the catalyst powder with or without a filler onto the
substrate under the influence of gravity.

3,915,746; October 28, 1975; H. Kohlmuller and K. Strasser
"Fuel Cell Battery Having an Improved Distributor Arrangement"

A fuel cell battery having an improved distributor arrangement, particularly
useful with fuel cell batteries of the filter press type design and using a liquid
electrolyte and at least one gaseous reactant in which the distributor is placed
approximately in the center of the fuel cell battery thereby permitting use of
the optimum material and resulting in a reduction in electrolyte leakage current.

3,926,675; December 16, 1975; H. Kohlmuller and G. Kohlmuller
"Process for Making Electrodes Containing Raney Nickel and a Thiocyanate"

A method is disclosed for preparing electrodes containing Raney nickel to be
employed for the electrochemical conversion of hydrazine, in which a thiocyanate is introduced into the electrodes, advantageously by impregnating the
electrodes with a thiocyanate solution. Electrodes prepared according to the
process generally contain a thiocyanate approximately in the range of between
about 5 and about 25% by weight upon the amount of Raney nickel therein,
and are useful as anodes in hydrazine/oxygen fuel cells.

3,935,028; January 27, 1976; K. Strasser and D. Hasenauer
"Fuel Cell Set and Method"

An improved fuel cell set for the generation of electrical energy through the
reaction of gaseous reactants and a liquid electrolyte which is circulated.

3,944,434; March 16, 1976; D. Groppel and G. Siemsen
"Hybrid Electrode for Metal/Air Cells"

A hybrid electrode for metal/air cells of the type having a hydrophilic layer
of nickel on the electrolyte side for the precipitation of oxygen, hydrophobic
layer of plastic on the gas side and a hydrophobic layer of carbon between
these two layers which contains an embedded metallic structure for use in
dissolving oxygen, in which the metallic structure of the carbon layer is
firmly connected to the nickel layer and protrudes at least partially on the
gas side beyond the carbon layer and into the plastic layer, the plastic layer
also having a metallic structure embedded therein which is firmly connected
to the part of the metallic structure protruding from the carbon layer.

3,960,598; June 1, 1976; H. Kohlmuller
"Filter Press-Type Fuel Cell Battery"

A fuel cell battery of the filter press-type which uses a liquid electrolyte and
at least one gaseous reactant, with the individual fuel cells which make up
the battery having gas and electrolyte chambers separated from each other by
asbestos diaphragms with the asbestos diaphragms having electrolyte-impermeable
gas-tight outer zones of increased thickness which are pressed together or
against separator sheets and contain supply ducts formed therein, in which the
width of the outer zone of increased thickness on the electrolyte side is made

smaller than the width of the same part on the gas side and where the means
such as a support frame or electrode causing separation in the electrolyte
chamber are sized to extend beyond the inner edge of the outer zone on the
gas side.

3,979,224; September 7, 1976; K. Strasser
"Fuel Cell Battery Consisting of a Plurality of Fuel Cells"

A fuel cell battery consisting of a plurality of fuel cells which are individually
enclosed by plastic frames for reacting gaseous reactants and a liquid electrolyte.

3,992,223; November 16, 1976; H. Gutbier
"Method and Apparatus for Removing Reaction Water from Fuel Cells"

Method of removing reaction water from a fuel cell includes bringing electrolyte
of the fuel cell in contact with a diaphragm having a given capillary depression
pressure and permeable to water vapor so that water vapor contained in the electrolyte diffuses through the diaphragm into a chamber containing a gas at a
given pressure, and passing the diffused water vapor to a location at which it
engages with a cooled condensation surface and condenses thereon, the electrolyte having a hydrostatic pressure equilibrated by at least one of the pressures exerted by the gas and the capillary depression in the diaphragm.
Apparatus for carrying out the foregoing method includes a gas chamber, a
chamber for electrolyte, means separating the chambers from one another
comprising a diaphragm permeable to water vapor contained in the electrolyte
a condensation surface located in the gas chamber for condensing water vapor
coming in contact therewith, and means connected to the gas chamber for
discharging therefrom water condensed by the condensation surface.

Siemens AG, Munich, Germany (with Varta AG)

3,544,377; December 1, 1970; E. Justi and R. Wendtland
"Oxygen-Hydrogen Storage Battery Cell"

Oxygen-hydrogen cell for storing electric energy by electrolysis of water and
recombination of electrolysis gas. The same cell is used for both steps. Valve
electrode covers are on the electrodes. Both nickel and silver oxygen electrodes
are required.

3,567,519; March 2, 1971; R. Wendtland and A. Winsel
"Depolarization Process"

A method of controlling the concentration polarization during the operation
of a fuel cell. Electrolyte is passed through the small pores of an electrode
from one face thereof in one electrolyte chamber to the other face in another
electrolyte chamber while reaction gas is passed through the coarser pores of
the electrode.

3,576,676; April 27, 1971; R. Wendtland and A. Winsel
"Galvanic Fuel Cell Battery and Process"

A process for controlling the heat of reaction and the concentration polarization during an electrochemical process which involves maintaining two currents of electrolyte flow through pairs of electrolyte chambers maintained
at different pressures. A portion of the electrolyte flows through the electrodes and out of a low pressure electrolyte chamber and another portion
of electrolyte flows directly into the chamber and out therefrom. The
battery for operating the process.

3,597,275; August 3, 1971; A. Winsel and R. Wendtland
"Process of Operating Fuel Cell"

An electrolysis or fuel cell which includes a plurality of porous electrochemical
reactive electrodes positioned alternately with porous electrically nonconductive

diaphragms; the pores of the diaphragms at the faces adjoining the electrodes have a median radius smaller than that of the pores at the faces of the electrodes. A special electrode for use in a fuel cell battery is also disclosed as well as a process for the operation of the electrolysis or fuel cell. The fuel or electrolysis cell does not require separate gas chambers for each cell of the gases.

3,660,166; May 2, 1972; A. Winsel
"Gas Diffusion Electrode"

Catalytic electrode having improved heat dissipating means. The electrode embodies a foraminous metallic web having deposited over major portions of its surfaces porous catalytic material framed within gasket material and essentially having sufficient web surface exposed to effectively remove heat from the electrode during its operation in a fuel cell or other exothermic catalytic processes.

Societe Anonyme pour l'Equipment Electrique de Vehicles S.E.V. Marshall, Issy Les Moulineaux, France

3,840,405; October 8, 1974; E.J.P. D'Ange d'Orsay
"Circulating Fuel Cell with Crenellated Electrode"

A circulating fuel cell or fuel reactor containing an anodic and a cathodic compartment separated by an ionically conducting membrane, each compartment containing an electrode for collecting electrical current, generated by reaction of a chemical body contained in an electrically conductive, circulating liquid within one compartment with another chemical body in the other compartment of the reactor.

Societe Generale de Constructions Electriques et Mecaniques Alsthom et Cie., Paris, France

3,690,954; September 12, 1972; B. Warszawski and H. Vandenberghe
"Fuel Cell and Electrolyte Supply System"

The marginal portions of thin, sheet-like electrodes and diaphragms are pierced to form, when the diaphragms and electrodes are assembled in alternately placed stacks, ducts and channels. The end plates holding the stack of electrodes together are formed with openings and ducts interiorly of the thickness thereof to interconnect the ducts and channels through the electrode-diaphragm stack in selected hydraulic circuits. Two distinct groups of ducts are provided, one in communication with the faces of the electrodes at alternate sides, to permit application of electrolyte charged with oxidizing or reducing-type reactants, respectively; the other group of ducts or channels is independent of the first group and noncommunicating with the first set of ducts to provide for intercommunication of the electrolyte itself.

3,746,578; July 17, 1973; B. Warszawski
"Fuel Cell Battery"

To enable use of a carrier electrolyte having a reactant therein fuel cell units may be stacked together, each fuel cell being formed of an embossed electrode having protuberances extending on at least one side, or possibly on alternate sides and separated from each other by semipermeable diaphragms. In the case of a battery the electrodes are thin sheets held within a plastic frame, embossed to have the protuberances extending by several tenths of a millimeter, for example, 0.5 mm, and the sides are preferably coated with catalysts. The marginal portions of the frames, the electrodes and diaphragms and spacer elements, if used, are formed with openings to provide the stack with two electrolyte supply and drain systems. Opposite faces of the electrodes are in communication with alternate supply systems of electrolyte charged

with reactant by fine ducts or microchannels, so that the electrolyte can wash
over the faces of the electrodes. The electrolyte applied to one system includes
a reducing agent, and to the other an oxidizing agent, to provide anodic and
cathodic electrode surfaces, respectively, separated by permeable diaphragms.
At least one of the additive agents may be a nonregeneratable material, soluble
in the electrolyte. The additive agents may also be emulsified nonsoluble gaseous
or liquid material.

3,755,000; August 28, 1973; P. Demange
"Electrodes and Fuel Cells"

An improved electrode structure wherein the electrode is a nonporous embossed sheet. The electrode surface contains a plurality of projections arranged
in bands. Each band occupies an area and the top and bottom of the band
are substantially parallel to one of the margins and substantially perpendicular
to the direction of contemplated fluid flow across the electrode. Each band
of projections is an array of parallel elongated ridges inclined to the direction
of fluid flow. The angle of the inclination is such that $d \sin \alpha = ne$, wherein
d is the distance between the top and bottom of the band, e is the distance
between two adjacent ridges, and n is a whole number. The angle of inclination must also fall between $15°$ and $30°$. The invention also includes a fuel
cell comprising a stack of the electrodes.

3,764,391; October 9, 1973; B. Warszawski, H. Vandenberghe and B. Verger
"Fuel Cell and Electrode Structure Therefor"

An assembly of stacked electrodes, alternating with semipermeable diaphragms
is built up, with the marginal portions of the electrodes and diaphragms
formed with flow duct openings to direct electrolyte and reactants to flow
lengthwise of the stack of electrodes and diaphragms and, by means of channels formed in the electrodes, to wash over the faces of the electrodes; the
electrode face is embossed with projections arranged in a plurality of adjacent
elementary groups, each in itself parallel to the marginal portions of the electrode, the projections themselves being formed in arrays of elongated ridges,
peak points, or the like, which may be continuous or discontinuous, and
which are inclined with respect to the marginal portions, the direction of inclination of adjacent groups being opposite each other so that the median
flow direction of electrolyte and reactant washing over the face of the
electrodes will be approximately perpendicular to the marginal portions
from which the flow emanates.

3,773,559; November 20, 1973; B. Warszawski, B. Verger and P. Domenjoud
"Hydraulic Type Voltage Control for a Battery of Series Connected Fuel Cells"

In a battery of several fuel cells connected in series and supplied with a common recycled electrolyte flow, the first cell is continuously supplied with an
adjustable flow of energizing reagent, the adjustment being over a range carrying the voltage across the terminals of the cell between minimum and
maximum working value. Each of the other cells, which may or may not be
supplied with reagent during the use of the battery, is shunted by a rectifier
that allows current to pass in the direction of the normal discharge current of
the battery as a whole, and the group of cells is supplied with energizing
reagent having a total flow subject to adjustment between the flow necessary
to obtain the minimum voltage at the terminals of the first cell and that
which provides the maximum voltage at the terminals of the battery, this
total flow being made up of a variable number of individual cell flows such
that at a given moment, ratio between the flow at that moment and the
maximum flow is the same for all the cells being energized. This system is
particularly applicable to motors for propelling land vehicles.

3,814,631; June 4, 1974; B. Warszawski, B. Verger and P. Demange
"Framed Electrodes Containing Means for Supplying or Draining Liquid Along
the Edge of an Electrode"

> A framed electrode having a flat frame around a thin sheet electrode. The
> frame contains supply orifices in one side and drainage orifices in another
> side. The supply orifices and preferably also the drainage orifices communi-
> cate with the surface of the electrode through a plurality of microchannels.
> A plurality of microchannels diverge from each of said orifices toward the
> side of the electrode. Preferably the microchannels diverging from a given
> orifice are symmetrically positioned. The microchannels are preferably
> separated from the side of the electrode toward which they lead by a fluid
> distribution area positioned therebetween. The fluid distribution area con-
> tains a plurality of spaced projections.

3,899,356; August 12, 1975; P. Groult, F. Hubert, J. Daunay and P. Bono
"Porous Electrode for a Fuel Cell and Method of Making Same"

> A porous electrode for a fuel cell comprising a catalytic layer comprised of
> a mixture of active carbon and Teflon, a stop layer comprised of Teflon de-
> posited on the catalytic layer, and current collector means including a grid
> compressed on the two preceding layers, so that it crosses through the stop
> layer and penetrates into a part of the thickness of the catalytic layer. The
> electrode has a porosity in the order of 50%.

3,902,916; September 2, 1975; B. Warszawski
"Rechargeable Electrochemical Generator"

> An electrochemical generator (e.g., a battery) which is rechargeable electro-
> chemically with liquid electrolyte having negative electrodes of zinc and the
> positive nonporous electrodes that are contacted on the same face by the
> electrolyte and an oxygen-containing gas. The generator or battery comprises
> a repetitive unit of groups of thin, plate-like components, each group having
> a total thickness that is preferably smaller than about 1 mm and comprising
> an impervious bipolar electrode, a microporous membrane, and means for
> maintaining an essentially constant spacing between said bipolar electrode and
> the membrane. These generators are capable of providing long service lives
> and/or a large specific power or output.

3,902,919; September 2, 1975; C. Hespel
"Electrochemical Cell Suitable for Operating in all Positions"

> A fuel cell for land vehicles, space craft and submarines is arranged in the
> center of a circular shell and supplied with hydrazine, hydrogen peroxide
> and potassium hydroxide by two outside tanks. The residual gases and a
> fraction of the potassium hydroxide coming from the cell are conveyed into
> the vicinity of the geometrical center of two auxiliary compartments of potas-
> sium hydroxide arranged inside the shell, these compartments comprising, also,
> tubes for discharging the gases and electrolyte, external of the shell, whose
> upstream inlet opens into the vicinity of the geometrical center, so that the
> volume of electrolyte in these compartments remains substantially constant,
> whatever the position of the cell may be.

3,925,099; December 9, 1975; P. Bono, P. Demange, C. Leclere and A. Tissier
"Fuel Cell with a Decarbonation Cycle, Consuming a Mixed Fuel"

> The invention consists in injecting a formate into the carbonated electrolyte
> going from the power section of a methanol-air primary battery to the decar-
> bonation section thereof, in a quantity such that formate ions thus generated
> jointly with the formate ions formed in the power section by oxidation of
> methanol ensure the decarbonation of the electrolyte in the decarbonation
> section of the battery and make it possible to obtain a maximum value of the
> ratio between the reaction surfaces of the power section and the decarbonation

section of the primary cell. More particularly, in certain cases, the injection of methyl formate makes it possible to obviate the separate injecting of methanol and of a formate.

4,002,493; January 11, 1977; B. Warszawski
"Fuel Cell Structure and System, More Particularly for a Carbon Fuel and Atmospheric Air"

This structure is constituted by several cells each comprising, more particularly an anode and a cathode separated by a corrugated membrane partly immersed in the electrolyte, as well as a condensor-exchanger device for the water drawn off by the combustive, that water being made to flow back towards the electrolyte by means of a plane part of the membrane coming and bearing against one of the walls of the condensor-exchanger.

Societe Les Piles Wonder, Saint-Ouen, France

3,836,398; September 17, 1974; M. Garcin and A. Gauthier
"Electrochemical Generators of the Metal-Air or Metal-Oxygen Type"

The generator comprises in combination a container in which are housed a gaseous double diffusion electrode constituting a first assembly with an obturator which carries and closes said container and an anode dry electrolyte assembly, removable and replaceable independently of the first assembly. The anode dry electrolyte assembly is U-shaped and is arranged in the container on both sides of the gaseous double diffusion electrode.

Sony Corp., Tokyo, Japan

3,592,698; July 13, 1971; H. Baba
"Metal Fuel Battery with Fuel Suspended in Electrolyte"

A metal fuel battery consisting of a liquid-containing chamber at least one wall of which is a gas diffusion positive electrode including a metal oxidation catalyst therein, a negative electrode positioned in the chamber and spaced from the positive electrode, a liquid electrolyte having a finely divided metal fuel powder suspended therein and means for circulating the electrolyte through the chamber.

3,758,342; September 11, 1973; H. Baba
"Metal Fuel Battery and System"

A metal fuel battery having a number of cells, each cell having at least one gas diffusion electrode (positive electrode) and one cathode collector electrode (negative electrode) periodically between which a liquid electrolyte having powder metal fuel dispersed therein is caused to flow in a downward direction. The battery is used in a system in which the metal fuel is replenished from time to time as is also the electrolyte.

Southern California Gas Co., Los Angeles, Cal. (with Consolidated Natural Gas Service Co. and Southern Counties Gas Co. of California)

3,589,942; June 29, 1971; F.B. Leitz, Jr. and D.K. Fleming
"Bipolar Collector Plates"

Southern Counties Gas Co. of California, Los Angeles, Cal. (with Consolidated Natural Gas Co. and Southern California Gas Co.)

3,589,942; June 29, 1971; F.B. Leitz, Jr. and D.K. Fleming
"Bipolar Collector Plates"

The Standard Oil Company, Cleveland, Ohio

3,657,015; April 18, 1972; F. Veatch, E.C. Milberger and R.D. Presson
"Hydrazine Fuel Cell and Process of Operating Said Fuel Cell"

A process for conversion of chemical energy directly into electrical energy
which comprises the steps of supplying an aqueous solution of hydrazine
to the interface between an anode and an aqueous electrolyte in a fuel cell,
supplying an oxidizing agent to the interface between a cathode and an
aqueous electrolyte in said fuel cell said anode and cathode being ionically
connected, and electrically connecting said anode and cathode through an
electrical load.

3,719,528; March 6, 1973; R.K. Grasselli and J.L. Callahan
"Fuel Cell Containing an Electrolyte Consisting of an Aqueous Solution of
Arsenic Acid"

Improved ion-containing and conducting medium for electrochemical reaction apparatus comprising phosphoric acid and/or arsenic acid and water.
The principles of this invention, for exemplary purposes, will be described
in reference to a fuel cell for directly converting chemical energy into
electrical energy, it being understood, however, that these principles are
applicable to other types of electrochemical reaction apparatus as well.

Stanford Research Institute, Menlo Park, Cal.

3,741,809; June 26, 1973; M. Anbar
"Methods and Apparatus for the Pollution-Free Generation of Electrochemical
Energy"

Methods and apparatus for the pollution-free generation of electrochemical
energy from coal, hydrocarbons or other carbonaceous fuels by a cyclic
operation wherein a metal such as lead is oxidized in an electrochemical cell
delivering electromotive force, with the resulting lead oxide then being reduced in a molten salt melt within a regeneration chamber by the addition
thereto of the fuel. The reduced lead metal is then returned to the cell for
the next cycle. A useful salt melt for employment in the regeneration
chamber comprises admixed alkali metal carbonates, and preferably a melt
of this character is employed both as the vehicle in which the lead is reduced
by the fuel as well as for the electrolyte in the cell. Passage of the lead and
lead oxide between the respective cell and regeneration zones can be continuous or intermittent. The total operation is carried out without formation of pollutant gases, and only CO_2, N_2 and H_2O vapors are vented. Any
sulfur present in the fuel is converted to lead sulfate and is removed from
the salt, usually by treatment of a sidestream.

AB St. Powercell, Helsingborg, Sweden

3,811,952; May 21, 1974; O.L. Siwersson and K.G. Tell
"Method and Apparatus for Supplying Entirely or Substantially Entirely Con-
sumable Metallic Material to a Magnetic Electrode in a Metal/Oxygen or
Metal/Air Cell

A method and apparatus for supplying substantially entirely consumable
metallic material to a magnetic electrode in a metal/oxygen cell which also
includes an oxidizing agent electrode and an electrolyte, comprises supplying the metallic material in the form of particles having ferromagnetic
properties and in such proximity to the magnetic electrode that the
particles are attracted by and retained to it.

Studebaker Corp., South Bend, Ind.

3,554,803; January 12, 1971; A.R. Poirier
"Fuel Cell Designed for Efficient Stacking"

A fuel cell battery having a plurality of unitized duplex fuel cells, each being
comprised of conventional fuel cell elements. The unitized duplex cells are
mounted in the battery such that each cell operates independently of all
other cells in the battery. A fluid fuel circulates inside each duplex cell and a
fluid reactant circulates between adjacent duplex cells, thereby permitting each
cell to be removed for repair or replacement without the disassembly of the
entire battery.

Teledyne Isotopes, Westwood, N.J.

3,879,218; April 22, 1975; F.J. Kellen and S.S. Tomter
"Apparatus for Controlling the Feeding of Reactant to a Fuel Cell"

Apparatus for controlling the feeding of reactant to a fuel cell having a valve
means for regulating the amount of reactant fed to the fuel cell, means
in operative relationship with the fuel cell for sensing concentration of the
reactant in the fuel cell, means for controlling operation of the valve means
in accordance with the concentration of reactant and means electrically
isolating, with respect to at least dc energy, the terminals of the fuel cell
from the controlling means and the sensing means to eliminate leakage of
electrical energy between the fuel cell terminals, the sensing means, and the
controlling means so that the voltage read-out from the sensing means is accurate and is not affected by any electrical leakage from the fuel cell.

Texas Instruments Inc., Dallas, Texas

3,598,656; August 10, 1971; I. Trachtenberg
"Fuel Cell Utilizing Apertured Metal Foil Electrodes"

A fuel cell having a pair of foraminous electrodes wherein one of the
electrodes includes apertured metal foil bonded to an expanded metal support.

3,615,845; October 26, 1971; F.L. Gray
"Fuel Cell Electrolyte Control"

An electrolyte control system for a fuel cell which has an anode and a cathode
spaced by an electrolyte carrying matrix wherein porous capillary conduits communicate from an electrolyte reservoir to points uniformly along the matrix.
For example, the porous capillary conduits can be positioned uniformly along
the matrix adjacent the cathode and thereby supply electrolyte directly to the
cathode, and to the anode and matrix at whatever rate the matrix will absorb
the electrolyte.

3,617,385; November 2, 1971; F.L. Gray
"Fuel Cell"

A fuel cell electrode subassembly for a multicell power package having at
least a pair of thin, porous, elongated electrodes, e.g., cathodes, which are
operatively spaced to receive an anode carrying a molded matrix and are
connected in series to a third thin, porous electrode, e.g., an anode, which
is aligned with the space between the first two electrodes hermetically
sealed with an electrically conductive material. Each subassembly can be
interfitted with other subassemblies to yield several series-connected fuel
cell arrays.

3,619,144; November 9, 1971; M.S. Bawa and J.K. Truitt
"Apparatus for Reforming Hydrocarbon Fuels to Produce Hydrogen"

Apparatus for reforming hydrocarbon fuels by admixing a hydrocarbon and
water to form a feedstream, contacting the feedstream with an inorganic salt

or mixture of inorganic salts in the molten state, and separating the gaseous effluent generated by this contact. After the hydrogen-containing effluent is removed from the molten salt, means are provided that the latter may be mixed with oxygen at a temperature and pressure sufficient to permit combustion between the oxygen and any residual carbon which may be retained in the salt, after which further means are provided to remove the gaseous reaction products of this combustion.

3,622,394; November 23, 1971; M.S. Bawa, H.R. Kroeger and J.K. Truitt
"An Electrolyte-Matrix Material for Use in Molten Carbonate Fuel Cells"
>This invention relates to fuel cells. More particularly it relates to an electrolyte-matrix material and the process of making the same for use in high-temperature molten carbonate fuel cells.

Textron, Inc., Providence, R.I.

3,840,407; October 8, 1974; N.-P. Yao, R.L. Oliver and H.N. Seiger
"Composite Porous Electrode"
>A composite porous electrode for use as a gas-diffusion electrode in a battery assembly is formed by coating one surface of a porous plaque of noncorrosive metal with a layer of hydrophobic material, for repelling aqueous electrolyte solution, and coating the opposite surface with a layer of electronically conductive material dispersed in a hydrophobic binder therefor. The second coating is formed of a plurality of successive, integral layers of increasing concentration of electronically conductive material. Principally, for larger sized electrodes the second coating is formed with a thickness differential from bottom to top to balance hydrostatic pressure differentials in the electrolyte. A battery assembly is provided in which a pair of porous electrodes are secured on opposite sides of a consumable electrode with aqueous electrolyte therebetween. The housing is formed with ducts and cavities to direct electrochemically reactive fluid, such as halogen gas, against the outer surfaces of the porous electrodes.

Toyota Chuo Kenkyusho KK, Japan

3,666,561; May 30, 1972; T. Chiku
"Electrolyte Circulating Battery"
>An electrolyte circulating battery having a plurality of cells electrically connected in series and supplied with circulating electrolyte, wherein there is provided a main electrolyte inlet passage and branched inlet passages leading therefrom to each cell, and branched outlet passages leading from each cell to a main electrolyte outlet passage, the branched inlet and outlet passages being made sufficiently long and of sufficiently small cross section to prevent passage of any substantial amount of internal current between cells. Gas, such as air and oxygen, is introduced into the branched inlet end and/or outlet passages through needles to form bubbles in the electrolyte, thereby further increasing the electric resistance in said branched passages.

AB Tudor, Stockholm, Sweden

3,684,950; August 15, 1972; O.K.-G. Nilsson
"Lead Storage Battery Cell Electrolyte Density Measuring Device and Method"
>The measurement of the charge in a lead-acid storage battery through the determination of the density of the sulfuric acid electrolyte by making two conductivity measurements of the electrolyte with the second measurement being made after a dilution or concentration of the sulfuric acid to thereby resolve an ambiguity in the conductivity-density relationship.

Tyco Laboratories, Inc., Waltham, Mass.

3,739,573; June 19, 1973; J. Giner
"Device for Converting Electrical Energy to Mechanical Energy"
> An electrochemical energy conversion device comprising a container, one or more electrochemical cells in said container adapted to produce a gas or to consume said gas according to the polarity of current passed by said one or more cells, polarity reversible means for passing current through said one or more cells, and pressure-responsive moveable mechanical means arranged for movement according to changes in the gas pressure in said container. Each cell comprises a gas producing and consuming electrode, a counter electrode, and an electrolyte contacted by both electrodes.

Union Carbide Corp., New York, N.Y.

3,553,029; January 5, 1971; K.V. Kordesch and W.G. Darland, Jr.
"Electrode with Carbon Layer and Fuel Cell Therewith"
> A fuel cell including a multilayer electrode comprising (a) a metal liquid repellent, gas permeable layer; (b) a liquid permeable, conductive layer and; (c) a middle layer containing carbon. Where the electrode is a fuel electrode, the middle layer is catalyzed with metals catalytically active in fuel cell anode reactions. The repellent can be a hydrophobic resin, such as polytetrafluoroethylene.

3,556,856; January 19, 1971; R.J. Elbert
"Three-Layer Fuel Cell Electrode"
> Electrodes which comprise (a) a porous metal layer having relatively fine pores forming the electrolyte side of the electrode (b) a center layer of porous metal having relatively coarse pores, one side of which is bonded to the fine pored metal layer, and (c) a gas permeable layer of plastic bonded active material which forms the gas side of the electrode and which is bonded to the other side of the coarse pored metal layer, at least a portion of such bonding taking place through penetration of the plastic bonded material into the coarse pores.

3,562,018; February 9, 1971; E.G. Munck and R.F. Hauser
"Battery Comprising Carbon Electrode Wetproofed with Polyterpene Resin"
> Gas permeable carbon electrodes for batteries are rendered repellent to battery electrolyte by polyterpene resin on at least the electrochemically active surfaces of the electrodes.

3,595,698; July 27, 1971; K.V. Kordesch
"Hydrazine Fuel Cell with Acrylic Acid Polymer Membrane"
> Alkali metal salts of ethylene-acrylic acid copolymers are useful as selective, ion permeable membranes for permeable membranes for hydrazine-air fuel cells and permit operation of the cell at high fuel concentrations.

3,629,009; December 21, 1971; R.J. Bennett
"Auxiliary Component Package for Oxygen-Metal Batteries"
> A high-rate oxygen-metal battery is provided with an auxiliary component package for the storage, release and pressure regulation of the oxygen gas. The package includes an oxygen storage tank supported on a basal member which also serves as a housing for all the remaining auxiliary components, i.e., an oxygen-release mechanism, a gas-pressure regulator and a circuit switch.

3,672,998; June 27, 1972; W.G. Darland, Jr.
"Extended Area Zinc Anode Having Low Density For Use in a High Rate Alkaline Galvanic Cell"
> For use in a high rate alkaline galvanic cell, an extended area anode compact

is provided composed of elongated forms of zinc such as zinc fibers or wool, in pressure-formed, multipoint physical contact throughout the body of the anode compact. The anode compact may also be formed from fabricated metal such as expanded zinc metal or screen. In forming the anode compact, the zinc fibers, wool or expanded zinc metal is compression molded to a controlled low bulk density of below 2.5 grams per cubic centimeter. To attain reasonable high electrode efficiencies on the order of 70% of theoretical and above at electrical current drains of about 250 amperes per square foot, the anode compact is formed to a low bulk density of from about 1 to 1.75 and preferably from about 1 to 1.50 grams per cubic centimeter. Optimum electrode efficiencies are attained if the bulk density of the anode compact is maintained within the range of from about 1.0 to 1.25 grams per cubic centimeter. Suitable means are provided in the cell employing the anode compact for maintaining its internal temperature at least at a minimum operating temperature required to discharge the cell at high current densities.

3,674,563; July 4, 1972; R.J. Bennett
"Oxygen-Depolarized Galvanic Cell"

Disclosed is a high performance oxygen-depolarized galvanic cell which is capable of supplying electric power at extremely high electrode current densities. The cell basically includes a porous, electrolyte absorbent, extended area anode made from an electrochemically active metal, a thin, porous fixed zone oxygen-depolarizable cathode, and a porous ionically permeable separator interposed between the anode and cathode and containing an alkaline electrolyte, the alkaline electrolyte being an aqueous solution of potassium hydroxide having a concentration of between about 10 and 14 normal.

3,783,026; January 1, 1974; K.V. Kordesch
"Air-Depolarized Cells Utilizing a Cyanate or Thiocyanate-Containing Electrolyte"

Air-depolarized cells are disclosed which have an improved aqueous electrolyte. The electrolyte, which is neutral, slightly acid, or slightly alkaline, contains: (a) salt such as a halide of an alkali metal, alkaline earth metal, aluminum, zinc, or ammonia, and (b) a cyanate or thiocyanate salt. The air-depolarized cells of the invention alleviate the problem of carbonation with attendant plugging of the pores in the cathode that is often encountered with air-depolarized cells utilizing strong alkaline electrolytes.

3,847,673; November 12, 1974; K.V. Kordesch and M.B. Clark
"Hydrazine Concentration Sensing Cell for Fuel Cell Electrolyte"

A cathodic sensing cell is provided which monitors hydrazine concentration in the electrolyte of a fuel cell. The hydrazine in the electrolyte reacts on the surface of a depolarized cathode lowering the open circuit voltage of the cathode proportional to the amount of hydrazine present. This effect is used to regulate the hydrazine concentration in an operating hydrazine fuel cell.

3,849,201; November 19, 1974; K.V. Kordesch
"Anode and Sensing Cell for Hydrazine Fuel Cell"

The disclosure describes a sensing cell which monitors hydrazine concentration in the fuel cell electrolyte by detecting the potential differences between (a) an anode which forms part of a complete circuit including a load and (b) a reference anode, both anodes being in contact with the electrolyte. Also described is a hydrazine fuel cell which includes the improved anode and/or the sensing cell.

**3,880,671; April 29, 1975; K.V. Kordesch and A. Kozawa
"Corrosion Inhibitor System for Alkaline Aluminum Cells"**

> The aluminum electrodes in alkaline aluminum galvanic cells are inhibited
> against corrosion while the cell is on open circuit or while the cell is being
> operated at very low discharge current, by an inhibitor system comprising
> an alkali metal citrate plus a lead compound, a tin compound, or both a
> lead compound and a tin compound.

**3,960,601; June 1, 1976; D.A. Schulz
"Fuel Cell Electrode"**

> Improved fuel cell electrodes produced by spinning a carbonaceous pitch
> having a mesophase content of from about 40% by weight to about 90%
> by weight to form carbonaceous pitch fiber; disposing staple lengths of the
> spun fiber in intimately contacting relationship with each other in a nonwoven
> fibrous web; heating the web produced in this manner in an oxidizing atmos-
> phere to thermoset the fibers to an extent which will allow the fibers to
> maintain their shape upon heating to more elevated temperatures; heating
> the web containing the thermoset fibers to a carbonizing temperature in an
> oxygen-free atmosphere so as to expel hydrogen and other volatiles; im-
> pregnating the web with a hydrophobic resin dispersion, compressing the
> impregnated web to remove excess fluid and form a thin, paper-like sheet;
> drying the compressed sheet, and heating it at a temperature sufficiently
> elevated to cause the hydrophobic resin impregnant to fuse and bond the
> fibers of the web together so as to increase the structural integrity and hy-
> drophobicity of the web; and then treating the web with a metallic catalyst
> to produce the desired electrode.

Union Oil Co. of California, Los Angeles, Calif.

**3,560,264; February 2, 1971; R.E. Biddick
"Fuel Cell with Electrolyte or Fuel Distributor"**

> In combination an electrolyte and fuel distributor means and fuel cell com-
> prising a plurality of electrodes separated by absorbent members, at least a
> portion of said plurality of electrodes having exposed surfaces and hydrophilic
> material secured in physical contact with said exposed surfaces whereby elec-
> trolyte collects on said hydrophilic material; said electrolyte and fuel distributor
> comprising a fluid impermeable closed-end conduit having a plurality of sub-
> stantially coplanar apertures; a fluid absorbent element coextensive with and
> occupying at least a portion of the enclosure formed by said conduit, siphon-
> fluid conducting means extending through each of said apertures, one end of
> said siphon means contacting said fluid absorbent element, the other end of
> said siphon means adapted to communicate at least one of electrolyte and
> fuel to said absorbent members in said fuel cell; and means provided in said
> conduit through which at least one of electrolyte and fuel can be supplied
> thereto.

**3,708,341; January 2, 1973; R.E. Biddick
"Electrode with Passageways and Weir-Shaped Electrolyte Collecting Means"**

> A porous fuel cell electrode having a gas surface and an electrolyte surface,
> a plurality of passageways therethrough in fluid communication between
> said gas surface and said electrolyte surface, a coating of hydrophobic ma-
> terial on said gas surface, a plurality of fluid collecting droplets of elec-
> trolyte formed on said gas surface, for channeling the accumulated elec-
> trolyte into said passageways and returning said electrolyte to said electrolyte
> surface of said electrode, said fluid collecting means being weir-shaped semi-
> conduits on said electrode passageway to form a static head of accumulated
> fluid.

U.S. National Aeronautics and Space Administration, Washington, D.C.

3,910,814; October 7, 1975; H. McBryar
"Reconstituted Asbestos Matrix"

> An asbestos matrix suitable for use in a fuel cell or electrolysis cell is pro-
> duced having a greater porosity and bubble pressure for a given thickness,
> improved homogeneity, and more uniform thickness than heretofore known.
> The matrix is produced by first shredding the asbestos; forming a slurry of
> the asbestos with water or a low boiling hydrocarbon, preferably an alcohol
> such as methanol; forming a mat by passing the slurry through a properly
> sized porous plaque having a piece of filter paper on top; drying the mat so
> produced; and rolling the mat so produced to yield desired thickness and
> surface finish.

U.S. Secretary of the Air Force, Washington, D.C.

3,891,461; June 24, 1975; Y. Harada and A.Z. Hed
"Chemical Protection of Asbestos"

> A method for protecting the asbestos fiber matrix of a hydrogen-oxygen
> type fuel cell from the degradative effects induced by a potassium hy-
> droxide fuel cell electrolyte which comprises adding potassium silicate in
> proper proportions to the potassium hydroxide electrolyte.

U.S. Secretary of the Army, Washington, D.C.

3,549,422; December 22, 1970; O.C. Wagner
"Rechargeable Metal-Air Battery"

> A rechargeable metal-air unit cell is provided in which a flat cadmium
> anode is positioned intermediate of and spaced from a pair of flat air
> cathodes and means provided to transport the electrolyte to the electrodes.
> The electrolyte transport means include an electrolyte reservoir adjoining
> the electrodes, each cell being in combination with an electrode separation
> system consisting of electrolyte absorbent material layers extending from
> the cathode and anode faces respectively into the reservoir; with layers of
> nonoxidizable membrane material positioned between the layers of elec-
> trolyte absorbent material.

3,717,506; February 20, 1973; R.E. Hopkins
"High Voltage Deposited Fuel Cell"

> This invention pertains to an improved fuel cell of the solid electrolyte
> variety and a method of manufacturing same. The improved fuel cell is
> especially adaptable to high voltage, high power applications in that a
> large number of planar cell assemblies may be stacked in a compact manner
> to afford a relatively high power density.

3,718,507; February 27, 1973; D. Linden
"Bi-Cell Unit for Fuel Cell"

> A bi-cell unit is provided for a fuel cell. The bi-cell unit is comprised of
> two electrode packs. Each of the electrode packs includes an anode, an
> air cathode, and a separator between the anode and the cathode. The two
> packs are assembled together with the anodes spaced apart and facing each
> other, the space between the anodes defining an anolyte chamber through
> which a liquid mixture of fuel and electrolyte is circulated. This invention
> relates in general to a bi-cell unit for a fuel cell and in particular, to a bi-
> cell unit for a hydrazine-air fuel cell.

3,745,127; July 10, 1973; W.J. Asher
"Composition of Matter Containing Carbon"

A new composition of matter has been made. It consists of porous carbon
with a very thin layer of alumino silicate on the pore walls. The alumino
silicate contains exchangeable sodium cations. Ion exchange can be used
to change adsorption properties or load catalyst elements. These materials
can be used for fuel cell electrodes, adsorbents or catalyst supports. This
invention relates to a new composition of matter suitable for use as the fuel
electrode of a hydrocarbon-air fuel cell, as an adsorbent, and as a support
for catalysts.

3,753,780; August 21, 1973; D.L. Fetterman
"Fluctuation Sensitive Fuel Cell Replenishment Control Means"

This invention relates, in general, to electrical energy source regulation
means and in particular, to means for replenishment of selected energy
sources, such as fuel cells, having a voltage-load function characteristic
curve which must be maintained at all load levels for optimum performance of the system. In this invention, electronic control means adapted
to sense minute fluctuations in the output voltage and load current serve
to actuate replenishment means, as appropriate, to maintain the voltage
level per each load point on the characteristic curve.

3,844,839; October 29, 1974; J. Perry, Jr.
"Fuel Cell Electrode"

A porous iron plaque impregnated with palladium black is used as the
anode of a hydrazine-air electrochemical cell. This invention relates in
general to fuel cell electrodes, and electrochemical cells containing the
electrodes, and in particular, to an anode for a hydrazine-air electrochemical cell and to a hydrazine-air electrochemical cell containing the anode.

3,948,681; April 6, 1976; H.J. Barger, Jr. and A.A. Adams
"Fuel Cell Utilizing Direct Hydrocarbon Oxidation"

A fuel cell utilizing direct hydrocarbon oxidation is disclosed. The fuel
cell comprises an anode, a cathode, a fuel compartment, an oxidant compartment, a compartmented electrolyte chamber and an electrolyte disposed
in the compartmented electrolyte chamber. The basic electrolyte utilized
is the monohydrate of trifluoromethyl sulfonic acid $CF_3SO_3H \cdot H_2O$.

4,000,003; December 28, 1976; B.S. Baker and R.N. Camp
"Fuel Cell-Secondary Cell Combination"

A hybrid fuel cell secondary battery system suitable for low power sensor
applications is provided. The system comprises in combination, a fuel cell,
a fully contained fuel and oxidant source for the fuel cell, a DC to DC
power processor for boosting the voltage output from the fuel cell, and a
nickel-cadmium battery in parallel with the output from the DC to DC
processor to sustain peak power drains.

U.S. Secretary of the Interior, Washington, D.C.

3,551,209; December 29, 1970; J.J. Alles, H.E. Ricks, W.D. Reed and J.B.
Carter
"Formation of Skeletal Metal Solid Electrolyte Fuel Cell Electrodes"

A porous nickel, iron or cobalt electrode is coated on a solid ceramic
electrolyte by (a) electrolessly depositing the metal on the electrolyte;
(b) electrodepositing a second layer of the metal, while alloyed with
molybdenum or tungsten, onto the electrolessly deposited layer; and
(c) heating the electrodeposited layer in the presence of water vapor to
drive off the molybdenum or tungsten as a volatile oxide.

U.S. Secretary of the Navy, Washington, D.C.

3,592,692; July 13, 1971; E.C. Jerabek
"Apparatus for Battery Cell Connections"

A method and apparatus are disclosed for making intercell connections in a metal-air battery wherein the metal electrode is removed and another inserted. The connection utilizes integral or added parts of the anode and cathode in a low impedance sliding contact. This contact is of low mass, impossible to misconnect, protected from corrosion, does not impede access of air to the air electrode and requires no other wiring to connect the individual cells of the battery in series.

3,592,694; July 13, 1971; H.B. Urbach, R.E. Smith, R.J. Bowen and D.E. Icenhower
"Hydrazine Fuel Cell Control System"

Apparatus for controlling hydrazine concentration in fuel cells. A hydrazine sensor comprises an elongated porous anode, guarded from the cathode by a porous separator   prevent cathode gas from contacting the anode. The current output of the sensor is fed into a comparator then into a hydrazine-feeding system which delivers fuel to the fuel cell.

3,623,914; November 30, 1971; W.N. Carson, Jr.
"Metal Anode Package"

A salt package is provided for use with mechanically rechargeable primary metal-air cells. The salt package is formed from a binder-salt mixture spread as a dry solid on a portion of the nonconductive webbing which serves as a cell separator and which surrounds the replaceable metal anode. The binder, which rapidly disintegrates when wetted, is selected to be nonreactive with cell components and to provide a durable solid when dry. Means for simultaneous insertion of the salt package and replaceable anode are provided.

3,645,794; February 29, 1972; R.P. Hamlen and E.G. Siwek
"Disposable Anode Package"

The disclosure describes a disposable anode package for use in mechanically rechargeable oxygen-depolarized metal-air batteries. The anode package includes a cellophane envelope which retains the flocculent discharged anode material. A pull tab at the top of the cellophane envelope allows the spent anode to be removed. Dependent on the thickness of the individual cell, nonreactive mesh or fiber separator materials are used to prevent the discharged flocculent matter from dropping to the bottom of the anode package.

3,736,187; May 29, 1973; J.H. Harrison, R.J. Bowen, H.B. Urbach and D.E. Icenhower
"Pressure Equilibrated Gas Fuel Cells and Method"

A self-pressurized fuel cell system for operation at deep sea ambient pressures includes a fuel cell having oxygen and fuel manifolds. Oxygen is produced and fed within the system to the oxygen manifold, and a fuel solution is produced and fed within the system to the fuel manifold. Pressure is maintained within a relatively flexible low strength housing surrounding the cell substantially the same as pressure ambient to the housing due to by-product gases produced by the cell being fed to the housing.

3,748,180; July 24, 1973; J.V. Clausi, M.B. Landau and G. Vartanian
"Fuel Cell System for Underwater Vehicle"

A fuel cell system for an underwater vehicle having at least one fuel cell module for supplying electrical power for said vehicle. A first tank containing hydrogen is connected to one reactant chamber of a fuel cell module and a second tank containing oxygen is connected to another reactant chamber of the fuel cell module. The hydrogen-product water output from the hydrogen

reactant chamber is condensed and separated and the water produced is
stored in the first tank containing hydrogen. The fuel cell module is placed
in a containment vessel which is pressurized with nitrogen. A catalytic
reactor is provided in the containment vessel to form water in the event
that there are simultaneous leaks in both the hydrogen and oxygen high
pressure supply lines. A pressure transducer is also provided in the containment vessel for shutting down the system if there is a high pressure leak
in either the hydrogen or oxygen line.

**3,795,544; March 5, 1974; J.V. Clausi, M.B. Landau, L.H. Otter and R.D.
Sawyer**
"Pressure Balanced Fuel Cell System for Underwater Vehicle"

A pressure balanced fuel cell system for an underwater vehicle having at least
one fuel cell module for supplying electrical power for said vehicle. A first
compartmented tank having a resilient separator is provided for supplying
fuel for an oxygen generation system and a second compartmented tank having a resilient separator is provided for supplying enriched anolyte to the fuel
cell. Pressure balance is maintained in the first compartmented tank by sea
water and in the second compartmented tank by the spent anolyte recovered
from the fuel cell module.

3,909,297; September 30, 1975; E.J. Zeitner, Jr. and J.H. Kennedy
"Lithium-Chlorine Battery Design"

A lithium-chlorine fuel cell having a porous graphite cathode through which
gaseous chlorine is diffused into an electrolyte-containing reaction chamber,
and a molten lithium-wettable metallic fiberous matrix anode which is capable of delivering molten lithium to said reaction chamber. A method for
activating the fuel cell. A battery having at least one of the fuel cells attached in series.

United Technologies Corp., Hartford, Conn.

3,553,023; January 5, 1971; T. Doyle
"Fuel Cell Gas Reversal Method and System"

A system for periodically reversing the direction of flow of a reaction gas
through a fuel cell to prevent a local drying-out of the cell in the region
adjacent the inlet end of the cell.

3,573,102; March 30, 1971; J.W. Lanc, J.H. Hirschenhofer and R.L. Gelting
"Fuel Cell Gas Manifold System"

A manifold system for distributing the flow of reactant gases through a fuel
cell to prevent a drying out of the cell. The fuel cell includes a fuel electrode, an oxidant electrode, and an aqueous electrolyte retentive matrix
located between the electrodes. The manifold system includes a gas
chamber for distributing the flow of at least one of the reactant gases over
the surface of its electrode in a circuitous path between the inlet and outlet of the chamber. The inlet and outlet ends of the chamber are located
closely adjacent to each other with respect to the path followed by the gas
flowing through the chamber so that the driest and wettest regions of the
cell near the inlet and outlet ends, respectively, of the chamber are close
together to prevent a drying-out of the cell.

3,576,677; April 27, 1971; S.J. Keating, Jr. and R.D. Sawyer
"Fuel Cell Process Air Control"

The air supply which provides oxygen to the cathode of a fuel cell utilizing
an aqueous potassium hydroxide electrolyte includes a constant volume
output pump which is fed by a portion of the partially expended air exhausting
from the cathode and also draws air from the atmosphere through a valve
which increases atmospheric air flow in response to increasing fuel cell

power plant load current, thereby to provide increased oxygen content and decreased water vapor pressure to the process air in response to increasing current loads.

3,585,077; June 15, 1971; E.I. Waldman
"Reformer Fuel Flow Control"

A fuel cell control apparatus is disclosed wherein the reformer feed flow is regulated responsive to the hydrogen consumption demands of the fuel cell. The basic mode of operation concerns the control of reformer feed flow as a function of fuel cell gross current and biasing the feed flow as a function of reformer temperature. Flow settings are achieved rapidly through gross current control and the reactor temperature provides a gradual trimming of the flow settings.

3,585,078; June 15, 1971; R.A. Sederquist and J.W. Lane
"Method of Reformer Fuel Flow Control"

A fuel cell control is disclosed wherein the method of regulating the reformer feed flow is responsive to the hydrogen consumption demands of the fuel cell. The basic mode of operation concerns the control of reformer feed flow as a function of fuel cell gross current and biasing the feed flow as a function of reformer temperature. Flow settings are achieved rapidly through gross current control and the reactor temperature provides trimming of the flow settings.

3,607,416; September 21, 1971; J.H. Sizer, Jr. and J.D. Giner
"Spinel-Type Electrodes, Process of Making and Fuel Cell"

A process of forming a fuel cell electrode comprised of a nickel oxide-cobalt oxide spinel surface is disclosed. The process consists of first forming a nickel oxide layer on a substrate, then impregnating it with a solution of cobalt salt and subsequently oxidizing the impregnated substrate to form the mixed oxide spinel-type electrode.

3,607,419; September 21, 1971; S.J. Keating, Jr.
"Fuel Cell System Control"

A fuel cell system is controlled responsive to the hydrogen partial pressure in the anode effluent and the reformer-reactor temperature. The control concerns regulating stream flow to a venturi mixer dependent upon the hydrogen partial pressure. Auxiliary fuel flow may be scheduled responsive to abrupt changes in gross current to eliminate the time lapse normally associated with flow of fuel from the reformer to the fuel cell. Supplemental fuel is supplied to the reformer-burner dependent upon reactor temperature.

3,615,839; October 26, 1971; R.A. Thompson, A.H. Levy and E.M. Hoyle
"Fuel Cell System with Recycle Stream"

A high-temperature integrated fuel cell system with high overall thermal efficiency is achieved by utilizing fuel cell waste heat as the source of energy for the fuel-reforming process. A hydrocarbon fuel is mixed with a recycle stream containing water vapor and the mixture is introduced into the anode chamber of a high-temperature molten carbonate fuel cell. Within the anode chamber is a suitable catalyst which, with the addition of fuel cell waste heat, produces a hydrogen-rich stream. Hydrogen reacts electrochemically at the anode. The anode effluent essentially contains hydrogen, carbon dioxide, and steam. Carbon dioxide and a portion of the water is removed from the anode effluent and is transferred to the process air supply since a relatively high concentration of carbon dioxide is desirable at the cathode in the molten carbonate system. The remaining moist anode stream is recycled to be mixed with the fresh fuel.

3,615,849; October 26, 1971; A. Hall
"Fuel Cell Dielectric Heat Transfer Medium"

A device is disclosed whereby a heat transfer medium is inserted between a fuel cell stack and an attached boiler to improve the heat transfer from the cell stack to the boiler. The medium consists essentially of a porous nonconductive matrix saturated with a dielectric grease.

3,615,862; October 26, 1971; H.A. Roth and W.R. Lasko
"Fuel Cell Electrodes"

Activated fuel cell electrodes are prepared by electrodeposition of a noble metal catalyst on a solid foil or porous electrode substrate. The electrodeposition is effected from an alkali metal hydroxide solution, preferably molten alkali metal hydroxide. The electrolysis is generally carried out at temperatures of about $300°$ to $500°F$ at current densities of about 5 to 150 ma/cm^3 by passing a direct electrical current to the electrode substrate through the alkali metal hydroxide electrolyte solution. Palladium-black is a preferred noble metal catalyst, and palladium or palladium-silver alloy foils and porous sintered nickel structures are the preferred electrode substrates.

3,634,139; January 11, 1972; C.A. Reiser
"External Reservoir and Internal Pool Fuel Cell System and Method of Operation"

A fuel cell module is disclosed for use with aqueous electrolyte fuel cells with provisions for accepting large electrolyte volume changes which occur as a result of changes in ambient conditions and/or power output. The module incorporates internal cell pools combined through an internal manifold system to an external electrolyte reservoir. The advantage of this particular arrangement lies in its ability to provide broad tolerance to electrolyte volume changes and is particularly suitable for use with compact fuel cells having lightweight electrodes.

3,649,360; March 14, 1972; D.P. Bloomfield and W.J. Olsson
"Combined Water Removal and Hydrogen Generation Fuel Cell Powerplant"

The disclosure concerns a fuel cell and a metal hydride bed within a casing. The fuel cell requires hydrogen for the fuel at the anode. Water vapor, the by-product of the electrochemical reaction, may be evolved through the anode and diffused to the metal hydride bed within the casing where the water vapor and metal hydride react to generate hydrogen. Hydrogen diffuses to the anode. This self-contained system requires no external controlling devices or moving parts. A feature of this system lies in the provision of supplementary water vapor through a specially tapered wick connected to an external water reservoir.

3,650,838; March 21, 1972; J.D. Giner and J.R. Moser
"Electrochemical Cell with Platinum-Ruthenium Electrode and Method of Using with Ammonia"

An improved method of, and an apparatus for, generating electrical energy is described comprising feeding ammonia to an anode comprising a major amount of platinum and a second metal of the platinum group in minor but effective amounts. Preferred percentages range from about 75 to 98 parts platinum and from about 25 to about 2 parts of the promoter metal on an atomic weight basis.

3,655,448; April 11, 1972; H.J. Setzer
"Hydrogen Generator Desulfurizer Employing Feedback Ejector"

Water which is to be mixed with hydrocarbon fuel for feedstock in a hydrogen generator has low pressure hydrogen diffused therein so as to permit a hydrogen rich fuel-water admixture to enhance desulfurization of the admixture, without fuel sulfur poisoning of a diffuser. The hydrogen is fed back serially or in parallel from any point downstream of a steam reforming reactor.

3,664,873; May 23, 1972; R.F. Buswell, P.G. Nicholas and W.J. Olsson
"Self-Regulating Encapsulated Fuel Cell System"

> A hydrogen and oxygen fuel cell powerplant is disclosed wherein the fuel
> cell is encapsulated within a sealed pressure tank for undersea applications.
> The system is self-regulating to compensate for variations in the venting
> and leaking of reactant gases. Hydrogen gas in the capsule is reacted or
> burned, and the resulting water vapor is condensed and collected in a tank
> at the bottom of the capsule. A simple method of controlling the reactant
> purge flow has been devised to include essentially a pressure regulator in the
> oxygen purge line and an orifice in the hydrogen purge line.

3,668,013; June 6, 1972; T.C. Franz
"Fuel Cell System with Pneumatic Fuel Flow Control"

> A control is disclosed for automatically metering fluid flow as a function
> of the difference between two pressures or of a single pressure. A restric-
> tion is disposed in the fluid outlet line so that the metered flow will depend
> upon pressure differential across the restriction. The pressure upstream of
> the restriction is regulated by a valve in the supply line activated by a com-
> puting diaphragm assembly consisting essentially of three spaced intercon-
> nected disc elements responsive to the difference between the two pressures
> or a single control pressure when one of the pressures is fixed.

3,671,317; June 20, 1972; E. Rifkin
"Method of Making Fuel Cell Electrodes"

> Inexpensive screen substrates for fuel cell electrodes are provided with a
> nonporous vinylidene fluoride corrosion resistant coating. Thereafter, a
> second, metal particle and porous polymer, coating is provided. Cure of
> the second coating must be compatible with the first coating and monitored
> to prevent the flow of the polymer in the second coating so that a porous
> structure with the desired hydrophilic properties and conductive properties
> preferred in electrode structure can be obtained.

3,677,823; July 18, 1972; J.C. Trocciola
"Fuel Saturator for Low Temperature Fuel Cells"

> Fuel cell waste heat supplies the energy to saturate a dry fuel gas supply
> stream with water vapor. The fuel saturator may be featured as an integral
> component of the fuel cell stack. The mixture is directed to a reformer
> where the fuel is steam reformed to produce a hydrogen rich stream which
> is utilized as the fuel for the electrochemical reactions occurring within the
> fuel stack.

3,694,310; September 26, 1972; R.C. Emanuelson, R.C. Stewart and R.W.
  Vine
"Fuel Cell Organic Fiber Matrix"

> A wettable organic fiber matrix is provided by coating phenolic resin fibers
> with a phenolic beater addition resin forming the matrix curing the beater
> addition resin, and making the matrix wettable by heating the matrix in
> air or in a partially inert atmosphere.

3,698,957; October 17, 1972; R.A. Sanderson
"Fuel Cell System Having a Natural Circulation Boiler"

> A natural circulation boiler for a fuel cell system is provided which both
> cools the fuel cell and provides steam for a steam reformer; the boiler is in
> contiguous heat exchange relationship with the fuel cell.

3,704,172; November 28, 1972; J.K. Stedman and R. Cohen
"Dual Mode Fuel Cell System"

> A dual mode fuel cell system is provided by the combination of a fuel cell
> power section with separate waste heat and electrolyte diluent removal sub-
> systems for open cycle operation and for closed cycle operation.

3,716,415; February 13, 1973; R.N. Gagnon, C.V. Banic and A.P. Grasso
"Fuel Cell Electrolyte Concentration Control"

A fuel cell electrolyte concentration control is provided by wet bulb and dry bulb temperature sensors disposed downstream of a fuel cell in a reactant gas recirculation conduit; excess electrolyte diluent removal means are provided and are controlled by control means as a function of the difference between the sensed dry bulb and wet bulb temperatures.

3,743,544; July 3, 1973; R.C. Stewart, Jr.
"Fuel Cell"

Means for sealing and unitizing a fuel cell are provided by a seal having an integral O-ring on its outer edge for sealing between two coolant plates, the seal having a step-shaped inner edge spaced from said O-ring by a supporting gasket, the inner edge being in mating contiguity with the edge of the matrix.

3,745,047; July 10, 1973; S. Fanciullo, R.N. Gagnon and W.S. Summers
"Proportional Action Electronic Fuel Control for Fuel Cells"

A proportional electronic fuel control controls the feed flow of fuel to the reformer of a fuel cell system in response to fuel cell current, reformer temperature and the position of the fuel flow control valve. The electronic control produces a proportional control signal which actuates solenoids in a digital manner to regulate the position of the fuel flow control valve. A deadband is provided in the electronic control to prevent continuous cycling of the control valve.

3,748,179; July 24, 1973; C.L. Bushnell
"Matrix-Type Fuel Cell with Circulated Electrolyte"

A compact, matrix-type electrochemical cell utilizing a circulating electrolyte is described. The cell comprises an anode, a cathode, a matrix containing an ion-conductive electrolyte between the anode and cathode, and a porous metal plate containing porous pins positioned adjacent to one of said anode and cathode in order that the pins of the plate are in contact with the anode or cathode over the limited surface area of the pins. Electrolyte circulated behind the porous plate floods the porous plate, porous pins, and electrode directly beneath the pins and makes contact with the electrolyte contained in the electrolyte matrix. As a result of the plurality of contact points, electrolyte is circulated into and from the matrix. The advantages of both a circulating electrolyte system and a matrix-type electrolyte system are realized.

3,761,316; September 25, 1973; J.K. Stedman
"Fuel Cell with Evaporative Cooling"

A fuel cell assembly utilizing the waste heat of a fuel cell to provide evaporative cooling of the cell is provided by a hydrophobic separator disposed in heat conducting relationship with the fuel cell. A coolant liquid is fed under pressure to a cavity on one side of the hydrophobic separator, and as vapor evolves from the coolant liquid, it passes through the hydrophobic separator to ambient.

3,765,946; October 16, 1973; L.H. Werner and J.C. Trocciola
"Fuel Cell System"

A fuel cell system comprising an anode, a cathode and an electrolyte in combination with a reformer system comprising a reformer and reformer gas separator is described. The reformer gas separator comprises a first membrane selectively permeable to a consumable reactant gas (hydrogen) and a second membrane spaced from said first membrane selectively permeable to impurities or detrimental gases (CO_2). A sweep gas is maintained at the downstream side of the second membrane reducing the partial pressure of the detrimental gas and effectively increasing the partial pressure of the reactant gas providing a more efficient system.

3,769,090; October 30, 1973; M. Katz and J.K. Stedman
"Electrochemical Cell"

> A compact electrochemical cell utilizing a circulating electrolyte in which the anode and cathode are very closely spaced, decreasing the internal resistance of the cell, is described. The cell comprises in spaced relation a reactant chamber, a gas (bubble) distributor, an electrolyte chamber, a first electrode which is flooded with electrolyte, an electrolyte chamber, a second electrode, and a second reactant chamber. The advantages of a circulating electrolyte are achieved while permitting the use of electrodes and electrode separators having low or zero bubble pressures and while maintaining the internal resistance of the cell low.

3,775,185; November 27, 1973; H.R. Kunz and M.A. Walsh
"Fuel Cell Utilizing Fused Thallium Oxide Electrolyte"

> A fuel cell system for the direct generation of electricity from a fuel and oxidant is described employing an electrolyte containing thallium oxide. This fuel cell system will operate efficiently using air as an oxidant and impure hydrogen or a hydrocarbon as the fuel at relatively low temperatures, i.e., $250°$ to $300°C$.

3,779,811; December 18, 1973; C.L. Bushnell and J.K. Stedman
"Matrix-Type Fuel Cell"

> A compact electrochemical cell is described comprising an anode, a cathode, a matrix containing an ion-conductive electrolyte between the anode and cathode and porous metal plates containing porous pins positioned adjacent each of said anode and cathode in order that the pins of the plates are in contact with said anode and cathode over the limited surface area of the pins. The electrolyte volume of the cell is controlled by electrolyte movement through the pins of the porous plate, thereby stabilizing the electrochemical performance of the cell.

3,801,372; April 2, 1974; R.H. Shaw
"Fuel Cell System Comprising Separate Coolant Loop and Sensing Means"

> A fuel cell system comprising a fuel cell stack, a pressurized reactant supply, and a circulating coolant, in which the coolant is circulated by a hydrostatic pump is described. The hydrostatic pump is created by introducing a gaseous reactant into one leg of a coolant loop, thereby decreasing the density of the coolant and providing a hydrostatic head to circulate the coolant. The pump has no moving parts, eliminating a source of potential mechanical failure.

3,801,374; April 2, 1974; G.H. Dews and R.W. Vine
"Graphite and Vinylidene Fluoride Structures for Fuel Cells"

> A corrosion resistant coolant and support plate molded from graphite powder and vinylidene fluoride is disclosed for use in acid electrolyte fuel cells. A limited amount of vinylidene fluoride is used to obtain the structural characteristics so that thermal and electrical conductivity may be maintained at a satisfactory level.

3,811,951; May 21, 1974; J.K. Stedman
"Venturi Tube Regulator for a Fuel Cell"

> Hydrogen is provided at constant pressure to a fuel cell through a Venturi tube. A highly concentrated solution of electrolyte is introduced into the cell from one bladder valve and diluted electrolyte is removed from the cell to a second bladder valve in proportion to the reduction of pressure at the throat of the Venturi tube.

3,839,091; October 1, 1974; D.P. Bloomfield, N.A. Hassett and J.K. Stedman
"Regenerative Fuel Cell"

> A regenerative fuel cell assembly is provided in which a fuel cell is integrated

with an electrolysis cell; the fuel cell assembly, including a water storage
matrix, is disposed adjacent to the electrolysis cell assembly, which also
includes a water storage matrix, and spaced therefrom by a hydrogen gas
passage.

3,847,672; November 12, 1974; J.C. Trocciola and M.A. Walsh III
"Fuel Cell with Gas Separator"

A fuel cell system comprising an anode, a cathode, and an electrolyte in
combination with a gas separator is described. The gas separator comprises a tile or block of salt in molten or solid state having opposed reactive surfaces. The salt is a mixture of $M_xO + M_xCO_3$ where M is an
element such as an alkali or alkaline earth metal. A gas stream containing
a fuel gas (hydrogen) and CO_2 is fed to one surface of the salt tile or block
at which surface the carbon dioxide is chemically taken up by the salt. A
sweep or stripping gas is maintained at the downstream surface of the salt
tile or block at which surface carbon dioxide is released, and carried away
by the sweep gas, for subsequent cycling to the cathode of the cell if
desired to reduce concentration polarization of the cell.

3,850,969; November 26, 1974; W.S. Summers and F.G. Charest
"Purge Timer for a Fuel Cell"

A purge valve for purging a fuel cell is automatically opened during a
purge cycle after the fuel cell has expended a preselected number of
ampere-hours. In one specific embodiment a plating current, proportional to the current provided by the fuel cell, causes platable material of
an electrolytic cell to be transferred from the anode to the cathode thereof
whereby the platable material on the cathode is proportional to the ampere-
hours that have been expended by the fuel cell. When substantially all of
the platable material has been transferred, the voltage of the electrolytic
cell rises, thereby initiating the purge cycle. In another embodiment a
current-to-frequency converter provides pulses to a binary counter in proportion to the current provided by the fuel cell whereby the accumulated
count in the counter is proportional to the expended ampere-hours. When
the accumulated count corresponds to the preselected number of expended
ampere-hours, the purge valve is automatically opened.

3,855,002; December 17, 1974; C.R. Schroll
"Liquid Electrolyte Fuel Cell with Gas Seal"

Escape of reactant gases from a fuel cell is prevented by sandwiching an
electrolyte-saturated matrix between a separator plate and an electrode,
utilizing the electrode for support and the electrolyte itself to provide a
wet capillary seal against the escape of gas.

3,859,138; January 7, 1975; S.T. Narsavage, R.W. Vine and R.C. Emanuelson
"Novel Composite Fuel Cell Electrode"

Electrode substrates comprising infusible, cured novolac fibers and silicon
dioxide fibers are fabricated into a suitable form. The substrate is made
electrically conductive by heat treatment in an inert atmosphere to render
the novolac fibers conductive. The substrates are relatively inexpensive,
have good resistance to acid and alkaline electrolytes, have low ohmic resistance, and when treated with an electrocatalyst provide a high performance
electrode.

3,859,139; January 7, 1975; G.H. Dews, D. Eichner and F.S. Kemp
"Novel Composite Fuel Cell Electrode"

Method of making a composite electrolyte matrix/electrode assembly comprising filling the pores of a porous hydrophilic matrix resistant to alkali
or acid electrolyte with a completely volatile filler such as water, solidifying
the filler, applying a catalytic mixture to the surfaces of the matrix while

the filler is in the solidified form, and thereafter removing the filler. The
filler, being present while the catalyst is applied to the matrix, prevents the
catalytic particles which are electrically conductive from impregnating the
matrix which can cause internal cell shorting. Further, a catalyst layer with
a uniform, flat surface is obtained.

3,867,206; February 18, 1975; J.C. Trocciola, C.R. Schroll and D.E. Elmore
"Wet Seal for Liquid Electrolyte Fuel Cells"

Escape of reactant gases from a fuel cell is prevented by sandwiching an
electrolyte saturated porous electrode end and an electrolyte saturated matrix
between surfaces which are thereby wet by the electrolyte, the electrolyte
providing a wet capillary seal against the escape of gas.

3,877,989; April 15, 1975; E.I. Waldman and J.R. Aylward
"Power System and an Electrochemical Control Device Therefor"

A transducer is provided which comprises a metal/gas/electrolyte electrochemical cell which delivers a high-gain change of gas pressure as a function of millivolt shunt voltage impressed upon its electrodes. The device
is simple, relatively inexpensive and reliable with a high degree of sensitivity
enabling its use, for example, in a fuel cell power plant to sense and use
load current to directly control the power plant.

3,878,296; April 15, 1975; R.W. Vine, W.J. Harrison and R.C. Emanuelson
"Reaction of Lithium Carbonate and Fibrous Aluminum Oxide to Produce Lithium Aluminate"

An electrolyte matrix comprising fibrous lithium aluminate for use in an
electrochemical cell comprising an alkali or alkaline earth carbonate electrolyte
is described. The matrix has good thermal cycle characteristics from ambient
to the operating temperature of the cell, and good electrolyte retention and
wicking properties.

3,880,670, April 29, 1975; B.H. Shinn
"Electrochemical Cell Separator Plate Means"

Separator plate means, disposed between two adjacent electrochemical cells,
have a thin layer of electrically and thermally conductive material disposed
on the surfaces thereof which are contiguous with the cells; the separator
plate means forms gas compartments with each adjacent cell and includes
coolant means therein; conducting means, such as pins, are disposed in the
separator plate means for providing an electrical path from one cell, through
the separator plate means, to the adjacent cell and for providing a heat conductive path from each of the cells to the coolant compartment.

3,905,832; September 16, 1975; J.C. Trocciola
"Novel Fuel Cell Structure"

A compact electrochemical cell comprising a pair of opposed electrodes; an
electrolyte matrix containing an aqueous, ion-conductive electrolyte between
the opposed electrodes; and an electrolyte reservoir positioned behind and
partially defined by at least one of said electrodes. The electrode partly
defining the reservoir has a continuous hydrophobic surface and select hydrophilic areas substantially uniformly distributed within the boundaries of the
electrode surface. The electrolyte volume of the cell is controlled by electrolyte movement between the electrolyte matrix of the cell and the reservoir
through the select hydrophilic areas of the electrode, thereby stabilizing the
electrochemical performance of the cell.

3,912,538; October 14, 1975; G.H. Dews and F.S. Kemp
"Novel Composite Fuel Cell Electrode"

A fuel cell electrode comprising a continuous carbon fiber substrate, a graphite-
hydrophobic polymer sublayer partially impregnated into said fiber substrate,

and a catalyst-hydrophobic polymer layer on said sublayer. The electrode has
good chemical stability and electrical conductivity; permits excellent control
of the reaction interface of the electrode; and permits a low-catalyst loading,
providing a relatively inexpensive electrode.

3,915,747; October 28, 1975; W.S. Summers and S. Fanciullo
"Pulse Width Modulated Fuel Control for Fuel Cells"

The injection of fuel into a fuel cell hydrogen generator is controlled by
the actuation of a fuel injector valve in response to a pulse width modulated
square wave as a function of fuel cell stack operation. The off time of the
fuel injector valve is fixed at a constant value by comparison of a ramp voltage produced by a first integrator with a constant reference voltage. The on
time of the fuel injector valve is variable and is determined by comparing a
ramp voltage from a second integrator with a variable voltage which is a function of fuel stack voltage and fuel cell gross current. A fail-safe circuit is incorporated to insure that the fuel injector does not remain in a continuous
on or off state if a malfunction occurs in the control circuitry.

3,923,546; December 2, 1975; M. Katz, S.W. Smith and D. Reitsma
"Corrosion Protection for a Fuel Cell Coolant System"

The internal coolant system of a fuel cell power plant utilizes a soluble salt
of a metal in the coolant fluid to inhibit the corrosion of those fuel cell
components that corrode due to shunt currents flowing through the coolant
fluid. In a preferred embodiment a soluble salt of iron is used.

3,932,197; January 13, 1976; M. Katz and A. Kaufman
"Method for Catalyzing a Fuel Cell Electrode and an Electrode So Produced"

A porous conducting particle, hydrophobic bonded, substrate supported electrode is prewetted with the electrolyte. A dc voltage is applied to the electrode to assist in the prewetting with the electrolyte. A soluble catalyst-
containing material is then introduced into the electrode and the catalyst
deposited within the electrode. By appropriate selection of the porous conducting particles and the catalyst-applying techniques, precise control of the
location of the catalyst can be obtained. If graphite materials are used as
the conducting particles, a catalyst-containing salt is allowed to dissolve in the
electrolyte in the prewetted electrode, and the catalyst-containing material is
reduced to the metal. If the reduction is done by reaction with a reducing
gas such as hydrogen, the catalyst will be deposited only in those regions of
the electrode at which there is an electrolyte-reactant gas interface which is
in electrical-conducting relationship with the substrate. Alternatively, extremely precise amounts of catalyst can be deposited within the electrode
structure by use of a solution of a compound of the catalyst whose wettability
with the hydrophobic material varies as the solution evaporates. By this
technique almost 100% of the catalyst can be deposited within the electrode
structure on the hydrophilic region, with virtually no losses in the hydrophobic material.

3,940,285; February 24, 1976; R.C. Nickols, Jr. and J.C. Trocciola
"Corrosion Protection for a Fuel Cell Coolant System"

The internal coolant system of a fuel cell power plant utilizes a gas in the
coolant fluid to inhibit the corrosion of those fuel cell components that
corrode due to shunt currents flowing through the coolant fluid. In a
preferred embodiment hydrogen gas is used.

3,945,844; March 23, 1976; R.C. Nickols, Jr.
"Electrochemical Cell"

An electrochemical cell comprising an anode, a cathode, and an electrolyte in
combination with support or coolant plates fabricated from a polymer/metal
composite, such as polysulfone/nickel, is described. The plates are relatively

inexpensive, easily fabricated, have good thermal conductivity, low electrical resistance, and good structural stability.

3,956,014; May 11, 1976; D.A. Landsman and E.I. Thiery
"Precisely-Structured Electrochemical Cell Electrode and Method of Making Same"

An electrochemical cell electrode, having clearly segregated and structured hydrophobic gas channels and electrolyte-filled catalyst channels is provided in which the size and shape of the various passages are such that they offer little resistance to the passage of reactants and products. The electrode consists of alternate layers of porous hydrophobic material and porous hydrophilic catalyst-containing material. The electrode is formed by first depositing the alternate layers of materials and then cutting the structure in a plane at right angles to the plane of deposition at a thickness equal to that of the end product. Electrodes of varying size can be constructed of panels so produced.

3,961,986; June 8, 1976; E.I. Waldman
"Method and Apparatus for Controlling the Fuel Flow to a Steam Reformer in a Fuel Cell System"

A fuel cell system includes an ejector for pumping steam and fuel into a steam reformer. Steam is the primary flow through the ejector and fuel is the secondary flow. The rate of steam flow is metered by a variable area orifice in the ejector. The fuel passes through a laminar restrictor in the conduit carrying the fuel to the ejector. A pressure regulator increases and decreases the pressure of the fuel supplied to the laminar restrictor by amounts equal to increases and decreases in the ejector back pressure thereby eliminating the ejector back pressure as a factor affecting the rate of fuel flow into the ejector. The laminar restrictor is designed to match the fuel flow to the ejector pumping characteristics so as to provide the desired steam to fuel ratio for the steam reformer for each operational mode of the fuel cell.

3,964,929; June 22, 1976; P.E Grevstad
"Fuel Cell Cooling System with Shunt Current Protection"

A fuel cell coolant system for use with a plurality of fuel cell stacks connected electrically in series is adapted to use a nondielectric coolant carried through each stack by electrically conductive tubes. Each stack also includes a plenum for distributing coolant to the tubes. To prevent short circuits across the stacks due to shunt currents in the coolant the plenums are electrically insulated from a grounded main coolant supply line, such as by dielectric hose connections, and each plenum is electrically connected to one end of its respective stack. By this invention the maximum driving potential for any shunt current is the potential drop across a stack rather than the potential drop from a stack to ground. In a preferred embodiment dielectric hoses are used to connect the plenums to the tubes in order to prevent short circuiting across cells within a stack.

3,964,930; June 22, 1976; C.A. Reiser
"Fuel Cell Cooling System"

A cooler for removing waste heat from a stack of fuel cells includes a plurality of tubes for carrying the coolant through the stack. The tubes are disposed adjacent the nonelectrolyte side of electrodes in the stack in grooves or passageways formed in the surface of plates which separate one cell in the stack from another. Since the tubes are exposed to the electrolyte used in the stack they must be made from or at least include a protective coating of material which is stable in the electrolyte. Preferably this material is also a dielectric to prevent shunt currents from passing into the tubes and coolant which may be water.

3,969,145; July 13, 1976; P.E. Grevstad and R.L. Gelting
"Fuel Cell Cooling System Using a Non-Dielectric Coolant"

A cooler for removing waste heat from a stack of fuel cells uses a nondielectric coolant which is carried in a plurality of tubes passing through one or more separator plates in the stack. Preferably the coolant is water so that heat removal is by evaporation of the water within the tubes by boiling. The tubes are electrically insulated from the cells by a coating of dielectric material such as polytetrafluoroethylene. In one embodiment of the present invention the cooler tubes are connected to the stack coolant supply conduits by dielectric hoses having a high length to diameter ratio to provide a several hundred thousand ohm impedance path in case of a flaw in the protective dielectric coating, in order that a short circuit of the stack does not occur.

3,973,993; August 10, 1976; D.P. Bloomfield and M.B. Landau
"Pressurized Fuel Cell Power Plant with Steam Flow Through the Cells"

A fuel cell power plant for producing electricity uses pressurized air and fuel in the cells. The air is compressed by compressor apparatus powered by waste energy in the form of hot pressurized gases including hot pressurized steam produced by the power plant. In one embodiment the compressor apparatus includes a turbine operably connected to a compressor, and hot pressurized gases produced by the power plant flow into the turbine thereby driving the compressor. The steam is generated by heat from the fuel cells, passes through the fuel cells adjacent the cathode electrode thereof, and is delivered into the turbine along with the other gases.

3,976,506; August 24, 1976; M.B. Landau
"Pressurized Fuel Cell Power Plant with Air Bypass"

A power plant for producing electricity uses fuel cells run on pressurized air and pressurized fuel. The air is compressed by compressor apparatus which is driven by waste energy in the form of hot pressurized gases produced in the power plant. In one embodiment the compressor apparatus includes a turbine operably connected to a compressor for driving the same. At part power, in order to maintain the pressure of the air being fed to the cells while reducing the amount of air fed to the cells, a portion of the compressed air is bypassed around the fuel cells, is increased in temperature using an auxiliary burner, and is delivered into the turbine to help drive the compressor. By doing this the mass flow of air through the cells may be reduced without reducing the mass flow of gases into the turbine thereby maintaining the speed of the compressor at part power.

3,976,507; August 24, 1976; D.P. Bloomfield
"Pressurized Fuel Cell Power Plant with Single Reactant Gas Stream"

A fuel cell power plant for producing electricity uses pressurized air and fuel in the cells. The power plant includes an autothermal reactor for processing the fuel and a compressor driven by a turbine for compressing the air used by the fuel cells. Pressurized effluent gases from the cathode side of the cell and pressurized fuel is delivered into the autothermal reactor and from the reactor passes into the anode side of the cells. Effluent gases from the anode side of the cells is delivered into the turbine thereby driving the compressor. A burner is used to increase the temperature of the gases entering the turbine. The burner is run on air and unburned fuel in the effluent gases from the anode side of the cells.

3,979,225; September 7, 1976; S.W. Smith and L.J. Bregoli
"Nitrogen Dioxide Regenerative Fuel Cell"

A fuel cell is disclosed in which the cathode is a gaseous diffusion fuel cell electrode operating with an acid electrolyte and nitrogen dioxide (NO_2) alone or with oxygen. The cathode half cell reaction produces nitric oxide (NO)

and water and the NO_2 is externally regenerated by reaction of NO with
oxygen to produce the nitrogen dioxide for reuse in the cell. When both
NO_2 and oxygen are used, oxidation of the NO formed back to the NO_2
occurs within the cathode itself so it is possible to get more than 100%
utilization of the NO_2 in the fuel cell.

3,979,227; September 7, 1976; M. Katz and A. Kaufman
"Method for Catalyzing a Fuel Cell Electrode and an Electrode So Produced"

A porous conducting particle, hydrophobic bonded, substrate supported electrode is prewetted with the electrolyte. A dc voltage is applied to the electrode to assist in the prewetting with the electrolyte. A soluble catalyst-containing material is then introduced into the electrode structure and the catalyst
deposited within the electrode. By appropriate selection of the porous conducting particles and the catalyst-applying techniques, precise control of the
location of the catalyst can be obtained. If graphite materials are used as the
conducting particles, a catalyst-containing salt is allowed to dissolve in the
electrolyte in the prewetted electrode, and the catalyst-containing material is
reduced to the metal. If the reduction is done by reaction with a reducing
gas such as hydrogen, the catalyst will be deposited only in those regions of
the electrode at which there is an electrolyte-reactant gas interface which is
in electrical-conducting relationship with the substrate. Alternatively, extremely precise amounts of catalyst can be deposited within the electrode
structure by use of a solution of a compound of the catalyst whose wettability with the hydrophobic material varies as the solution evaporates. By
this technique almost 100% of the catalyst can be deposited within the electrode structure on the hydrophilic region, with virtually no losses in the hydrophobic material.

3,981,745; September 21, 1976; J.K. Stedman
"Regenerative Fuel Cell"

A regenerative fuel cell assembly is provided in which a fuel cell is integrated
with an electrolysis cell. In a preferred embodiment the fuel cell assembly
and electrolysis cell assembly are spaced apart and have their hydrogen electrodes in facing relationship; a water transport matrix is disposed in the space
between the hydrogen electrodes, and a water storage matrix is disposed
adjacent the oxygen electrode of the fuel cell. During operation of the
fuel cell the water storage matrix holds the water produced by the fuel cell;
during electrolysis cell operation the water in the water storage matrix passes
to the water transport matrix which carries it to the electrolysis cell where
it is consumed.

3,982,961; September 28, 1976; A.P. Grasso
"Fuel Cell Stack with an Integral Ejector for Reactant Gas Recirculation"

A stack of fuel cells operating on gaseous reactants includes an ejector in
integral heat exchange relationship with the stack for recirculating one of
the reactant gases through the cells of the stack. The recirculating reactant
is continuously heated by waste heat from the cells as it recirculates thereby preventing condensation of water from the recirculating reactant gas and
thereby maintaining the dew point constant from the time the reactant gas
leaves the cells until it is mixed with fresh reactant in the ejector. The fresh
reactant gas is preheated prior to being introduced to the ejector so that
there is no condensation throughout the entire loop. The recirculation rate
relative to the amount of fresh reactant can be controlled to regulate the
dew point at the entrance to the cells to best advantage. By this invention
flooding of the electrodes or drying of the electrodes does not occur.

3,982,962; September 28, 1976; D.P. Bloomfield
"Pressurized Fuel Cell Power Plant with Steam Powered Compressor"

A fuel cell power plant for producing electricity uses pressurized air and fuel in the cells. A compressor is driven by a turbine operably connected thereto and provides compressed air to the cells. The turbine is driven by a working fluid in a hot, pressurized, vapor state. Energy to convert the working fluid into this state is waste energy produced by the power plant, such as stack waste heat. In one embodiment the power plant includes a steam reforming reactor and a reactor burner. Effluent gases from the anode side of the cell are used in the reactor burner. The working fluid is water and the turbine is driven by steam which is condensed to the liquid state after passing through the turbine. The liquid water is reconverted to steam by passing it into heat exchange relationship with the stack and it is then delivered again into the turbine. Part of the steam may be used in the steam reforming reactor. Preferably the effluent gas from the reactor burner and the effluent gases from the cathode side of the cells is delivered into an air turbine for generating electrical power in addition to the electrical power produced in the fuel cells.

3,990,912; November 9, 1976; M. Katz
"Electrolyte Regeneration in a Fuel Cell Stack"

In a fuel cell stack utilizing an alkali metal electrolyte, the electrolyte is distributed in parallel between the electrodes of a plurality of fuel cells and is then fed to a regenerator cell which converts carbonate ions to molecular CO_2 gas which is discharged from the cell. Regeneration is effected through the establishment of a hydroxyl ion gradient within a regenerator cell. The regenerated electrolyte is then returned to the fuel cells. In this manner a carbonate buildup in the cell is prevented.

3,994,748; November 30, 1976; H.R. Kunz and C.A. Reiser
"Method for Feeding Reactant Gas to Fuel Cells in a Stack and Apparatus Therefor"

A flow scheme for feeding a reactant gas to the cells of a fuel cell stack wherein the cells are connected electrically in series. For example, the fuel gas is passed in parallel over a portion of each fuel electrode and thereupon into a mixing manifold which directs the exhausted gases in parallel over a different portion of each fuel electrode and thereupon into another manifold. This is continued, depending upon the stack configuration, with the exhausted gases passing back and forth in parallel over different portions of the fuel electrodes and exhausting into a manifold until the fuel gas has covered the entire fuel electrode of each cell in the stack. This reduces the harmful effect of a blockage within the reactant gas chamber of a cell and also reduces the harmful effect caused by a maldistribution of current in one of the cells in the stack.

4,001,041; January 4, 1977; M.C. Menard
"Pressurized Fuel Cell Power Plant"

A fuel cell power plant for producing electricity uses pressurized reactants in the cells. In one embodiment air for the fuel cells is compressed in a compressor driven by a turbine which is powered by waste energy produced in the power plant in the form of a hot pressurized gaseous medium. The power plant includes fuel conditioning apparatus comprising a steam reforming reactor and a reactor burner to provide heat for the steam reforming reactor. Effluent gases from the anode side of the cells are delivered into the reactor burner and from the reactor burner are combined with effluent gases from the cathode side of the cells. The combined gases are used to drive the turbine. Water to produce steam for the steam reforming reactor is recovered from the effluent gases from the anode side of the cells before

they are delivered into the reactor burner and from the effluent gases from
the cathode side of the cells.

4,001,042; January 4, 1977; J.C. Trocciola, D.E. Elmore and R.J. Stosak
"Screen Printing Fuel Cell Electrolyte Matrices Using Polyethylene Oxide as the Inking Vehicle"

A matrix for retaining the electrolyte in a fuel cell is applied to the surface
of one or both the electrodes by screen printing using an inking vehicle which
is an aqueous solution of polyethylene oxide. This method produces a very
thin, continuous and uniform matrix layer and is well suited to production
operations.

4,002,805; January 11, 1977; E.I. Waldman
"Method and Apparatus for Controlling the Fuel Flow to a Steam Reformer in a Fuel Cell System"

A fuel cell system includes an ejector for pumping steam and fuel into a
steam reformer. Steam is the primary flow through the ejector and fuel is
the secondary flow. The rate of steam flow is metered by a variable area
orifice in the ejector. The fuel passes through a fuel control valve before
reaching the ejector. A pressure regulator maintains a constant pressure
drop across a variable area orifice in the valve. The fuel flow is thus
metered by the valve orifice area and not by the ejector pumping characteristics. Ejector back pressure is also eliminated as a factor affecting the
amount of fuel pumped by the ejector. Means are provided to simultaneously
control the orifice areas of the ejector and the fuel valve in order to maintain
a desired ratio of steam to fuel for the steam reformer for each operational
mode of the fuel cell. In a preferred embodiment this means is a mechanical
linkage connecting the ejector and fuel valve to a common actuator.

4,004,947; January 25, 1977; D.P. Bloomfield
"Pressurized Fuel Cell Power Plant"

A fuel cell power plant for producing electricity uses pressurized reactants
in the cells. In one embodiment air is the oxidant and is compressed in
a compressor driven by a turbine. The turbine is powered by waste energy
produced in the power plant in the form of a hot pressurized gaseous medium.
For example, effluent gases from both the anode and cathode sides of the
cells is delivered into the turbine which in turn drives the compressor. In a
preferred embodiment the effluent gases from the anode side of the cells
is first delivered into a burner for providing heat to a steam reforming
reactor, and the effluent gases from the burner are delivered into the turbine. In another embodiment, in addition to effluent gases delivered from
the anode side of the cells into the burner, a portion of the effluent gases
from the anode side of the cells is also delivered into the steam reforming
reactor to provide steam for the fuel processing.

4,017,663; April 12, 1977; R.D. Breault
"Electrodes for Electrochemical Cells"

A novel electrode for use in an electrochemical cell is provided by a
pyrolytic carbon coated, hydrophobic polymer impregnated carbon paper
having a catalyst-hydrophobic polymer layer applied thereon.

4,017,664; April 12, 1977; R.D. Breault
"Silicon Carbide Electrolyte Retaining Matrix for Fuel Cells"

In a fuel cell utilizing an acid electrolyte, such as H_3PO_4, the electrolyte
retaining matrix is made from silicon carbide. The silicon carbide has
been found to be virtually inert to H_3PO_4, at fuel cell operating temperatures and provides all the other necessary and desirable matrix properties.
This matrix is expected to have a life of at least 40,000 hours under normal
fuel cell operating conditions.

Universal Oil Products Co., Des Plaines, Ill.

3,881,957; May 6, 1975; R.H. Hausler
"Electrochemical Cell Comprising a Catalytic Electrode of a Refractory Oxide
and a Carbonaceous Pyropolymer"

> An electrochemical cell having a catalytic electrode which is comprised of a
> refractory oxide having a surface area from 1 to 500 square meters per gram,
> and a carbonaceous pyropolymer forming at least a monolayer of said refractory
> oxide, said electrode having a conductivity at room temperature of from 10^{-8}
> to 10^{0} inverse ohm-centimeters.

Varta AG, Frankfurt, Germany

3,544,382; December 1, 1970; D. Spahrbier, K. Wandschneider and R. Eckardt
"Fuel Cell Device for Continuous Operation"

> In the liquid supply conduit system used for supplying gas precursors in liquid
> form to the catalytically active surface of gas diffusion electrodes there is
> provided a pressure pipe or rising main which is used to regulate the flow of
> such liquids to such electrodes during continuous operation thereof under
> various electrical load conditions.

3,546,019; December 8, 1970; H.H. von Doehren and H.A. Schultze
"Fuel Cell and Fuel Cell Battery"

> A fuel cell including an electrolyte chamber with electrolyte which includes
> a solid cellular material with intercommunication cells which is capable of
> taking up the excess volume of electrolyte formed during the reaction.

3,553,026; January 5, 1971; A. Winsel
"Method of Removing Water of Reaction During Fuel Cell Operation"

> A process and a fuel cell battery comprising porous gas diffusion electrodes
> which is operated to remove water of reaction by passing one or both gas
> mixtures containing gaseous reactants and inert gas through a first group or
> groups of interconnected gas spaces of electrodes of equal polarity, dehumidify-
> ing a portion or all of the effluent gas mixture containing gaseous reactant,
> recycling dehumidified gas to the first group of gas spaces and passing a por-
> tion of the effluent gas from the first group of gas spaces to a second group
> of gas space connected in series for gas flow, and venting the inert gases
> from the last gas space in the series with a minimum loss of the reactant
> gases.

3,562,019; February 9, 1971; D. Spahrbier
"Reserve Fuel Cell Battery"

> Reserve fuel cell battery with liquid oxygen and hydrogen gas yielding mate-
> rials and electrolyte stored therein.

3,573,038; March 30, 1971; M. Jung and H.H. von Doehren
"Porous Formed Catalyst Body and Process for its Production"

> Catalyst powders, such as platinum, palladium or Raney metals, are rendered
> insensitive to air or oxygen by treatment with an oxygenated chlorine, bromine
> or iodine compound, such as potassium or sodium iodate, chlorate or bromate.
> The powder obtained is essentially hydrogen-free, nonpyrophoric and reactivatable.
> The powder may be molded by heat and pressure to a formed porous body.

3,576,732; April 27, 1971; K. Weidinger and A. Kalberlah
"Cast Electrical Batteries and Process for Their Production"

> An electrical device for the production or use of electrical energy which
> has a cast casing enclosing a plurality of positive and negative gas diffusion
> electrodes compactly arranged in an alternating sequence of polarity and in
> such a way that a portion of the surfaces of all of the electrodes of one, or
> of both, polarities project on at least one side of the arranged electrodes in

a uniform manner beyond recessed surfaces of all of the electrodes of the opposite polarity. The casing is cast from synthetic resin and has continuous electrolyte supply canal means therein for supplying electrolyte to, and removing electrolyte from, the electrodes; the casing also has holes for supplying operating gas to and removing such gas from, the electrodes. Process for fabricating the electrical device which includes compactly arranging the electrodes with removable spacing elements and cores therebetween, casting the resulting compact structure in a suitable casting resin, removing the spacing elements and cores to provide internal electrolyte chambers and conduits and then drilling holes through the casing and into the electrodes to provide gas inlet and exit openings for operating gases.

3,591,421; July 6, 1971; H.A. Schultze and D. Spahrbier
"Porous Electrode Having Lyophobic Material Affixed to the Walls of the Pores"

A porous gas diffusion electrode is made up of a single layer of porous material having two opposite faces and pores extending from one face to the opposite face of the electrode. These pores open through the two opposite faces. The walls of the pores are provided with particulate lyophobic material. At the wall area of the pores near their openings at one face of the electrode, the particulate lyophobic material is densely affixed to the wall. The density of the distribution of the particulate lyophobic material decreases along the length of the pores to essentially no coverage adjacent the pore openings at the opposite face of the electrode.

3,594,232; July 20, 1971; D. Spahrbier
"Device for the Automatic Adjustment of the Supply of Liquids"

The supply of liquid, which may be catalytically decomposed with the evolution of gas, to a main decomposing catalyst body, is regulated by the pressure of gas generated from such liquid on an auxiliary catalyst body. The main catalyst body may be in a gas diffusion electrode.

3,620,844; November 16, 1971; E. Wicke and A. Kuessner
"System for the Activation of Hydrogen"

A fuel cell comprising an electrode designed for the electrochemical utilization of gaseous hydrogen, said electrode comprising a metallic body of a heavy metal lattice capable of absorbing and diffusing the hydrogen, and at least a large portion of one side of said body being in intimate contact with a dissimilar metal hydrogenation catalyst layer, consisting of a metal hydride, the catalyst of said layer being selected from the group consisting of uranium hydride, titanium hydride, thorium hydride, cerium hydride and zirconium hydride and the metal of said lattice being selected from the group consisting of tantalum, tantalum alloys, palladium and palladium-silver alloys, and a conduit delivering the gaseous hydrogen to said one side of the metallic body.

3,649,467; March 14, 1972; A. Winsel and H.-J. Schwartz
"Vaporization Apparatus for the Removal of Heat and Concentration of Electrolyte in a Fuel Cell Battery"

Process and equipment for recovering, in liquid form, vaporizable components from compositions by bringing the composition into contact with one side of a porous wall through which the component to be recovered is caused to permeate as a vapor, and the vapor is subsequently condensed on a second wall and recovered therefrom in liquid form. The second wall may also be porous and the vapors of the vaporizable components may be allowed to pass through the second wall before being recovered in condensed form.

3,666,562; May 30, 1972; D. Sprengel
"Fuel Cell with Control System and Method"

A control system is provided for improving the venting of accumulated inert gas from electrochemical energy conversion means such as fuel cell batteries. The venting period is initiated by a drop in cell potential but is independent of cell potential recovery time.

3,668,011; June 6, 1972; H. Grune and A. Winsel
"Galvanic Cell Battery with Gas Diffusion Electrodes"

Galvanic cell battery having a plurality of gas diffusion electrodes of one or both polarities wherein the gas chambers of all the electrodes of one polarity are arranged in a plurality of groups such that, with respect to the flow of operating gas through the gas chambers, all the gas chambers in one group thereof are connected in parallel with one another and all the groups of gas chambers of the electrodes of the same polarity are connected together in series with one another and the number of gas chambers in the respective groups decreases continuously from the first to the last of the series of groups.

3,668,012; June 6, 1972; M. Jung and H.H. von Doehren
"Electrode for Electrochemical Devices and Method of Its Manufacture"

An electrode especially suitable for use in electrochemical cells, like in a fuel cell with alkaline electrolyte in which the catalytically active metal is Raney metal, Raney iron or Raney cobalt coated with particulate copper, mercury, silver, or alloy or a mixture thereof. The fuel cell comprises conventional elements, an alkaline electrolyte and said electrode. The electrodes have improved catalytic activity, especially improved load capacity and an improved rest potential.

3,770,509; November 6, 1973; A. Winsel, S. Guenter, G. Horst and K. Weidinger
"Duplex Gas Diffusion Electrodes with Gas Diffusion Means"

A multilayered gas diffusion unitary electrode body for an electrochemical apparatus, fuel cell elements and electrolysis cell, said gas diffusion electrode having two catalytically active layers and comprising two electrode parts, each part having at least two layers, said parts being bonded to each other into a unitary construction by a bonding or connecting middle layer which joins the respective larger pore layers of the respective electrode parts, said bonding layer comprising a resin or resin coated metal foil or web. A fuel cell battery comprising such multilayered electrodes. A method for bonding the electrode to give the structure described.

3,785,870; January 15, 1974; A. Winsel
"Method and Apparatus for the Removal of Carbon Dioxide from Gas Mixtures and Fuel Cell Combination"

Carbon dioxide is removed from a gaseous mixture containing carbon dioxide and hydrogen by washing the gas mixture with a liquid in an absorber where the carbon dioxide is absorbed, and desorption thereof in a desorber. The absorber and desorber contain a biporous element having a pore system of high capillary pressure and a pore system of low capillary pressure.

3,811,949; May 21, 1974; M. Jung
"Hydrazine Fuel Cell and Method of Operating Same"

An amalgamated metal, like nickel, iron or cobalt to use hydrazine electrochemically, or mixtures and alloys, an electrode, a fuel cell and battery and methods of using same.

3,817,792; June 18, 1974; D. Spahrbier
"Arrangement of a Fuel Cell Battery"

A fuel cell battery comprised of a plurality of separate battery units or elements, coordinated to each battery unit a unit storage tank for liquid fuel-electrolyte mixture and a forwarding system for the circulation of fuel-electrolyte mixture coordinated to each battery unit and unit storage tank combination. All unit storage tanks are arranged separately of each other, such as in vertically stacked fashion, and connected by overflow tubes to permit cascading of the liquid fuel-electrolyte mixture successively through the unit storage tanks by gravity. Fresh fuel-electrolyte mixture is fed from a feed tank by a metering device to the uppermost of the unit storage tanks.

3,843,410; October 22, 1974; D. Spahrbier
"Fuel Cell Power Station"

A self-contained source of well regulated electric power employing a set of fuel cell batteries, whose electrolyte-fuel mixtures are replenished by a cascade supply arrangement automatically controlled to maintain predetermined concentration distributions within the batteries. Reactant gas for the batteries is obtained by decomposition of a stored liquid. Heat exchanger means are provided to absorb the heat of decomposition and to stabilize the operating temperature of the batteries.

3,977,902; August 31, 1976; H. Sauer and D. Spahrbier
"Fuel Cell System"

An assembly of fuel cells, enclosed in a container, is immersed in a tank holding the electrolyte. Operating gas is used to produce electrolyte circulation through the cells.

Varta AG, Frankfurt, Germany (with Siemens AG)

3,544,377; December 1, 1970; E. Justi and R. Wendtland
"Oxygen-Hydrogen Storage Battery Cell"

3,567,519; March 2, 1971; R. Wendtland and A. Winsel
"Depolarization Process"

3,576,676; April 27, 1971; R. Wendtland and A. Winsel
"Galvanic Fuel Cell Battery and Process"

3,597,275; August 3, 1971; A. Winsel and R. Wendtland
"Process of Operating Fuel Cell"

3,660,166; May 2, 1972; A. Winsel
"Gas Diffusion Electrode"

Volkswagen AG, Wolfsburg, Germany

3,737,344; June 5, 1973; K.V. Benda, H. Binder, W. Faul and G. Sandstede
"Process for Increasing the Activity of Porous Fuel Cell Electrodes"

A process for increasing the activity of porous electrodes which contain tungsten carbide having the composition WC_{1-x}' where $0 \leqslant x \leqslant 0.2$, and which are used in connection with the anodic oxidation of carbon monoxide, hydrogen or hydrogen-containing fuels in fuel cells that utilize an acid electrolyte. The process involves placing at least one surface of at least one of the untreated tungsten carbide electrodes in an alkaline electrolyte containing a reducing agent. A potential of up to +850 mv, measured with relation to a hydrogen electrode in the same electrolyte is applied to the tungsten carbide electrode. The potential is maintained for a period of up to 20 hours. The invention also includes the activated tungsten carbide electrode.

Westinghouse Electric Corp., Pittsburgh, Pa.

3,558,360; January 26, 1971; E.F. Sverdrup, A.D. Glasser and D.H. Archer
"Fuel Cell Comprising a Stabilized Zirconium Oxide Electrolyte and a Doped Indium or Tin Oxide Cathode

A fuel cell having an electrolyte composed of a ridged ceramic material that conducts current by diffusion of oxygen ions at about 1000°C, having a fuel electrode on one side of the electrolyte and an air electrode on the opposite side thereof, and the air electrode being composed of one of the materials including indium oxide and tin oxide which is doped to be an electronic conductor having a resistivity of less than 10^{-3} ohm-centimeters.

3,616,408; October 26, 1971; W.M. Hickam
"Oxygen Sensor"

An oxygen solid electrolyte cell is provided having a pair of longitudinal holes. The interior of each of the holes is coated with a porous platinum coating which provides the electrodes of the cell. Each of the holes is adapted to be connected to a separate source of gas to permit flow of different gases over the electrodes within the holes.

3,658,996; April 25, 1972; R. Frumerman and J.D. McAdoo, Jr.
"System for the Removal of Hydrogen from Nuclear Containment Structures"

A system for the direct combustion of hydrogen within a reactor containment structure to prevent the formation of an explosive atmosphere therein. A blower produces a predetermined flow of containment hydrogen to a combustion chamber and additional amounts of hydrogen and oxygen are added to facilitate complete burning so as to minimize the quantity of hydrogen remaining within the containment. A separate system cools and condenses the gases leaving the chamber so as to prevent stratification and localized temperature increases.

3,671,323; June 20, 1972; Y.L. Sandler
"Hydrophobic Layers for Gas Diffusion Electrodes"

A gas diffusion electrode having a gas entrance side and an electrolyte contacting side for use with a liquid electrolyte and a gas in an electrochemical cell, comprises a coherent porous body, containing an electrical conductor and a hydrophobic outer layer on the gas entrance side, the hydrophobic outer layer comprising cloth material impregnated with wet proofing polymer.

3,682,707; August 8, 1972; Y.L. Sandler
"Stabilization of Silver Catalyst in an Air Diffusion Electrode"

An electrochemical cell, comprising an anode, a gas diffusion cathode containing silver catalyst, and an aqueous electrolyte between the anode and cathode to which is added at least 0.01 mol of either carbonate, phosphate or chloride ions per liter of electrolyte, is used to cause current to flow externally through a load circuit connecting the anode and cathode.

3,733,221; May 15, 1973; Y.L. Sandler and D.D. Durigon
"Gas Diffusion Electrode"

A gas diffusion electrode for use with a liquid electrolyte in an electrochemical cell, which electrode includes a porous electrical conductor between two porous layers. The first porous layer on the liquid electrolyte side is composed of a conducting material and a binder. The second layer of a thickness of from 3 to 7 mils is on the gas side and contains a conducting material, a binder and a catalyst; and it is less than about one-half as thick as the first layer and more hydrophobic, whereby gas diffusion rates are optimized and the catalyst is employed in a highly efficient manner.

3,770,508; November 6, 1973; Y.L. Sandler
"Cathodic Activation of Gas Diffusion Electrodes"

A method of activating a gas diffusion electrode and operating a metal-gas cell containing an anode, a gas diffusion cathode having a gas fed into one side, and an electrolyte between the electrodes and permeating the other opposite side of the cathode, with a current flowing through a load circuit connecting the anode and cathode; by reducing the rate of gas flow

to the gas diffusion cathode, at least once for at least about 2 minutes, to
less than 5 cm^3/min for each square centimeter of exposed gas diffusion
electrode area, so as to make the potential of the gas diffusion cathode more
cathodic.

3,799,811; March 26, 1974; R.N. Sampson and J. Chottiner
"Hydrophobic Mats for Gas Diffusion Electrodes"

A gas diffusion electrode having a gas entrance side and an electrolyte contacting side, for use with a liquid electrolyte and a gas in an electrochemical
cell, comprises a coherent porous catalyzed body, containing an electrical
conductor and a hydrophobic outer layer on the gas entrance side. The hydrophobic outer layer comprising a mat containing fibrillated very high molecular weight polyethylene.

3,877,994; April 15, 1975; J. Chottiner
"Catalytic Paste and Electrode"

A catalytic paste is disclosed which contains 3 to 55% finely-divided carbon,
2 to 10% binder, 0.4 to 6% catalyst, 30 to 70% water, and 0.01 to 3% water-
soluble alkali-resistant thickener which, in a 2% aqueous solution, has a vis-
cosity of at least 100 centipoises. The thickener is preferably a nonionic
cellulosic ether. A conducting plaque is filled with the catalytic paste to form
an electrode.

3,925,100; December 9, 1975; E.S. Buzzelli
"Metal/Air Cells and Air Cathodes for Use Therein"

An air cathode for use in metal/air cells comprising a hydrophobic layer
laminated to hydrophilic layer wherein the hydrophilic layer comprises a
metal current collector integrally molded into a composition comprising:
(1) an oxygen absorption/reduction carbon; (2) 50 to 70% by weight
manganese dioxide; (3) at least one of polytetrafluoroethylene and fluorinated ethylene propylene; and (4) Ag-Hg catalyst. The hydrophobic
layer preferably comprises a sheet of porous, unsintered, completely fibrillated polytetrafluoroethylene which is laminated to the hydrophilic layer.
A metal/air cell is provided which includes said air cathode together with
a spaced apart anode selected from iron, zinc, cadmium, or like metal, in
an alkali hydroxide electrolyte.

3,930,094; December 30, 1975; R.N. Sampson and J. Chottiner
"Hydrophobic Mats for Gas Diffusion Electrodes"

A gas diffusion electrode having a gas entrance side and an electrolyte
contacting side contains an electrical conductor and a hydrophobic outer
layer on the gas entrance side, the hydrophobic outer layer comprising
a mat containing fibrillated very high molecular weight polyethylene.

3,935,027; January 27, 1976; C.J. Warde and A.D. Glasser
"Oxygen-Reduction Electrocatalysts for Electrodes"

Gas electrodes which contain a hydrophilic layer are disclosed. The hydrophilic layer is adapted to provide a gas-liquid interface therein; and includes an oxygen absorption/reduction material selected from the group
consisting of $CaMnO_3$ and $Ca_{(1-x-z)}MnO_3 \cdot Na_x Y_z$ and a current collector.
A hydrophobic layer can then be laminated to the hydrophilic layer.

Re 28,792; April 27, 1976; R.J. Ruka and J. Weissbart
"Electrochemical Method for Separating O_2 from a Gas; Generating Electricity; Measuring O_2 Partial Pressure; and Fuel Cell"

A fuel cell and method of operation of same is detailed for measuring oxygen
pressure, separating oxygen from a gas, or generating electrical energy. The
fuel cell has first and second electrodes of selected materials with a solid parti-
tion of a solid electrolyte between the electrodes. The solid electrolyte consists of a solid solution of selected oxides having a high degree of oxygen ion
conductivity.

Yardney International Corp., New York, N.Y.

3,551,208; December 29, 1970; Z. Stachurski
"Cell with Displaceable Electrode"

> This invention is directed to an electrochemical cell or battery in which one
> of the electrodes is adapted to be displaced relative to the other electrode
> during normal operation of the cell or battery. The invention also includes
> a method of operating such cells so that at predetermined intervals the spac-
> ing between the two electrodes comprising the cell is modified.

3,594,233; July 20, 1971; A. Charkey and F.P. Kober
"Rechargeable Gas-Polarized Cell"

> Cell with an external gas electrode enclosing a plurality of reversible metal
> electrodes, the latter electrodes being separated from one another and from
> the gas electrode by inert conductor screens serving as auxiliary recharging
> electrodes; the auxiliary electrode or electrodes proximal to the gas electrode
> may be separated therefrom by a permeable inert dielectric spacer storing a
> reserve quantity of liquid electrolyte.

3,594,234; July 20, 1971; M. Lang and R. Di Pasquale
"Air Depolarized Fuel Cell"

> An air depolarized fuel cell with a metal electrode in a gas-permeable envelope
> electrode of microporous carbon-Teflon mixture, containing an alkaline elec-
> trolyte, is surrounded by a shell of microporous Teflon closely spaced from the
> envelope electrode by a small clearance designed to receive products of inter-
> action of carbon dioxide with the electrolyte.

3,600,230; August 17, 1971; Z. Stachurski and R. Di Pasquale
"Gas-Depolarized Cell with Hydrophobic-Resin-Containing Cathode"

> A gas electrode for use in a gas-depolarizable current-generating cell comprises
> a unitary structure formed of (1) a conductive apertured current-collecting
> member, i.e., a metallic grid or screen, (2) a porous conductive layer consist-
> ing of a hydrophobic resinous material and a network of conductive material
> in fibrous form in intimate contact with one surface of the apertured current-
> collecting member, and (3) a porous catalytically active layer consisting of a
> hydrophobic resinous material containing particles of catalytically active mater-
> rial dispersed therein, in intimate contact with the outer surface of the hydro-
> phobic conductive layer.

3,647,550; March 7, 1972; F.P. Kober and H.B. West
"Disposable Electrochemical Cells and Method of Operating Same"

> An electrochemical cell with an air-depolarizable cathode, a consumable anode
> of a base metal (e.g., zinc or aluminum) and an interelectrode separator laden
> with the anhydride of an aqueous electrolyte, such as sodium chloride, is
> activated by the introduction of a volume of water sufficient for only a frac-
> tional discharge of the stored electric energy. Upon prolonged idleness after
> activation, the depletion of the residual water supply by chemical action and
> evaporation halts the interaction between the stored electrolyte and the
> anode material, thereby conserving the remaining energy and inhibiting cor-
> rosion.

3,650,839; March 21, 1972; M. Lang and R. Di Pasquale
"Fuel Cell Battery"

> Several fuel cells, each comprising a metal electrode in a gas-permeable
> envelope electrode with a catalytically effective inner surface, are juxta-
> posed to form a battery and are separated by solid spacers which are car-
> ried on an interposed partition or project directly from confronting outer
> surfaces of adjoining envelope electrodes to form channels for the admis-
> sion of air or oxygen to the respective cells.

3,834,944; September 10, 1974; E. Dennison
"Multi-Cell Metal-Fluid Battery"

> An improved, multi-cell metal-fluid battery comprises a container having a side-wall and endwalls defining a hollow interior within which are stacked a plurality of cells electrically connected in series, each cell containing a pair of electrodes and electrolyte activated upon contact with a fluid. Fluid flow control means are provided which distribute moisture to the electrolyte and an oxygen-containing gas to the cells to sustain the reaction in the cells. The fluid flow control means include means of separating each of the cells in the stack to provide a transverse fluid flow passageway therebetween, circulation means such as a blower activated by a motor, means of spacing the stack of cells inwardly of the sidewall to define a peripheral fluid flow passageway, central openings in the cells defining a return passageway, fluid supply means and fluid cooling means such as a cooling core.

NONASSIGNED PATENTS

D.C. Avampato, R.N. Dibella and R.E. Vasas

3,811,950; May 21, 1974
"Biochemical Fuel Cell and Method of Operating Same"

> A chemical fuel cell is presented wherein electrical energy is generated by the reaction of enzymes, substrates and nucleotides, the reaction being, in part, similar to parts of the tricarboxylic acid cycle in the human body.

C. Berger

3,567,666; March 2, 1971
"Separation Means"

> This invention relates to a novel composite product which is employed as a separation means in reverse osmosis, gas mixtures and galvanic devices. The efficient separation mode is a result of the channel structure of the polymer coupled with the uniform porosity of the particulate within the polymer.

3,607,410; September 21, 1971
"High Power Density Galvanic Cell"

> This invention relates to a novel element in galvanic cells, applicable to fuel cells and batteries. The inclusion of hollow porous capillaries increases greatly the electrode surface area in a given volume yielding substantially larger power densities than conventional systems.

H. Binder, A. Köhling and W. Kuhn

3,854,994; December 17, 1974
"Gas Electrodes"

> A method of making a porous electrode having an electrically conductive electrocatalytically active layer and a contiguous gas permeable hydrophobic layer comprising filtering a first suspension comprising essentially polytetrafluoroethylene powder in propanol-(1), isopropanol, butanol, or dichloromethane, and withdrawing liquid therefrom to form a first porous layer comprising a matted and damp polytetrafluoroethylene filter cake, placing a metal screen on the filter cake, filtering a second suspension comprising essentially carbon powder, graphite fibers, and polytetrafluoroethylene powder in propanol-(1), isopropanol, butanol, or dichloromethane, by withdrawing liquid therefrom through the screen and the first layer to form a second porous layer including the screen on the first layer, drying the two-layer structure, and heating it to at least about 330°C, and preferably not more than about 380°C, in a nonoxidizing atmosphere. Each filtering typically is carried out by applying suction to each suspen-

sion through a filter medium, preferably in at least two steps with the
suction rate increasing from each step to the next. The metal screen
typically is made of nickel, silver, gold, platinum, or corrosion resistant steel.

M. Bonnemay, G.R. Bronoël, J.-B. Donnet, J. Lahaye and J.A. Sarradin

3,870,565; March 11, 1975
"Cathode for the Reduction of Oxygen and a Rechargeable Cell Comprising
Said Cathode"

An electrochemical catalyst is placed on a support so as to constitute an air
cathode for the reduction of oxygen which is primarily intended for use in a
reversible or rechargeable fuel cell. The catalyst consists of finely-divided and
nonporous carbon of the carbon black type constituted by a chain of substantially spherical particles and having a specific surface within the range of
50 to 600 m^2/g, the carbon black having been subjected to a thermochemical
treatment in a chlorine atmosphere at a temperature within the range of $1500°$
to $3000°$C.

W.C. England

3,988,905; November 2, 1976
"Reversible Mechanical-Thermal Energy Cell"

A reversible mechanical-thermal energy cell comprising: reversible intake and
exhaust passages respectively leading to and from reversible rotatably connected rotary intake and exhaust volumetric periodically vanishing displacement devices of unequal rates of volumetric displacement, said rotary intake
and exhaust volumetric displacements flowably connected by a reversible com-
pression-expansion conduit, said conduit being in thermal communication
with a thermal energy reservoir for containing matter subject to thermal
change; and a compressible-expandible fluid reversibly traversing from said
intake to said exhaust passage via said volumetric displacements and said
conduit, said fluid subject to volumetric, pressure and thermal change in
said conduit. Applicable in a range including thermal energy storage and
retrieval, heating, cooling, cooking, refrigerating, and of fixed installation or
portable.

A.C. Erickson

3,615,838; October 26, 1971
"Fuel Cell Unit with Novel Fluid Distribution Drain, and Vent Features"

A fuel cell is provided with reactant and electrolyte gaskets each defining a
central area having a fluid distributor located therein. The electrolyte gasket
is provided with notched or apertured fingers to uniformly distribute fluid
while the reactant frames may be similarly constructed or provided with
stacked screens having offset apertures.

L.L. Fehrenbacher

3,634,113; January 11, 1972
"Stabilized Zirconium Dioxide and Hafnium Dioxide Compositions"

A type C mixed rare earth oxide solid solution is used to eliminate the mono-
clinic phase of zirconium dioxide and thus produce a stable refractory. The
type C solid solution consists primarily of oxides of dysprosium, erbium,
ytterbium, and holmium with small amounts of thulium, terbium, and lu-
tetium. This solid solution is applicable to hafnium dioxide and to mixtures
of hafnium dioxide with zirconium dioxide as well.

J.B. Fox

3,853,628; December 10, 1974
"Fuel Cell"

A cell having a magnetic anode about which is retained metallic fuel particles with an electrolyte solution of iron II and iron III chlorides. Graphite cathodes with finned outer surfaces provide increased surface area and may be of perforate construction.

H. Ito

3,575,728; April 20, 1971
"Electric Cells Using Urea Monosulfonic Acid as Electrolyte"

The invention provides an electric cell in which urea monosulfonic acid is used as the electrolyte. The other parts of the cell, excepting the electrolyte, involve the following three types in combination of the elements; the first one comprising carbon as a positive electrode, zinc as a negative electrode and manganese dioxide or air as a depolarizer; the second one comprising iron or nickel as a positive electrode, zinc as a negative electrode and air as a depolarizer; and the third one comprising copper as a positive electrode, zinc as a negative electrode and cupric oxide as a depolarizer.

C.I. Johnsen

3,751,302; August 7, 1973
"Generating Alternating and Direct Electric Currents by Modified Fuel Cells"

The activity of a fuel cell is improved by using secondary electrodes to intensify ion-transport between fuel cell electrodes and rotating of the fuel cell to produce oscillating electric pulsations.

3,847,670; November 12, 1974
"Fuel Cells Using Electromagnetic Waves"

This invention relates to methods and means for invigorating electrochemical activities in producing fuel cells by causing advancing electromagnetic waves to impinge on migrating ionized oxidants flowing through one fuel cell passage whereby additional masses of ions are transported through electrolytic barrier while radiant energy accompanying said waves are also absorbed by entities in the other fuel cell passage; said transported ions reacting with entities of reactants while radiant energy absorption effects infusion and surcharging of energy content of entities to higher, excited energy levels by diversion of much of the energy liberated by exothermic reaction of said ions with reactants; said surcharging effecting emission of electrons to electrodes additionally enhancing electrical output of fuel cells. Oxygen-lean air will be produced as described when oxidant used is air. Recovery of liberated electrons from associated magnetic fields may also be effected.

D.S. Justice

3,827,912; August 6, 1974
"Magnetopolypile Tubular Battery and Method of Making Same"

A three-sixteenth inch inside diameter tube of one-half inch length suitable as an anode and a tubular cathode of the same dimension are joined by tubular insulation to form an electric cell. These cells are similarly joined to form a battery. The battery may be electrically connected to increase its voltage and amperage so that a three foot battery would be very slim and an eight inch diameter cylinder would house well over one hundred such batteries. In this method and at light weight a quick and conservative experimental instrument is at hand for eventual use as a power pack in an electric car.

W.P. Krostewitz

3,615,948; October 26, 1971
"Concentration Fuel Cell"

The present invention provides a fuel cell comprising a fuel chamber and oxidizer chamber, an electrolyte chamber having electrolyte therein, the electrolyte chamber being hermetically sealed from the first two chambers by a first closed electrode separating the electrolyte chamber from the oxidizer chamber, and by a second closed electrode separating the fuel chamber from the electrolyte chamber. Each of the electrodes are composed of a material selectively permeable to only one of the cell reactants and substantially impermeable to the others, their impurities, and reaction products. Said electrodes are formed by vapor deposition on a porous metal base of an easily removable metal coating which is selectively etched away.

R.W. MacCarthy

3,825,445; July 23, 1974
"Electrochemical Cell with Catalyzed Acid Peroxide Electrolyte"

A high energy density electrochemical cell or fuel cell having an anode containing lithium, calcium, magnesium, zinc, or aluminum with an acid electrolyte fuel containing a peroxide catalyzed by a salt of one or more metals selected from iron, cobalt, nickel and copper. The rate of reaction may be moderated by concentration of salts of one or more metals high in the electromotive series or by a primary alcohol. A preferred cell comprises a carbon cathode, a magnesium anode, and an aqueous electrolyte containing, in mols per liter, about 4 hydrochloric acid, 0.6 hydrogen peroxide, 1.2 magnesium chloride, 0.02 ferric chloride, and 0.02 cupric chloride.

S. Makishima, H. Hirai, K. Tomiie and T. Kudo

3,684,578; August 15, 1972
"Fuel Cell with Electrodes Having Spinel Crystal Structure"

A fuel cell is disclosed for use with high temperature. The cell is composed of electrodes having a spinel crystal structure made by sintering fine powders onto a solid electrolyte. The fine powders are composed of oxides of at least one metal of variable valency and of oxides of at least one metal selected from the group consisting of alkali metals, alkaline earth metals, rare earth metals, Zn, Cd, Ag, Al, Ga, In, Sn, Pb, As, Sb, Bi and Ta. The novel cell is able to produce electricity efficiently by using practical fuels and various kinds of oxidizers.

A.M. Mayo

3,692,585; September 19, 1972
"Fuel Cell Battery"

Battery is made very light, compact, and efficient by making fuel cells units very thin and generally flat and rectangular. Cell units are stacked in facewise relation to make box-like structure and electrically connected in series for working voltage. Each unit comprises fuel and oxidant electrode layers spaced by permeable insulating layer to provide electrolyte passage. Impervious, conductive barrier layer between each pair of adjacent cell units and in contact with both. In one form barrier layer has protuberances on each side contacting units to provide fuel and oxidant gas passages. Insulating layer has wettable surface and maintains electrolyte in contact with electrodes for high efficiency. It is about 0.010 inch or less thick to produce minimum ion path. Electrode layers are about 0.005 inch thick or less. All elements are held in tight facewise contact for compactness and also to cause all elements to provide structural support for each other.

D.M. Moulton and W. Juda

3,625,768; December 7, 1971
"Method of Operating Fuel Cell with Molten-Oxygen-Containing Electrolyte and Non-Porous Hydrogen-Diffusing Nickel Electrode"

Nonporous nickel is used as a hydrogen-diffusion electrode in an elevated-temperature, molten, oxygen-containing electrolyte. Hydrogen ions formed at the electrode-electrolyte interface protect the electrode from oxidation, and the potential of the electrode is adjusted to maintain the freedom from oxidation.

L. Nanis and F.R. McLarnon

3,663,300; May 16, 1972
"Foam Electrolyte, Fuel Cell, and Method for Increasing Effectiveness of Fuel Cells"

The introduction of one or more ingredients of a chemical reaction into a foam atmosphere for the purpose of promoting such reaction. With respect to the application of the present invention to fuel cells, the electrolyte in one or both the anode and cathode cabinets of a fuel cell is modified with the addition of a foaming agent so that the electrodes of the cell are in contact with foamed electrolyte.

P. Patin

3,740,269; June 19, 1973
"Process and Equipment for Regulating the Output of a Set of Fuel Cells"

The output of fuel cells fed by a forced flow of each reactant dissolved or emulsified in a dilute electrolyte with a constant concentration is regulated by electrically connecting the cells in series and continuously supplying each electrolyte charged with the corresponding reactant to the first cell and supplying the other cells with each electrolyte with its reaction at a low rate of flow and progressively supplying said other cells in parallel at a relatively high rate of flow of each electrolyte with its reactant as soon as the loss of charge through said other cells reaches predetermined values.

E. Petix

3,682,705; August 8, 1972
"Replaceable Electrochemical Cell Module"

A lightweight metal/air or metal/oxygen cell is described in which a consumable anode is inserted in and is removable from a cell module comprising an envelope cathode having a pocket for receiving the anode, and a nonconductive frame in which the cathode is mounted. The cell module is provided with an electrical wiring system such that contemporaneously with the replacement of the anode the proper electrical connections are made externally of the cell. The replaceable anode has an associated gripping member which supports an electrical connector constructed and arranged to mate with an electrical connector mounted on the cell module. These connectors are themselves electrically connected to the appropriate electrodes to establish, when mated, the desired connections for battery operation. The electrical connector mounted on the cell module is readily detachable therefrom, without need for special tools and without destroying any physical characteristic of the cell module, to permit rapid replacement of a single defective cell module in a string of cells used in a battery configuration.

M. Prigent and C. Dezael

3,692,649; September 19, 1972

"Process for Removing Carbon Dioxide from Gases"

A process for producing oxygen and extracting carbon dioxide from a gas or gaseous mixture by extraction with a basic solution, evolving the carbon dioxide and oxygen in the anode compartment of an electrolytic cell, and passing the depleted base to the cathode compartment of the cell for regeneration.

E. Reimers

3,865,630; February 11, 1975

"Electrochemical Cell Having Heat Pipe Means for Increasing Ion Mobility in the Electrolyte"

A molten salt battery wherein is sunk through the battery case a heat pipe into the electrolyte for the purpose of heating the electrolyte. The heat pipe's low thermal time constant makes possible rapid temperature control of the electrolyte without regard to the heat sink effect of the battery case of electrodes.

P.D. Richman

3,669,751; June 13, 1972

"Electric Battery Comprising a Fuel Cell Hydrogen Generator and Heat Exchanger"

An electric battery of very high capacity per pound. The preferred embodiment is a self-balancing system having a hydrogen-oxygen fuel cell, a circulating KOH electrolyte and a hydrogen generator in which an Si-Al mixture is reacted with the electrolyte to produce hydrogen and an insoluble aluminum silicate, thus taking up the water generated in the fuel cell. In the fuel cell the electrodes preferably have a hydrophobic surface on the gas side and a hydrophilic bubble barrier on the electrolyte side.

P. Roy and J.L. Krankota

3,865,709; February 11, 1975

"Carbon Activity Meter"

A carbon activity meter utilizing an electrochemical carbon cell with gaseous reference electrodes having particular application for measuring carbon activity in liquid sodium. The electrolyte container is electroplated with a thin gold film on the inside surface thereof, and a reference electrode consisting of CO/CO_2 gas is used.

S. Ruben

3,544,375; December 1, 1970

"Rechargeable Fuel Cell"

A rechargeable fuel cell utilizing an oxidizable and ionizable gas, such as hydrogen, methane or hydrazine as its anodic agent and a rechargeable or electrochemically reoxidizable metal oxide as its cathode. On discharging, oxidizable gas is supplied to the porous anode, and preferably by assistance of a catalyst, is combined with oxygen derived from the cathode, producing water. On charging, the said water is decomposed to hydrogen and oxygen of which the oxygen component serves to reoxidize the cathode. As the net water content is balanced between discharge and charge cycle, the cell of the invention can be continuously operated without the necessity of replacing its electrolyte.

D. Stanimirovitch, J.F. Laurent and P.R.J. Verrez

3,589,944; June 29, 1971
"Fuel Cells and Their Method of Operation"

A fuel cell adapted for operation with a reducing gas and an oxidizing gas and
a circulating liquid aqueous electrolyte constructed of porous gas-diffusion
positive and negative electrodes with the major, generally planar, surfaces of the
electrodes of opposing polarity in contact with such electrolyte and spaced apart
at a distance less than the maximum thickness of the ionic diffusion layer of
such electrolyte, e.g., about 0.05 to about 0.5 mm, with the oxidizing and reducing gases being supplied to their respective electrodes at the surface thereof
opposite such major surface. Such a fuel cell is operable at a temperature in
the range of from ambient to about $70°C$ and at gas overpressures less than
about $1\ kg/cm^2$ and insufficiently greater than the pressure on the electrolyte
to cause the gases to pass entirely through the electrodes with substantially
reduced polarization than if the spacing between the electrodes' major surfaces
was greater than stated.

C. Vanleugenhaghe

3,615,847; October 26, 1971
"Manufacture of Electrodes for Fuel Cells"

The electrochemical performance of a fuel cell electrode consisting of a laminated or compressed mixture of carbon, a catalyst and a fluorocarbon polymer as binder is improved by contacting the electrode with a metal reacting
with the fluorocarbon polymer and selected from the class of the alkali
metals and the alkaline earth metals at a temperature sufficient to cause reaction between the metal and the polymer, the reaction being confined to the
surface without degrading the bulk of the polymer.

M.N. Yardney and N. Kohen

3,632,449; January 4, 1972
"Method of Making and Operating a Gas-Depolarized Cell"

A battery of the metal/gas-electrode type includes a gas-depolarizable cathode
forming at least one pocket or passage open to the outside by way of a slot
(or a pair of opposite slots) in the cell housing. The cathode is in fluid-tight
contact with the slotted housing and subdivides the interior of the housing
into a gas passage or compartment and a surrounding electrolyte compartment.
Anode plates are disposed in the electrolyte compartment and may be interconnected to form a unit detachable from the cathode upon withdrawal of
the electrode assembly from the housing. For this purpose the housing may
be split into separably interfitting parts. The depolarizing gas circulates
through the interior of the housing by thermal convection.

3,682,706; August 8, 1972
"Gas Depolarized Cell"

Abstract identical to 3,632,449.

S. Zaromb

3,554,810; January 12, 1971
"Metal-Oxygen Power Source"

A power source comprises a metal anode and an oxygen depolarized cathode
at least partly enclosing a battery cell compartment, and a reversibly collapsible electrolyte storage reservoir, said collapsible reservoir being adapted to
control the volume of electrolyte contained in said battery cell compartment
and thereby to control the current density at the surface of said anode.

3,788,899; January 29, 1974
"Electrochemical Power Generation Apparatus and Methods"

A power source comprises an electrochemical battery cell including an anode
and an oxygen or hydrogen peroxide depolarized cathode, a liquid electrolyte
solution, electrolyte circulation means, an electrolyte container, and means
to produce changes in the volumes of electrolyte in said container and in said
battery cell and thereby control the current density at the surfaces of said
anode and cathode. The anode may be of a consumable metal type, especially
of aluminum. The volume changes may be effected by expanding or collaps-
ing a reversibly expansible pocket in said electrolyte container. When several
battery cells are connected in series, electrical current leakage paths through
the electrolyte circulation means are blocked by a system of valves which
sequentially and intermittently connect one of said cells at a time to said
circulation means. Heat exchange apparatus may be included comprising
a vacuum distillation means whereby impure water is converted into distilled
water, preferably by utilizing the heat generated in said power source.

NOTICE

Nothing contained in this Review shall be construed to constitute a permission or recommendation to practice any invention covered by any patent without a license from the patent owners. Further, neither the author nor the publisher assumes any liability with respect to the use of, or for damages resulting from the use of, any information, apparatus, method or process described in this Review.

ENERGY FROM BIOCONVERSION OF WASTE MATERIALS 1977

by Dorothy J. De Renzo

Energy Technology Review No. 11
Pollution Technology Review No. 33

One of the chief gaseous products of the anaerobic decomposition of organic matter is methane, CH_4. This is how natural gas was formed in prehistoric times along with other fossil fuels.

By applying this principle today in environmentally acceptable fashion it is possible to bioconvert municipal solid sewage, animal manure, agricultural and other organic wastes into substitute natural gas (95% CH_4). In its simplest essentials the process consists of loading the material into a digester (a closed tank with a gas outlet). Given favorable thermal and chemical conditions, the appropriate biological processes will then take their course.

The bioconversion of waste materials to methane provides at least partial solutions not only to the energy problem, but also to the solid waste disposal problem. The harvesting of heretofore undesirable vegetations, such as algae, water hyacinths, and kelp as "energy crops" offers unconventional opportunities for supplementary utilization of natural resources.

This book describes practical methods for the bioconversion of waste matter. It is based on reports of academic and industrial research teams working under government contracts. A partial and condensed table of contents follows here. Chapter headings and important subtitles are given.

1. SOURCES OF WASTE MATERIALS
Suitability & Characteristics
Quantities & Availability
Agricultural Crop Residues
Forests
Urban Wastes
Rural Wastes
Strictly Animal Wastes
Industrial Wastes
Cost Considerations

2. MECHANISMS & PATHWAYS
Anaerobic Decomposition Processes
Terminal Dissimilation of Matter
Degradation of Cellulose
Bacteria and Protozoa
Cellulases & Other Enzymes

Controlling Factors in
    Methane Fermentations
Acetate Utilization
Trace Organics
Effect of Temperature Changes

3. SOLID WASTE & SEWAGE SLUDGE
U. of Illinois Studies
Pfeffer-Dynatech Anaerobic
    Digestion System
Supporting Studies
Addition of Coal to Sludge
Synergistic Methane Production

4. METHANATION OF URBAN TRASH
Digester Feed Preparation
Digester Design
Gas Production
Gas Scrubbing Technology
Desirable Gas Characteristics

5. ANIMAL WASTE DIGESTION
Oregon State U. System
Animal Waste Management
Berkeley Conversion Studies
Digestion for Disposal
Digestion plus Photosynthesis
Dept. of Agriculture Study
Other Studies on Animal Waste Digesters
    Useable for Energy Purposes

6. INDUSTRIAL WASTES UTILIZATION
Petrochemical Wastewaters
Distillery Slops Digestion
Rum Distillery Slops
Process Flow Sheets
Design Criteria
Economic Analysis
Use of Biogas in the Sugar Industry
Winery Waste Treatment

7. METHANE FROM ENERGY CROPS
Integrated Conversion Systems
Digestion of Algae
University Studies
Single Stage vs. Two Stages
Mariculture Investigations
Conclusions and Drawbacks
Storage Difficulties

Note: Each chapter is followed by bibliographic reference lists in order to provide the reader with easy access to further information on these timely topics.

ISBN 0-8155-0656-2

223 pages

PRACTICAL TECHNIQUES FOR SAVING ENERGY IN THE CHEMICAL, PETROLEUM AND METALS INDUSTRIES 1977

by Marshall Sittig

Energy Technology Review No. 12
Chemical Technology Review No. 90

The above-captioned industries were selected for coverage in this book because they are the three largest consumers of energy in the U.S. economy. They constitute an interacting group: they use many common raw materials and are simultaneously feeding products and by-products to one another.

Practical thinking about industrial energy conservation requires interceptive calculations of such material transfers to produce positive energy savings.

The major conservation approaches are arranged as follows:

1. Waste Utilization
2. Process Integration
3. Process Modification
4. Design Modification
5. Maintenance and Insulation
6. Market Modification

The three major industries are subdivided into 39 individual processing industries. The following discussions and proposals are presented for practically every processing industry:

1. Process Technology Involved
2. Major Energy Conservation Options to 1980
3. Goal Year (1980) Energy Use Targets
4. Some Projections beyond 1980 to 1990

INDIVIDUAL PROCESSING INDUSTRIES

Introduction & Analytical Procedures
Metals Industry—General
Steel Industry
Aluminum Industry
Iron Foundries
Copper Industry
Ferroalloys
Non-Ferrous Foundries
Steel Foundries
Other Primary Non-Ferrous Metals
Non-Ferrous Processing Industry
Miscellaneous Metal Products
Secondary Non-Ferrous Smelting
  and Refining Industry

Primary Zinc Industry
Primary Lead Industry
Chemical and Allied Products
Alkalis and Chlorine
Industrial Gases
Inorganic Pigments
Other Industrial Inorganic Chemicals
Plastic Materials Industry
Synthetic Rubber Industry
Rayon and Cellulose Acetate Industry
Synthetic Fibers Industry
Drug Industry
Soaps, Detergents, & Toiletries
Paint Industries
Gums and Wood Chemicals
Coal Tar Chemicals, Dyes & Pigments
Aliphatic Organic Chemicals
Nitrogen Fertilizer Industry
Phosphate Fertilizer Industry
Fertilizer Mixing Industry
Agricultural Chemicals
Adhesives and Sealants
Explosives and Allied Industries
Printing Inks
Carbon Black
Fatty Acids & Allied Industries
Petroleum & Coal Products

As can be seen from this large and impressive table of contents, entries have been arranged in an encyclopedic manner whenever possible.

To be found at the end of the book is a complete and detailed list of reports and references cited throughout the work. In this list titles are complete with their publishers and other sources of procurement and are never abbreviated.

Under the Energy and Conservation Act, the U.S. Federal Energy Administration was required to set energy conservation targets for the most energy-intensive manufacturing industries. Goals to be attained by Jan. 1, 1980 were established by the end of 1976. Major contributions to the manuscript are acknowledged from the energy target documents, particularly from those on the chemical, petroleum, and metals industries.

ISBN 0-8155-0657-0   524 pages

HOW TO SAVE ENERGY AND CUT COSTS IN EXISTING INDUSTRIAL AND COMMERCIAL BUILDINGS 1976

An Energy Conservation Manual

**by Fred S. Dubin, Harold L. Mindell
and Selwyn Bloome**

Energy Technology Review No. 10

This manual offers guidelines for an organized approach toward conserving energy through more efficient utilization and the concomitant reduction of losses and waste.

The current tight supply of fuels and energy is unprecedented in the U.S.A. and other countries, and this situation is expected to continue for many years. Never before has there been as pressing a need for the efficient use of fuels and energy in all forms.

Most of the energy savings will result from planned systematic identification of, and action on, conservation opportunities.

Part I of this manual is directed primarily to owners, occupants, and operators of buildings. It identifies a wide range of opportunities and options to save energy and operating costs through proper operation and maintenance. It also includes minor modifications to the building and mechanical and electrical systems which can be carried out promptly with little, if any, investment costs.

Part II is intended for engineers, architects, and skilled building operators who are responsible for analyzing, devising, and implementing comprehensive energy conservation programs. Such programs involve additional and more complex measures than those in **Part I.** The investment is usually recovered through demonstrably lower operating expenses and much greater energy savings.

A partial and much condensed table of contents follows here:

PART I

1. **PRINCIPLES OF ENERGY CONSERVATION**
2. **MAJOR OPPORTUNITIES**
   Heating + Insulation
   Cooling + Insulation
   Lighting
   Hot Water
3. **ENERGY LOADS**
   Building Load
   Distribution Loads
   Equipment Efficiencies
   Building Profile vs. Energy Consumption
   Detailed Conservation Opportunities
4. **HEATING & VENTILATING**
   Primary Energy Conversion Equipment
   Building & Distribution Loads
   Energy Reducing Opportunities

5. **HOT WATER SAVINGS**
6. **COOLING & VENTILATING**
   Reducing Energy Consumption
   used for Cooling
7. **DISTRIBUTION & HVAC SYSTEMS**
8. **COMMERCIAL REFRIGERATION SYSTEMS**
9. **LIGHTING & HEAT FROM LAMPS**
   Fluorescents Turnoff vs.
   Replacement Cost & Labor
10. **POWER FOR MACHINERY**
APPENDIXES: COST VS. CONSERVATION

PART II

11. **ENERGY MANAGEMENT TEAMS**
12. **MECHANICAL SYSTEMS BACKGROUND**
13. **HEATING AND VENTILATION**
    Continuous Temperature Control
    Reduce Resistance to Air Flow
    Boiler & Burner Selection
14. **HOT WATER SYSTEMS**
15. **COOLING & VENTILATION**
    Condensers vs. Towers
    Losses Through Floors
    Air Shafts & Fenestration
16. **COMMERCIAL REFRIGERATION**
17. **HVAC & HEAT RECLAMATION**
18. **HEAT RECLAMATION EQUIPMENT**
19. **LIGHTING**
    Wattage Reduction
    Peak Load Reduction
    Non-Uniform Lighting
20. **POWER FOR MACHINERY**
    Correct Power Factors
    Exchange of Oversized Motors
    and Transformers
21. **CENTRAL CONTROL SYSTEMS**
    480/240 Volts Preferred to 120 V
22. **ALTERNATIVE ENERGY SOURCES**
    Solar Energy
    Methane Gas
23. **ECONOMIC ANALYSES & COSTS**
24. **APPENDIXES & REFERENCES**

Much of the technology required to achieve energy savings is already available. Current research is providing refinements and evaluating new techniques that can help to curb the waste inherent in yesteryear's designs. The principal need is to get the available technology, described here, into widespread use.

ISBN 0-8155-0638-4

725 pages